Topics in Current Genetics

16

Series Editor: *Stefan Hohmann*

Marja Makarow • Ineke Braakman (Eds.)

Chaperones

With 43 Figures, 1 in Color; and 3 Tables

Springer

Professor Dr. Marja Makarow
Institut of Biotechnology
P.O. Box 56
00140 Helsinki University
Helsinki
Finland

Professor Dr. Ineke Braakman
Universiteit Utrecht
Faculty of Science
P.O. Box 80083
3508 TB Utrecht
The Netherlands

The cover illustration depicts pseudohyphal filaments of the ascomycete *Saccharomyces cerevisiae* that enable this organism to forage for nutrients. Pseudohyphal filaments were induced here in a wild-type haploid MATa Σ1278b strain by an unknown readily diffusible factor provided by growth in confrontation with an isogenic petite yeast strain in a sealed petri dish for two weeks and photographed at 100X magnification (provided by Xuewen Pan and Joseph Heitman).

ISSN 1610-2096

ISBN 978-3-642-06902-4 e-ISBN 978-3-540-32581-9

Springer-Verlag is a part of Springer Science + Business Media
springeronline.com

Printed in Germany

Cover Design: Design & Production, Heidelberg
Printed on acid-free paper - 39/3152-YK - 5 4 3 2 1 0

Topics in Current Genetics publishes review articles of wide interest in volumes that centre around a specific topic in genetics, genomics as well as cell, molecular and developmental biology. Particular emphasis is placed on the comparison of several model organisms. Volume editors are invited by the series editor for special topics, but further suggestions for volume topics are highly welcomed.

Each volume is edited by one or several acknowledged leaders in the field, who ensure the highest standard of content and presentation. All contributions are peer-reviewed. All papers are published online prior to the print version. Individual DOIs (digital object identifiers) make each article fully citable from the moment of online publication.

Subscribers to the print version of *Topics in Current Genetics* receive free access to the online version. An online-only license is also available.

Editorial office:
Topics in Current Genetics
Series Editor: Stefan Hohmann
Cell and Molecular Biology
Göteborg University
Box 462
40530 Göteborg, Sweden
Phone: +46 733 547297
FAX: +46 31 7732595
E-mail: editor@topics-current-genetics.se
Website: http://www.topics-current-genetics.se

Foreword

Ineke Braakman and Marja Makarow

Molecular chaperones interact with virtually every newly synthesized protein. Their role is not limited to this, as an increasing number of protein-protein interactions are found to be mediated by molecular chaperones. They reside in large complexes, in every cellular compartment, and to some extent even outside cells. These proteins are of interest to a large number of scientists, not only to those interested in protein biosynthesis, but also in relation to protein transport, organelle biogenesis, and cell stress. Molecular chaperones have entered textbooks in Cell Biology and Biochemistry, and many undergraduate and graduate students are taught about the chaperone cycles of substrate binding and release and the co-chaperones that regulate a chaperone's activity.

Whereas excellent reviews on molecular chaperones are published, they often focus on the latest results without reiterating the basics. These reviews are important, but often too advanced to be useful for students and scientist who want to enter the chaperone field. Goal of this issue was to assemble a collection of reviews on molecular chaperones that would be both timely and basic, which would make them an excellent entrance for novices into the field and suitable for teaching purposes. The reviews together do not cover the complete field of chaperones but instead focus on one particular chaperone or one particular organelle or process. This still allows a both broad and deep coverage of each topic, but also gives an impression of the wealth of processes molecular chaperones are involved in.

We feel to have succeeded in the task we set ourselves, and thank all authors for their contributions. We hope that you will enjoy reading this issue as much as we enjoyed planning and assembling it.

Braakman, Ineke

Professor of Cellular Protein Chemistry, University of Utrecht, The Netherlands
i.braakman@chem.uu.nl

Makarow, Marja

Professor of Applied Biochemistry and Molecular Biology, University of Helsinki, Finland
marja.makarow@helsinki.fi

Table of contents

Regulation of the heat shock response by heat shock transcription factors 1
Ville Hietakangas and Lea Sistonen 1
Abstract 1
1 Preface 1
2 Transcriptional regulation of the heat shock response in bacteria 2
3 Regulation of the eukaryotic heat shock response via heat shock elements 3
4 Heat shock factors constitute a conserved family of transcriptional regulators 4
5 HSF1 – the prototypical heat shock transcription factor 7
5.1 Modular structure of HSF1 7
5.2 Rapid and transient HSF1 DNA-binding activity in response to stress 8
5.3 Complex regulation of HSF1 transcriptional activity by post-translational modifications 8
5.4 Interactions between HSF1 and the transcriptional machinery 11
5.5 Negative and positive regulation of HSF1 by interacting proteins ... 12
5.6 Role of HSF1 in normal development and physiology 14
6 HSF2 – a cooperative modulator of HSF1? 15
7 HSF3 – an avian-specific regulator of heat shock genes 17
8 HSF4 – regulator of eye development 18
9 Novel functions for HSFs in transcription 18
9.1 Broad repertoire of HSF target genes 18
9.2 Regulation of longevity by HSF1 20
9.3 HSF1-dependent transcription of satellite III repeats previously considered as heterochromatin 21
10 Conclusions and perspectives 22
Acknowledgements 23
References 23

The unfolded protein response unfolds 35
Maho Niwa 35
Abstract 35
1 Endoplasmic reticulum (ER): the journey to secretion 35
2 The unfolded protein response (UPR) pathway 37
2.1 Yeast UPR 37
2.2 Non-conventional HAC1 mRNA splicing 37
2.3 Are there additional RNA substrates for yeast Ire1p? 39
2.4 UPR specific transcription factors in yeast 39
3 Mammalian UPR 40
3.1 IRE1 40

3.2 ATF6 42
3.3 PERK 43
3.4 Transcriptional output of three UPR signaling branches 46
4 Phospholipids and the UPR 47
5 UPR signaling arm specific components identified to date 48
5.1 IRE1 signaling branch 49
5.2 PERK signaling branch 49
5.3 ATF6 signaling branch 50
6 BiP associates with the luminal domains of IRE1, PERK, and ATF6 50
7 Time dependent shift of the UPR response 51
8 Physiological roles of the UPR 52
9 Conclusions 53
Acknowledgements 54
References 54

Hsp104p: a protein disaggregase 65
Johnny M. Tkach and John R. Glover 65
Abstract 65
1 Hsp104p and thermotolerance in yeast 65
2 *In vitro* reconstitution of Hsp104p refolding activity 68
3 Hsp100 structure and function 70
4 Mechanism of protein disaggregation 73
5 Organization of the bichaperone network 76
6 Yeast prions and Hsp104p 77
7 Implications for protein aggregation disease 82
8 Final Remarks 84
References 84

Folding of newly synthesised proteins in the endoplasmic reticulum 91
Sanjika Dias-Gunasekara and Adam M. Benham 91
Abstract 91
1 The scope of protein folding in the ER 91
2 Entry into the ER 92
3 ER protein sorting 93
4 Signal peptide cleavage 93
5 The proline problem 95
6 Folding of ER glycoproteins 96
7 BiP 100
8 Disulfide bond formation 100
9 Introducing more ER chaperones 104
10 Folding of specialised proteins in the ER 104
11 Techniques, model systems and what's next 106
Acknowledgements 107
References 107
Abbreviations 116

Quality control of proteins in the mitochondrion 119
Mark Nolden, Brigitte Kisters-Woike, Thomas Langer, and Martin Graef... 119
Abstract 119
1 Stability of mitochondria.. 119
2 Molecular chaperone proteins and mitochondrial proteolysis 121
3 ATP-dependent proteases of mitochondria 122
3.1 Mitochondrial Lon proteases 124
3.2 Mitochondrial Clp proteases 126
3.3 AAA proteases in the inner membrane 127
4 Quality control of inner membrane proteins.. 131
4.1 Substrate recognition by AAA proteases 132
4.2 Substrate dislocation during proteolysis by AAA proteases 133
4.3 Additional components of the quality control system.................... 136
5 Regulation of quality control systems of mitochondria........................ 136
References 137

Chaperone proteins and peroxisomal protein import 149
Wim de Jonge, Henk F. Tabak, and Ineke Braakman.................................. 149
Abstract 149
1 Introduction 149
2 Peroxisomes 150
2.1 Formation, maintenance, and function.. 150
2.2 General properties of protein import.. 153
2.3 Proteins involved in protein import 153
2.4 Formation of peroxisomes 155
2.5 Recycling receptor import model.. 156
2.6 Docking and translocation of peroxisomal proteins........................... 158
2.7 Folding state and import of peroxisomal proteins.............................. 160
3 Involvement of Hsp70 in peroxisomal protein import 162
3.1 Hsp70 family introduction... 162
3.2 *In vivo* roles of Hsp70.. 165
3.3 Hsp70 and import of proteins into organelles 166
3.4 Hsp70 and peroxisomal protein import... 168
4 Involvement of Hsp90 in peroxisomal protein import 170
4.1 Hsp90 family introduction... 170
4.2 *In vivo* roles of Hsp90.. 170
4.3 Hsp90 and import of proteins into organelles 171
4.4 Hsp90 and peroxisomal protein import... 172
5 Concluding remarks 172
References 173

Proteasomal degradation of misfolded proteins 185
Robert Gauss, Oliver Neuber, and Thomas Sommer..................................... 185
Abstract 185
1 Introduction 185
2 Recognition and degradation of aberrant proteins in the cytosol 186

2.1 Hsp70 and Hsp90 chaperones – protein folding and re-folding...... 186
2.2 The ubiquitin-proteasome system.. 189
2.3 Some *U-box* ligases link chaperones to the ubiquitin-proteasome system.. 191
3 Protein degradation from the ER.. 195
3.1 Quality control in the ER and the unfolded protein response......... 195
3.2 The ER-associated protein degradation pathway........................ 195
3.3 Selection and recruitment of aberrant proteins........................ 196
3.4 Dislocation of terminally misfolded proteins........................ 200
3.5 Ubiquitination of aberrant ER-resident proteins........................ 202
3.6 Do E3 ligases play a central role in the ER degradation system?... 202
3.7 Driving-force of dislocation.. 204
4 Cdc48p/p97 – chaperoning poly-ubiquitinated proteins.................. 204
5 Diseases and toxins – what can go wrong in protein degradation?........ 206
5.1 Diseases associated with protein degradation........................ 207
5.2 Viruses and AB-toxins.. 208
5.3 Pharmacological chaperones – a new approach in fighting folding diseases.. 209
6 Conclusions.. 211
Acknowledgement.. 211
References.. 212

Template-induced protein misfolding underlying prion diseases 221
Luc Bousset, Nicolas Fay, and Ronald Melki.. 221
Abstract.. 221
1 Prion diseases.. 221
2 Formulation of the prion hypothesis.. 222
3 The mammalian prion PrP.. 222
3.1 Identification.. 222
3.2 Structure.. 222
3.3 Folding properties.. 223
3.4 Function.. 223
3.5 Cellular processing.. 224
4 The prions in yeast and fungi.. 224
4.1 Genetic criteria for the prions in yeast and fungi........................ 225
4.2 Characteristics.. 225
4.3 Structural features.. 228
5 Properties of the fibrillar forms of prion proteins........................ 229
6 Soluble oligomeric forms of the prion proteins........................ 231
7 Mechanistic models for prion propagation.. 231
8 Maintenance and inheritance.. 232
9 *In vitro* assembly process of prions proteins........................ 234
10 Prions and misfolding diseases, unquestioned issues, and unanswered questions.. 237
11 Conclusions and perspectives.. 239
References.. 239

The Hsp60 chaperonins from prokaryotes and eukaryotes **251**
M. Giulia Bigotti, Anthony R. Clarke, and Steven G. Burston 251
Abstract 251
1 The Group I chaperonins 251
1.1 Introduction 251
2 Structure of the Group I chaperonins 254
2.1 GroEL structure 254
2.2 GroES structure 255
2.3 The structure of the GroEL-ATP complex 255
2.4 The structure of the GroEL-GroES complexes 256
3 Interaction between Group I chaperonins and protein substrate 257
4 Allostery and asymmetry in nucleotide binding to GroEL 257
5 Reaction cycle of the Group I chaperonins 259
5.2 Encapsulation and the initiation of protein folding 261
5.3 Priming the complex for the release of GroES and polypeptide substrate 262
5.4 Ejection of the substrate and GroES from the *cis* ring 262
6 The Group I chaperonin-assisted protein folding reaction 263
7 The Group II chaperonins 264
7.1 Introduction 264
8 Group II chaperonin subunit composition and organization 266
9 Structure of the Group II chaperonins 267
10 Nucleotide-induced structural rearrangements in the Group II chaperonins 269
11 Allostery in the Group II chaperonins 271
12 Interaction between the Group II chaperonins and protein substrates .. 273
13 Future perspectives 274
References 275

Index **285**

List of contributors

Benham, Adam M.
Department of Biological Sciences, University of Durham, South Road, Durham, DH1 3LE, England.
Adam.benham@durham.ac.uk

Bigotti, M. Giulia
Department of Biochemistry, University of Bristol, School of Medical Sciences, Bristol BS8 1TD, UK

Bousset, Luc
Laboratoire d'Enzymologie et Biochimie Structurales, CNRS, 91198 Gif-sur-Yvette Cedex, France.
Present address: EMBL, 6 rue Jules Horowitz, BP181, 38042 Grenoble Cedex 9, France.

Braakman, Ineke
Department of Cellular Protein Chemistry, Utrecht University, Padualaan 8, 3584 CH, Utrecht, The Netherlands
i.braakman@chem.uu.nl

Burston, Steven G.
Department of Biochemistry, University of Bristol, School of Medical Sciences, Bristol BS8 1TD, UK
s.g.burston@bristol.ac.uk

Clarke, Anthony R.
Department of Biochemistry, University of Bristol, School of Medical Sciences, Bristol BS8 1TD, UK

de Jonge, Wim
Department of Cellular Protein Chemistry, Utrecht University, Padualaan 8, 3584 CH, Utrecht, The Netherlands. Current adress: Institute of Information and Computing Sciences, Utrecht University, Padualaan 14, 3584CH Utrecht, The Netherlands.

Dias-Gunasekara, Sanjika
Department of Biological Sciences, University of Durham, South Road, Durham, DH1 3LE, England.

Fay, Nicolas
Laboratoire d'Enzymologie et Biochimie Structurales, CNRS, 91198 Gif-sur-Yvette Cedex, France.

Gauss, Robert
Max-Delbrück-Centrum für Molekulare Medizin, Robert-Rössle-Str. 10, 13025 Berlin, Germany

Glover, John R.
Department of Biochemistry, University of Toronto, 1 King's College Circle, Room 5302, Medical Sciences Building, Toronto, Ontario, Canada M5S 1A8
john.glover@utoronto.ca

Graef, Martin
Institut für Genetik, Universität Köln, Zülpicher Str. 47, 50674 Köln, Germany

Hietakangas, Ville
Turku Centre for Biotechnology, P.O. Box 123, FIN-20521 Turku, Finland. Current address: EMBL Heidelberg, Meyerhofstrasse 1, D-69117 Heidelberg, Germany

Kisters-Woike, Brigitte
Institut für Genetik, Universität Köln, Zülpicher Str. 47, 50674 Köln, Germany

Langer, Thomas
Institut für Genetik, Universität Köln, Zülpicher Str. 47, 50674 Köln, Germany
Thomas.Langer@uni-koeln.de.

Melki, Ronald
Laboratoire d'Enzymologie et Biochimie Structurales, CNRS, 91198 Gif-sur-Yvette Cedex, France.
melki@lebs.cnrs-gif.fr

Neuber, Oliver
Max-Delbrück-Centrum für Molekulare Medizin, Robert-Rössle-Str. 10, 13025 Berlin, Germany

Niwa, Maho
Division of Biological Sciences, Section of Molecular Biology, University of California, San Diego, 9500 Gilman Drive, La Jolla, California, 92093-0377
niwa@ucsd.edu

Nolden, Mark
Institut für Genetik, Universität Köln, Zülpicher Str. 47, 50674 Köln, Germany

Sistonen, Lea
Turku Centre for Biotechnology, P.O. Box 123, FIN-20521 Turku, Finland
lea.sistonen@btk.fi

Sommer, Thomas
Max-Delbrück-Centrum für Molekulare Medizin, Robert-Rössle-Str. 10, 13025 Berlin, Germany
tsommer@mdc-berlin.de

Tabak, Henk F.
Department of Cellular Protein Chemistry, Utrecht University, Padualaan 8, 3584 CH, Utrecht, The Netherlands, and Laboratory of Cell Biology, University Medical Center Utrecht and Center for Biomedical Genetics, 3584 CX Utrecht, The Netherlands

Tkach, Johnny M.
Department of Biochemistry, University of Toronto, 1 King's College Circle, Room 5302, Medical Sciences Building, Toronto, Ontario, Canada M5S 1A8

Regulation of the heat shock response by heat shock transcription factors

Ville Hietakangas and Lea Sistonen

Abstract

The heat shock response is characterized by a rapid and robust increase in heat shock proteins upon exposure to protein-damaging stresses. This evolutionarily conserved cellular protection mechanism is primarily regulated at the level of transcription. In bacteria, heat shock-induced transcription is regulated by the activation of σ^{32} factor, whereas eukaryotes utilize heat shock transcription factors (HSFs) that bind to specific heat shock elements (HSEs) within the promoters of their target genes. Unlike yeasts, nematodes, and fruit flies, which have a single HSF, vertebrates and plants have an entire HSF family. In addition to stress-induced activation, some members of the HSF family are also activated under non-stressful conditions, including development and differentiation. The activity of HSFs is under post-translational control, requiring trimerization, DNA binding, and hyperphosphorylation. The interplay between different family members and other interacting proteins adds further complexity to HSF-mediated transcription. Here, we summarize the current knowledge of the transcriptional regulation of the heat shock response, highlighting recent advances in exploring the multi-faceted nature of heat shock transcription.

1 Preface

Maintenance of cellular protein homeostasis requires the proper function of molecular chaperones. Molecular chaperones comprise a great variety of proteins that are essential for protein folding and translocation across cellular compartments, assembling multi-protein complexes, preventing aggregation, and directing misfolded and short-lived proteins to degradation by the proteasome. Protein-damaging stresses, including heat shock, induce the expression of a subgroup of molecular chaperones, called heat shock proteins (Hsps), which consist of several protein families designated by their molecular weight, such as the Hsp90, Hsp70, Hsp60, and the small Hsp (sHsp) families. Upon stress, the Hsps prevent protein unfolding and aggregation, thereby maintaining the critical cellular structures and functions and protecting against apoptotic or necrotic cell death. The expression of Hsps is regulated by multiple mechanisms, among which the transcriptional regulation is most prominent. The transcriptional regulation of the heat shock response is conserved throughout the eukaryotes, where the different members of the HSF family share homologous functional domains and act mainly as transcriptional activators. In addition to stress stimuli, some HSFs respond to other signals and have distinct targets from the classical heat shock genes. Consequently, HSFs are involved in a number of vital physiological processes, such as development, differentiation, and regulation of longevity. The central role of HSF-mediated transcription in several cellular processes underlines the importance of understanding the functions of HSFs in normal physiology and disease conditions.

2 Transcriptional regulation of the heat shock response in bacteria

The function of Hsps and the presence of heat-shock inducible transcription are strikingly well conserved throughout evolution. Despite certain analogies, however, the transcriptional regulatory mechanisms are distinct in prokaryotes and eukaryotes. The regulation of the bacterial heat shock response is best characterized in *Escherichia coli* where the expression of Hsps, including DnaK (prokaryotic Hsp70), DnaJ (Hsp40), GrpE, GroEL (Hsp60), and GroES (Hsp10), is regulated by the product of the *rpoH* gene, i.e. the stress-inducible σ^{32} subunit of RNA polymerase (Fig. 1A; Grossman et al. 1984; for review see Arsène et al. 2000). Under non-stressful conditions, σ^{32} is maintained at a low level due to its rapid turnover, but upon exposure to heat shock, the concentration of σ^{32} is greatly increased through enhanced synthesis and increased stability, resulting in preferred transcription of σ^{32}-dependent heat shock genes (Straus et al. 1987; for review see Yura and Nakahigashi 1999; Arsène et al. 2000).

The σ^{32}-mediated transcription is controlled by a negative feedback system. The accumulating DnaK-DnaJ-GrpE chaperone machinery binds to σ^{32} and inhibits its activity (Tilly et al. 1983; Straus et al. 1990; Tomoyasu et al. 1998). Moreover, binding of DnaK-DnaJ to the σ^{32} promotes its degradation by the ATP-dependent metalloprotease FtsH (Tatsuta et al. 1998). Therefore, availability of DnaK/DnaJ is a direct sensor of cellular stress and a regulator of heat shock transcription (Tomoyasu et al. 1998). In addition to the central role of σ^{32}-mediated regulation of the heat shock response, which is well conserved in Gram-negative bacteria (for review see Yura and Nagahigashi 1999), distinct regulatory mechanisms have been characterized in other bacteria. For example, in the Gram-positive bacteria *Bacillus subtilis*, the HrcA repressor regulates major Hsps by binding to negatively acting CIRCE elements (Fig. 1B; for review see Yura and Nagahigashi 1999). Interestingly, the folding of HrcA is facilitated by the GroE chaperonin system. In response to increased protein damage, the GroE folding machinery is occupied and folding of the HrcA repressor is stalled (Mogk et al. 1997). Thereby this regulatory system also relies on the availability of Hsps, providing a direct sensing mechanism for protein misfolding.

A.

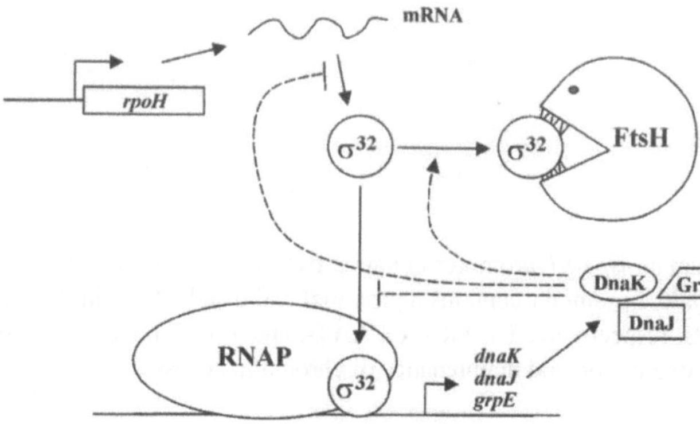

B.

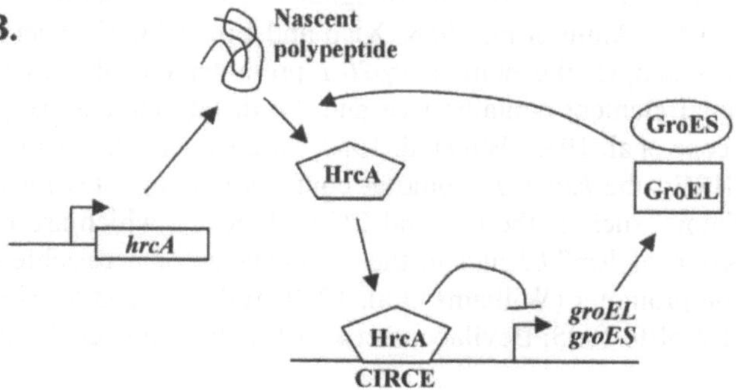

Fig. 1. Regulatory mechanisms of the prokaryotic heat shock response. A. In Gram-negative bacteria, the heat shock response is regulated by the σ^{32} factor. Accumulation of the heat shock proteins DnaK, DnaJ, and GrpE provides a negative feedback mechanism by inhibiting σ^{32} translation and association with the core RNA polymerase and by promoting σ^{32} degradation by FtsH. B. Regulation of the heat shock response by CIRCE elements and the HrcA repressor. Negative feedback regulation is provided by GroE chaperonins, which facilitate the folding of HrcA.

3 Regulation of the eukaryotic heat shock response via heat shock elements

Protein-damaging conditions have profound effects on eukaryotic transcription. While transcription is generally inhibited, the transcription of genes encoding Hsps can be increased more than 100-fold upon heat shock (Gilmour and Lis 1985). This robust activation of gene expression was initially reported by Ferruccio Ritossa (1962), who observed induction of specific chromosome puffs in the polytene chromosomes of *Drosophila buschii* larval salivary glands accidentally

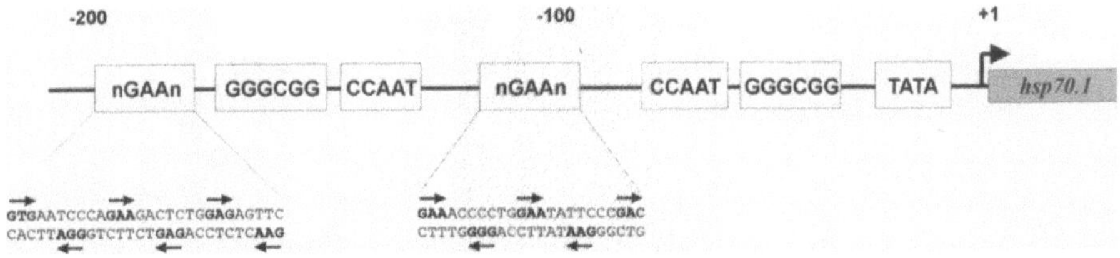

Fig. 2. The human *hsp70.1* promoter contains two stress-responsive heat shock elements. The human *hsp70.1* promoter contains a proximal and distal HSE with five and six inverted nGAAn repeats, respectively. The GC-, CCAAT-, and TATA-boxes are involved in constitutive *hsp70.1* expression and maintenance of chromatin accessibility.

exposed to elevated temperatures. Ten years later, Tissieres and coworkers discovered the stress-induced synthesis of Hsps (Tissieres et al. 1974), and after another decade, the heat shock element (HSE), a specific DNA sequence responsible for the transcriptional activation of heat shock genes, was identified (Pelham 1982). The consensus HSE consists of contiguous inverted pentameric repeats of the sequence nGAAn (Amin et al. 1988; Xiao and Lis 1988; Kroeger and Morimoto 1994). For example, the human *hsp70.1* promoter contains two HSEs, of which the proximal element contains five and the distal element six pentameric units (Fig. 2; Greene et al. 1987; Wu et al. 1987; Abravaya et al. 1991a, 1991b). In addition to the HSEs, the *hsp70.1* promoter contains binding sites for other transcriptional regulators, such as the GC and CCAAT boxes, which are involved in the basal expression of *hsp70.1* and in the maintenance of accessible chromatin architecture of the promoter (Williams et al. 1989; Williams and Morimoto 1990; Landsberger and Wolffe 1995; Bevilacqua et al. 1997; Imbriano et al. 2001).

4 Heat shock factors constitute a conserved family of transcriptional regulators

The eukaryotic transcription factors that bind to HSEs and activate transcription are called heat shock transcription factors or HSFs (for review see Wu 1995; Pirkkala et al. 2001). In yeast and fruit fly, a single HSF is responsible for HSE-mediated transcription (Sorger and Pelham 1988; Wiederrecht et al. 1988; Clos et al. 1990), but in vertebrates, the HSF family consists of several members sharing functional and structural properties, such as domains involved in DNA binding and oligomerization (Fig. 3). In mammals, three HSF family members, HSF1, HSF2, and HSF4, have been found, whereas avian species have an additional heat shock factor, HSF3 (Rabindran et al. 1991; Sarge et al. 1991; Schuetz et al. 1991; Nakai and Morimoto 1993; Nakai et al. 1997). The diversity of mammalian HSFs is increased by alternative splicing of HSF1, HSF2, and HSF4 (for review see Pirkkala et al. 2001). Despite the family name, not all heat shock factors are stress responsive. Among the vertebrate HSFs, only HSF1 and the avian-specific HSF3 have conclusively been shown to be activated by stress and to be essential for the

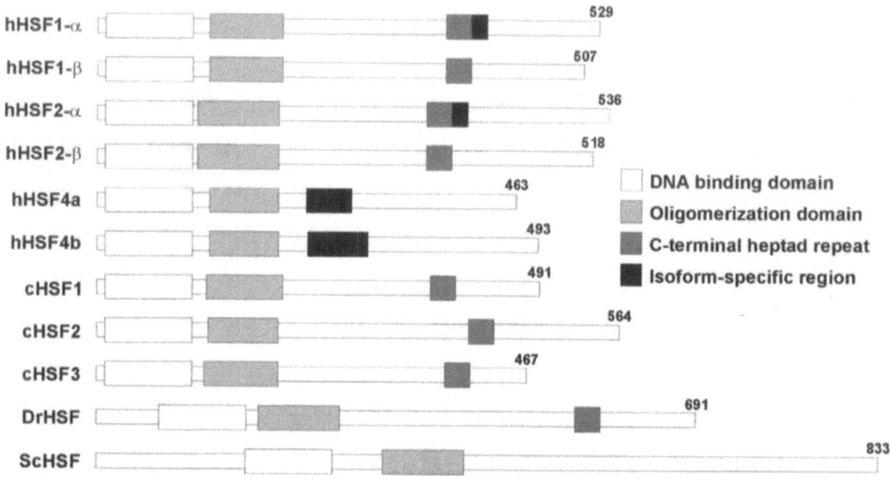

Fig. 3. Common structural features of heat shock transcription factors. A schematic presentation of conserved domains of HSFs (h, human; c, chicken; Dr, fruit fly; Sc, budding yeast). All HSFs have a DNA-binding domain and N-terminal heptad repeat involved in trimerization. HSF4 and ScHSF lack the C-terminal heptad repeat, which negatively regulates trimerization. Mammalian HSF1, HSF2, and HSF4 are expressed as two alternatively spliced isoforms. The number of amino acids in each HSF molecule is indicated.

transcriptional regulation of the heat shock response (Nakai et al. 1995; McMillan et al. 1998; Tanabe et al. 1998; Xiao et al. 1999). It has been postulated that the HSFs that are refractory to stress stimuli would regulate transcription under other circumstances, such as during development and differentiation, and possibly have target genes distinct from the classical heat shock genes (for review see Pirkkala et al. 2001). Recent studies, however, indicate that there is more interplay between the classical and non-stress-responsive HSFs than originally anticipated (Alastalo et al. 2003; He et al. 2003). In addition to the transcriptional activators of the HSF family, the HSF4a isoform lacks transcriptional activity and has been suggested to function as a repressor of transcription (Nakai et al. 1997; Tanabe et al. 1999).

Although heat shock transcription in plants is under the control of HSFs, the regulatory mechanisms differ substantially from those in animal species. A striking property of plant HSFs (Hsfs) is their great number and complexity. For example, in *Arabidopsis thaliana*, 21 genes coding for Hsfs have been found (for review see Nover et al. 2001). Based on structural features, these Hsfs can be assigned to three classes (A, B, and C) and 14 groups. Although the functional significance of the multitude of plant Hsfs is not fully understood, studies on tomato, which to date provides the best characterized Hsf system, have revealed novel regulatory principles that diverge between plant and animal HSFs. Unlike the stress-responsive HSFs in animals, which are mainly regulated post-translationally, certain tomato Hsfs display stress-inducible expression (for review see Nover et al. 2001). In addition, Hsfs are regulated hierarchically, as tomato

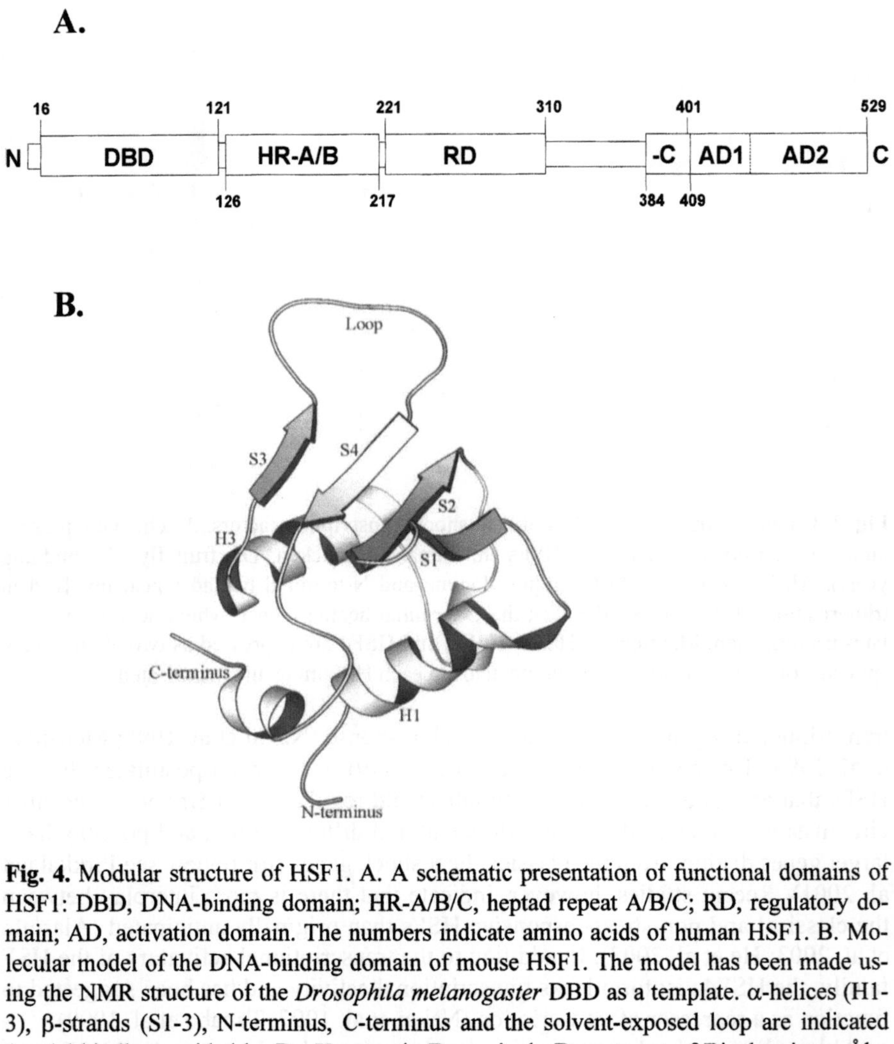

Fig. 4. Modular structure of HSF1. A. A schematic presentation of functional domains of HSF1: DBD, DNA-binding domain; HR-A/B/C, heptad repeat A/B/C; RD, regulatory domain; AD, activation domain. The numbers indicate amino acids of human HSF1. B. Molecular model of the DNA-binding domain of mouse HSF1. The model has been made using the NMR structure of the *Drosophila melanogaster* DBD as a template. α-helices (H1-3), β-strands (S1-3), N-terminus, C-terminus and the solvent-exposed loop are indicated (model kindly provided by Dr. Konstantin Denessiouk, Department of Biochemistry, Åbo Akademi University, Turku, Finland).

HsfA1 has been shown to be the master regulator of the heat shock response, mediating the stress-inducible expression of Hsps as well as that of HsfA2 and HsfB1 (Mishra et al. 2002). In contrast to the class A Hsfs, the class B and C Hsfs have not been shown to act as transcriptional activators *per se*. Recently, however, tomato HsfB1 was reported to enhance the promoter recognition and transcriptional competence of class A Hsfs, thereby providing evidence for a cooperative function between two Hsf classes in plants (Bharti et al. 2004).

5 HSF1 – the prototypical heat shock transcription factor

Mammalian HSF1 is the structural and functional ortholog of the single HSF in yeast and fruit fly, and cannot be replaced by any other member of the HSF family (McMillan et al. 1998; Xiao et al. 1999). HSF1 is a strong transcriptional activator with functional domains distributed in a modular manner, which is typical for most transcription factors. Inactive HSF1 exists as a monomer both in the nucleus and cytoplasm. Upon activation, HSF1 forms a trimer, concentrates in the nucleus and undergoes post-translational modifications, including phosphorylation and sumoylation. As HSF1 regulates a plethora of target genes and is activated by numerous stimuli, its activity is regulated by multiple mechanisms.

5.1 Modular structure of HSF1

The functional domains of HSF1 have been thoroughly mapped by mutational analyses. Deletion analysis of *S. cerevisiae* HSF defined the DNA-binding domain (DBD) (Wiederrecht et al. 1988), which was subsequently revealed to be the most conserved domain across the HSF family (Fig. 4A; for review see Pirkkala et al. 2001). The crystal structure of the DBD of *Kluyveromyces lactis* HSF (Harrison et al. 1994) and the solution structures of the DBDs of *K. lactis* and *D. melanogaster* (Damberger et al. 1994; Vuister et al. 1994a) place the HSF DBD in the group of helix-turn-helix motifs, consisting of three alpha helices, four beta strands, and a solvent-exposed loop (Fig. 4B). The DBD recognizes the inverted nGAAn repeats of the HSE through contacts in the major groove mainly by helix 3, which also makes several contacts with the phosphate backbone (Vuister et al. 1994b; Littlefield and Nelson 1999). As expected, helix 3 is rich in positively charged residues and comprises the most invariant region in the HSF sequence. Unlike in many other DNA-binding domains with a similar structure, the DBD loop of HSFs does not interact with DNA, but rather forms an interface between two adjacent HSFs, and therefore, possibly contributes to HSF trimerization or cooperativity between the trimers (Littlefield and Nelson 1999).

The trimerization domain is composed of three arrays of hydrophobic heptad repeats (HR-A/B) and resides C-terminal to the DBD (Sorger and Nelson 1989). Through hydrophobic interactions, the heptad repeats form a helical coiled-coil structure known as the leucine zipper. Spontaneous HSF leucine zipper trimerization is suppressed by another hydrophobic heptad repeat (HR-C) located near the C-terminus (Rabindran et al. 1993; Zuo et al. 1994). Since a mutation in HR-C leads to constitutive trimerization and DNA-binding activity of HSF1, HR-C is likely to form an intramolecular coiled-coil with HR-A/B, thereby preventing the intermolecular coiled-coil formation. Interestingly, in *S. cerevisiae* and *K. lactis* HSFs, which display constitutive DNA-binding activity, the HR-C is poorly conserved and its disruption does not affect the oligomerization status (Chen et al. 1993).

The bipartite activation domain of mammalian HSF1 (AD1, AD2) is located at the C-terminus of the molecule, and it contains several acidic and hydrophobic residues (Green et al. 1995; Shi et al. 1995). In the absence of stress stimuli, the function of the ADs is repressed by the regulatory domain (RD), which is not yet structurally solved but is localized between amino acids 220-310 in human HSF1 (Green et al. 1995; Newton et al. 1996). Although the RD has not been shown to affect DNA binding or subcellular localization of HSF1, it negatively regulates HSF1 transactivation capacity in reporter assays. This repressive activity of the RD is relieved upon stress, providing direct evidence for multi-level regulation of HSF1 activity (Green et al. 1995; Newton et al. 1996).

5.2 Rapid and transient HSF1 DNA-binding activity in response to stress

As supported by several studies, activation of HSF/HSF1 is a multi-step process, where the induction of DNA binding and the increase in transcriptional activity can be separated (for review see Holmberg et al. 2002). In fission yeast, fruit fly and vertebrates, the DNA-binding activity of HSF/HSF1 is under stringent control (Wu 1984; Mosser et al. 1988; Gallo et al. 1991; Sarge et al. 1993; Westwood and Wu 1993). While dimerization is known to be involved in activation of most transcription factors, HSF1 is activated by a transition from monomer to trimer, which causes a 10^4-fold increase in HSE-binding activity (for review see Wu 1995). An exception to this rule is the budding yeast HSF that has been indicated to exist constitutively as a trimer capable of binding to the HSEs also under non-stressful conditions (Sorger et al. 1987; Jakobsen and Pelham 1988).

HSF1 DNA binding occurs very rapidly. Kinetic analyses in *D. melanogaster* have shown that HSF is recruited to the *hsp70* promoter within seconds following heat shock, and HSF levels are saturated on the promoter in less than two minutes (Boehm et al. 2003). Therefore, it is obvious that HSF1 DNA binding needs to be regulated post-translationally. In cell-free systems, HSF1 acts as a direct sensor for certain stresses, as purified mammalian HSF1 and fruit fly HSF are able to acquire DNA-binding activity by undergoing homotrimerization in response to heat shock and H_2O_2 treatment (Goodson and Sarge 1995; Zhong et al. 1998). Recently, Ahn and Thiele (2003) demonstrated that the trimerization of mammalian HSF1 depends on two cysteine residues in the DNA-binding domain which form disulfide bonds in a redox-regulated manner. Mutation of the cysteines prevents HSF1 activation *in vivo,* suggesting that direct sensing of the intracellular redox status by intra- or intermolecular disulfide bonds can also influence the oligomerization status of HSF1.

5.3 Complex regulation of HSF1 transcriptional activity by post-translational modifications

Right after its cloning, yeast HSF was reported to display retarded electrophoretic mobility due to stress-inducible phosphorylation (Sorger and Pelham 1988). Similarly, mammalian HSF1 also undergoes inducible phosphorylation (hyperphosphorylation) upon activation by stress (Larson et al. 1988; Sarge et al. 1993). Several lines of evidence suggest that phosphorylation is involved in determining the stress-inducible transactivation capacity of HSF/HSF1. For example, *S. cerevisiae* HSF appears to display constitutive DNA-binding activity and the increased phosphorylation coincides with its transcriptional activation (Sorger et al. 1987; Sorger and Pelham 1988; Sorger 1990). Moreover, the inducible phosphorylation of mammalian HSF1 correlates temporally with *hsp70* transcription rather than with HSF1 DNA binding (Kline and Morimoto 1997), and the DNA binding of HSF1 can be induced by sodium salicylate, which neither activates *hsp70* transcription nor increases HSF1 phosphorylation (Jurivich et al. 1992, 1995). The heat shock response can also be modulated by pharmacological compounds which inhibit or stimulate phosphorylation-mediated signaling pathways (Holmberg et al. 1997; Xia and Voellmy 1997).

Phosphoamino acid analyses have revealed that mammalian HSF1 is predominantly phosphorylated at serine residues, albeit low levels of threonine phosphorylation have been detected (Chu et al. 1996; Cotto et al. 1996; Kline and Morimoto 1997; Xia and Voellmy 1997; Holmberg et al. 2001). Phosphopeptide mapping experiments have shown several tryptic phosphopeptides, suggesting that more than ten residues are phosphorylated altogether (Chu et al. 1996; Kline and Morimoto 1997; Xia et al. 1998). The initial efforts to map the critical amino acids involved in HSF1 activation led to the characterization of three serine residues, S303, S307, and S363, which all are proline-directed sites and prominently phosphorylated both in the presence and absence of stress (Chu et al. 1996; Knauf et al. 1996; Kline and Morimoto 1997). Surprisingly, phosphorylation of these sites is not needed for HSF1 activation. In fact, serines 303 and 307 display a repressive activity under non-stressful conditions, which is overridden by stress-induced activation of HSF1 (Chu et al. 1996; Knauf et al. 1996; Kline and Morimoto 1997). A recent study by Wang and coworkers (2003) has linked HSF1 phosphorylation to regulation of nucleocytoplasmic localization, showing evidence that HSF1 binds to the scaffold protein 14-3-3ε, which negatively regulates HSF1 by anchoring it in the cytoplasm. 14-3-3ε appears to bind specifically to phosphorylated serines 303 and 307, suggesting that the repressive effect reported for S303/307 phosphorylation could be partially explained by cytoplasmic sequestration.

In contrast to the repressive phosphorylation of S303, 307, and 363, stress-inducible phosphorylation of serine 230 in the N-terminal part of the regulatory domain has been shown to positively contribute to HSF1-mediated transcription (Holmberg et al. 2001). Therefore, it is plausible that phosphorylation regulates HSF1 activity both negatively and positively, and perhaps the ratio between repressive and activating phosphorylation determines the magnitude of HSF1-mediated transcription (Fig. 5; for review see Holmberg et al. 2002). As many of the phosphorylation sites remain to be characterized and target gene-specific regulation of HSF1 activity has not been systematically analyzed, our understanding of phosphorylation-mediated regulation of HSF1 is far from conclusive. Regarding the numerous HSF target genes (see Section 9.1), a recent report on budding yeast HSF elegantly demonstrates that phosphorylation can indeed regulate HSF activity in a target gene-specific manner (Hashikawa and Sakurai 2004).

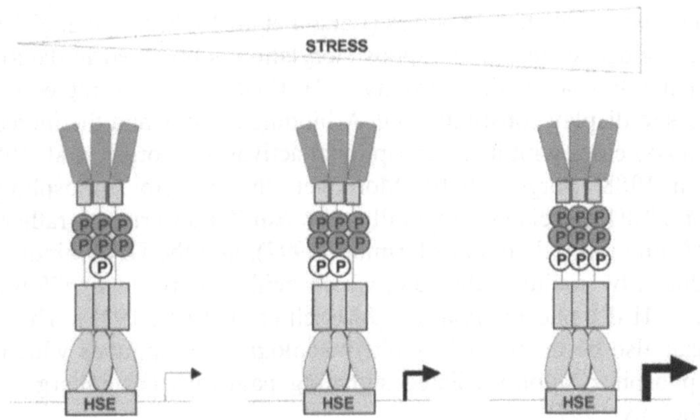

Fig. 5. Regulation of HSF1 activity by phosphorylation. HSF1 is a phosphoprotein containing both negatively (gray) and positively (white) acting phosphorylation sites. Upon stress, phosphorylation of the positively acting sites is increased, which potentiates the transactivation capacity of HSF1.

Several kinases have been suggested to target the known serine residues on HSF1 (for review see Holmberg et al. 2002). As serines 303, 307, and 363 are followed by a proline residue, MAP kinases are putative candidates, and indeed, MAPK/ERK has been reported to phosphorylate serines 303 and 307 or only serine 307, after which serine 303 can be phosphorylated by glycogen synthase kinase-3 (GSK-3) (Chu et al. 1996, 1998; Knauf et al. 1996; Kline and Morimoto 1997). Serine 363 can be phosphorylated by PKC isoforms alpha and zeta as well as by JNK (Chu et al. 1998; Dai et al. 2000), and S230 serves as a target for Ca^{2+}/calmodulin-dependent kinase II (CaMK II) (Holmberg et al. 2001). Since most evidence for kinases involved in phosphorylating HSF1 originates from *in vitro* phosphorylation reactions and overexpression of active kinases, the physiological relevance of the suggested kinases needs to be established. To this end, experiments utilizing knockout cell lines, kinase knockdown by RNA interference or specific inhibitors are warranted to unravel the signaling pathways, including both kinases and phosphatases that are involved in the regulation of HSF1. Considering the multitude of HSF1 phosphorylation sites, it is plausible that different phosphorylation-mediated pathways regulate HSF1 depending on the stress stimuli. This hypothesis is supported by a recent report on *S. cerevisiae* HSF, which is activated by phosphorylation by Snfl protein kinase upon glucose starvation but not upon heat shock (Hahn and Thiele 2004).

In addition to phosphorylation, HSF1 undergoes stress-inducible SUMO-1 modification on the well-conserved lysine residue 298 within the regulatory domain (Hong et al. 2001; Hietakangas et al. 2003). Hong et al. (2001) indicated that sumoylation would be needed for acquisition of HSF1 DNA-binding activity, however this was not supported by a subsequent study (Hietakangas et al. 2003). Although the regulatory roles of sumoylation and phosphorylation are still enigmatic, these post-translational modifications seem to occur in concert, i.e. one modification can affect another. According to Hietakangas and coworkers (2003), phosphorylation of serine 303 is a prerequisite for sumoylation of lysine 298, and inhibition of S303 phosphorylation prevents transient colocalization of HSF1 and SUMO-1 in nuclear stress bodies (see Section 9.3). The concurrent regulation of HSF1 activity by several post-translational modifications is likely to be essential for the fine-tuning of heat shock transcription in response to specific stimuli.

5.4 Interactions between HSF1 and the transcriptional machinery

The molecular basis for the inducible transcription of heat shock genes has been under intense investigation, and the *hsp70* promoter has provided one of the best models for studies of basic mechanisms of transcriptional regulation, especially elongation by RNA polymerase II (Pol II) (for review see Lis 1998). As originally shown in *D. melanogaster*, the uninduced *hsp70* promoter is constitutively occupied by Pol II, which is transcriptionally engaged but paused after transcription of a short stretch of RNA containing 21-35 nucleotides (Rougvie and Lis 1988; Rasmussen and Lis 1993; Brown et al. 1996). The C-terminal domain (CTD) of the paused Pol II is hypophosphorylated, and must be hyperphosphorylated before Pol II is actively involved in transcriptional elongation (for review see Svejstrup 2004). Phosphorylation of Ser2 in the CTD by the Cdk9-CyclinT1 heterocomplex called P-TEFb appears to be crucial for heat shock gene regulation, since P-TEFb is rapidly recruited to the heat shock loci upon stress induction, and its artificial recruitment to the *hsp70* promoter is sufficient to activate transcription in the absence of stress (Lis et al. 2000). However, the effect of P-TEFb-mediated CTD phosphorylation may be more complex than originally anticipated. Using flavopiridol, a specific P-TEFb kinase inhibitor, it was recently demonstrated that inhibition of Ser2 phosphorylation leads to failure of the 3' end processing of *hsp70* and *hsp26* mRNAs (Ni et al. 2004). The unprocessed mRNA is rapidly degraded, possibly by the exosome, which is also recruited to the Pol II elongation complex (Andrulis et al. 2002). These findings emphasize the tight coordination of transcription and mRNA processing that has emerged as a central regulatory feature of gene expression (for review see Proudfoot et al. 2002).

Only a few examples of direct connections between HSF1 and the components of chromatin-modifying complexes and transcriptional coactivators have been reported (Fig. 6). The SWI/SNF chromatin-remodeling complex interacts with the activation domain of human HSF1, and appears to mediate the changes in chromatin architecture necessary for proper Pol II elongation (Sullivan et al. 2001; Corey et al. 2003). In fruit fly, HSF has been shown to recruit the Mediator coactivator complex to the heat shock loci by directly interacting with the dTRAP80 subunit (Park et al. 2001). Moreover, the coactivator ASC-2 was recently shown to interact with the activation domain of human HSF1 and promote HSF1-mediated transcription (Hong et al. 2004). It remains to be elucidated how the recruitment of these coactivators is linked to the maturation of Pol II into the elongating complex. In the context of transcription-coupled 3' end processing of mRNA, the recent finding of symplekin, which is known to form a complex with polyadenylation factors (Takagaki and Manley 2000), as a partner protein for HSF1 is intriguing

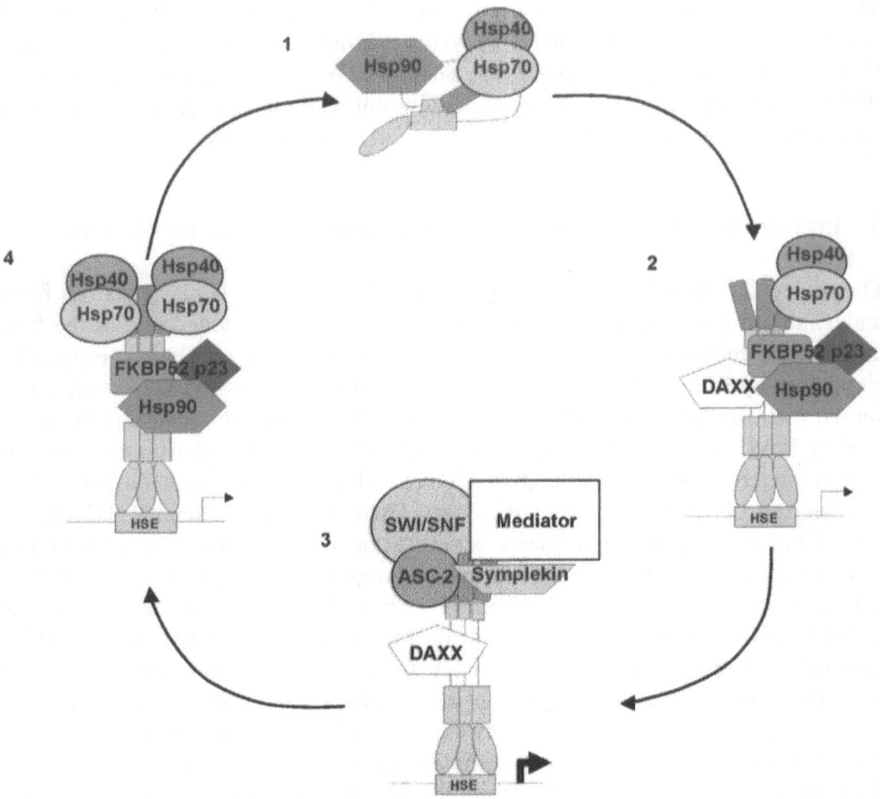

Fig. 6. Complex regulation of HSF1 activity by interacting proteins during various phases of the heat shock response. A hypothetical presentation of proteins and protein complexes interacting with HSF1 under non-stressful conditions (1), at the onset of transcription (2), at maximal transcription (3), and at the attenuation phase (4) of the heat shock response. For the abbreviations see text.

(Xing et al. 2004). Inhibition of the interaction between HSF1 and symplekin strongly reduced *hsp70* mRNA polyadenylation, indicating that HSF1 could be involved in cotranscriptional mRNA processing.

5.5 Negative and positive regulation of HSF1 by interacting proteins

HSF1 is an extremely potent transactivator and its constitutive activity would be harmful for the cells and the organism as a whole. Therefore, HSF1 is under negative control by interacting proteins, including its own targets such as Hsp70 and Hsp90 (Fig. 6; for review see Morimoto 1998). Several studies indicate that Hsp90 binds to HSF1. *In vitro* experiments have shown that immunodepletion of Hsp90 from cell lysates or inhibition of Hsp90 chaperone activity by geldanamycin leads to activation of HSF1 DNA binding (Zou et al. 1998). Geldanamycin also activates HSF1 *in vivo* and Hsp90 binding to HSF1 is reduced when HSF1 is activated by stress (Zou et al. 1998). The nature of the HSF1-interacting Hsp90 complex appears to depend on the oligomerization status of HSF1; while Hsp90 alone associates with monomeric HSF1, the activation domain of the HSF1 trimer interacts with the Hsp90-immunophilin (FKBP52)-p23 complex, suggesting that different Hsp90 chaperone complexes can regulate both the stress-sensing and the attenuation of HSF1 (Guo et al. 2001). The association of HSF1 with the Hsp90-immunophilin-p23 complex has also been demonstrated in *Xenopus laevis* oocytes (Ali et al. 1998; Bharadwaj et al. 1999). Microinjection of neutralizing antibodies against Hsp90 and p23 into the oocytes is sufficient to activate HSF1, and immunotargeting of Hsp90 and its interacting partners (e.g. Hip, Hop, p23, FKBP51, and FKBP52) delays HSF1 attenuation, providing further evidence for the importance of the Hsp90 chaperone machinery in HSF1 regulation (Bharadwaj et al. 1999).

Hsp70 and its cochaperone Hdj1/Hsp40 have been shown to interact with the HSF1 trimers by binding to the activation domain and interfering with its function (Abravaya et al. 1992; Baler et al. 1992; Shi et al. 1998). Since high levels of Hsp70 accumulate as a result of HSF1 transcriptional activity, this interaction likely plays a role in negative feedback at the attenuation phase of the heat shock response (Fig. 6). In cells expressing constitutively high levels of Hsp70, HSF1, and Hsp70 form a complex under normal conditions, and the complex disassembles upon heat stress, possibly contributing to full HSF1 activity (Alastalo et al. 2003). When Hsp70 is overexpressed in the absence of stress, the heat-inducible DNA binding and phosphorylation of HSF1 are not affected, but its transcriptional capacity is severely impaired (Shi et al. 1998). This suggests that titration of Hsp70 away from HSF1 is not the initial stimulus triggering the stress-inducible trimerization of HSF1, but rather regulates the magnitude of transcription by masking the activation domain from the transcriptional machinery. In conclusion, the negative feedback by Hsps comprises a conserved regulatory mechanism of heat shock gene transcription both in prokaryotes (see Section 2) and eukaryotes.

In contrast to Hsps, CHIP, another cochaperone of Hsp70, was recently reported to activate HSF1-mediated transcription (Dai et al. 2003). CHIP associates in the same complex with HSF1, probably via Hsp70, but the exact mechanism of its activating function is not known. Interestingly, CHIP-deficient mice are hypersensitive to environmental stress, displaying apoptosis in multiple organs when exposed to stress (Dai et al. 2003). Moreover, DAXX, a ubiquitous nuclear protein with anti- and pro-apoptotic properties, was identified as a mediator of HSF1 activation by directly interacting with trimeric HSF1, and thereby counteracting the repressive function of Hsps (Boellman et al. 2004).

As presented in Figure 6, HSF1 interacts with multiple protein complexes, including components of the transcriptional and mRNA processing machinery as well as molecular chaperones, during its activation-inactivation cycle. These interactions are highly dynamic and more transient than those occurring in the assembly and disassembly of intracellular hormone receptors and their accessory proteins (for review see Morimoto 2002). The multitude of HSF1 partner proteins and the transient nature of interactions may reflect the need for extraordinarily strict and rapid control of this potent transactivator under life-threatening conditions.

5.6 Role of HSF1 in normal development and physiology

In the past two decades, many molecular aspects of the HSF-mediated heat shock gene expression have been examined using yeast and fruit flies as well as mammalian and avian cell cultures. More recently, *in vivo* studies in complex multicellular organisms, employing targeted gene disruption or transgenic approaches, have started to unravel the numerous physiological functions of HSFs, some of which seem unrelated to Hsps. The versatile physiological roles of vertebrate HSFs are not surprising when considering that in both budding yeast and fission yeast, HSF is essential for cell growth and viability even in the absence of stress (Sorger and Pelham 1988; Wiederrecht et al. 1988; Gallo et al. 1993). In *D. melanogaster*, HSF is essential for oogenesis and early larval development independently of heat shock gene expression (Jedlicka et al. 1997). In adult flies, HSF is dispensable for cell growth and viability but essential for survival under extreme heat stress (Jedlicka et al. 1997).

In contrast to fruit fly HSF, mouse HSF1 is not required for embryonic development, although *hsf1*$^{-/-}$ mice display high prenatal lethality due to defects in the chorioallantoic placenta (Xiao et al. 1999). Disruption of *hsf1* also causes growth retardation and female infertility (Xiao et al. 1999). The adult HSF1-deficient mice and embryonic fibroblasts derived from these animals (*hsf1*$^{-/-}$ MEFs) display neither classical heat shock response nor thermotolerance, demonstrating that the classical functions of HSF1 cannot be replaced by any other HSF (McMillan et al. 1998; Xiao et al. 1999). Consequently, *hsf1*$^{-/-}$ MEFs are extremely sensitive to heat stress-induced apoptosis (McMillan et al. 1998). Moreover, the adult mice also are more sensitive to stress, as they display increased mortality in response to the bacterial endotoxin lipopolysaccharide (LPS), corresponding to increased production of the proinflammatory cytokine TNF-α (Xiao et al. 1999).

The infertility of *hsf1*$^{-/-}$ females is due to the requirement of HSF1 as a maternal factor (Christians et al. 2000). The *hsf1*$^{-/-}$ females have normal ovaries and reproductive tracts, and the embryos produced by *hsf1*$^{-/-}$ females can initiate early development but are arrested before the blastocyst stage, even when the *hsf1*$^{-/-}$ females have been mated with wild type males (Christians et al. 2000). It is notable that in the defective embryos derived from *hsf1*$^{-/-}$ females, the Hsp expression in the absence of stress is not markedly affected, suggesting that HSF1 mediates these physiological effects independently of heat shock gene expression (Xiao et al. 1999; Christians et al. 2000). In contrast, in the heart of adult HSF1-deficient mice, the basal expression of Hsp25, αB-crystallin, and Hsp70 is downregulated (Yan et al. 2002). Furthermore, oxidative stress, measured as a decreased glutathione/glutathione disulfate ratio and an increase in oxidative damage of mitochondrial proteins, was detected, which is likely to be due to the reduced Hsp levels.

HSF1 also affects the reproductive capacity of males. Expression of constitutively active HSF1 induces apoptosis of pachytene spermatocytes in testis (Nakai et al. 2000; Izu et al. 2004). As the activation of HSF1 has a markedly lower temperature threshold in testis than in other tissues, HSF1 may function as a quality control factor at certain stages of spermatogenesis (Sarge et al. 1995; Nakai et al. 2000; Izu et al. 2004). The pro-apoptotic function of HSF1 in spermatocytes might partially explain why thermal insult of the testis reduces fertility (Mieusset and Bujan 1995).

6 HSF2 – a cooperative modulator of HSF1?

Simultaneously with mammalian HSF1, another HSF family member, HSF2, was discovered (Sarge et al. 1991; Schuetz et al. 1991). Purified HSF2 protein was shown to possess affinity towards HSEs and stimulate HSE-dependent transcription *in vitro*, but in contrast to HSF1, it could not be activated by heat shock either *in vitro* or *in vivo* (Sarge et al. 1991; Sistonen et al. 1992). Electromobility shift analyses (EMSA) revealed constitutive HSF2 DNA-binding activities in mouse embryonal carcinoma cells, testis, blastocysts and post-implantation stage embryos, and inducible DNA binding in human K562 erythroleukemia cells undergoing hemin-mediated erythroid differentiation (Sistonen et al. 1992; Mezger et al. 1994; Murphy et al. 1994; Sarge et al. 1994; Rallu et al. 1997). Based on these observations, HSF2 was considered a development- and differentiation-related transcription factor clearly distinct from the *bona fide* stress-responsive HSF1. However, recent studies, as discussed below, are likely to change this view by providing evidence for HSF2 acting in concert with HSF1, possibly as a context-dependent modulator of HSF1-mediated transcription.

Although HSF2 shares only 35% of amino acid homology with HSF1, it contains the domains characteristic to HSFs, such as the conserved DNA-binding domain DBD and the N- and C-terminal heptad repeats HR-A/B and HR-C (see Fig. 3). Based on *in vitro* and *in vivo* DNA footprinting experiments, HSF2 binds to HSEs in a different manner than HSF1 (Sistonen et al. 1992; Kroeger et al. 1993). In addition, recombinant HSF1 and HSF2 proteins display specific binding properties and cooperativity between trimers (Kroeger and Morimoto 1994), suggesting that HSF2 might regulate a distinct set of target genes with slightly different HSEs. Experiments performed with HSF1/HSF2 chimeras have revealed that the DNA-binding specificity is determined by the loop within the DBD (Ahn et al. 2001; see Fig. 4B). Since HSF2 containing the loop of HSF1 adopts the capacity to be activated by heat stress, it is plausible that both the DNA-binding specificity and stress-sensing capacity of HSFs are determined by the same structure (Ahn et al. 2001). The transactivating capacity of HSF2 is weak when compared to HSF1, and the transcriptional activation domain is dispersed; in addition to the major AD in the C-terminus, another AD between HR-A/B and HR-C has been proposed (Yoshima et al. 1998). The function of the ADs is negatively regulated by adjacent regions, but HSF2 transactivation capacity has not been shown to respond to any stimulatory signal (Yoshima et al. 1998).

The regulation of HSF2 differs fundamentally from that of HSF1. No phosphorylation of HSF2 has been reported and its expression levels vary greatly in different tissues and cell types. As analyzed by EMSA, the expression level of HSF2 correlates with its ability to bind HSEs in differentiating K562 cells and during mouse embryogenesis (Sistonen et al. 1992; Rallu et al. 1997; Pirkkala et al. 1999). The hemin-induced upregulation of HSF2 is mainly due to increased mRNA stability rather than altered transcription or protein stability (Pirkkala et al. 1999; Nykänen et al. 2001; Päivi Östling and Johanna Ahlskog, unpublished observations). Unlike HSF1, which is a stable protein, HSF2 has a short half-life (approx. 1 hour), and it can be stabilized by proteasome inhibitors (Kawazoe et al. 1998; Mathew et al. 1998; Pirkkala et al. 2000), suggesting that HSF2 is subjected to ubiquitin-mediated proteasomal degradation. Although gel filtration and sedimentation analyses have indicated that HSF2 undergoes a dimer-to-trimer transition in hemin-treated K562 cells (Sistonen et al. 1994), it is not known whether HSF2 DNA-binding activity is regulated by any mechanism other than elevated expression. A putative regulatory mechanism is SUMO-1 modification, which occurs within the DBD loop and has been shown to induce DNA-binding activity *in vitro* (Goodson et al. 2001). Whether sumoylation is involved in the regulation of HSF2 DNA-binding activity *in vivo* remains to be elucidated.

HSF2 exists as two isoforms, HSF2-α and HSF2-β, which are formed by alternative splicing of exon 11 (Fiorenza et al. 1995; Goodson et al. 1995; see Fig. 3A). The isoforms may have distinct biological activities, since they are differentially expressed in adult tissues, during embryogenesis and spermatogenesis, and in differentiating K562 cells (Goodson et al. 1995; Leppä et al. 1997; Alastalo et al. 1998). Moreover, HSF2-α displays a more prominent transactivation capacity in reporter assays (Goodson et al. 1995), whereas overexpression of HSF2-β interferes with the hemin-induced activation of HSF2 in K562 cells (Leppä et al. 1997). Further studies will be needed to determine whether the HSF2 isoforms possess distinct regulatory functions *in vivo*.

For more than a decade, HSF2 was considered functionally unrelated to HSF1-mediated regulation of heat shock genes. Although HSF2 is active and abundantly expressed during the post-implantation phase of mouse embryogenesis and at certain stages of spermatogenesis and heart development, HSF2 activity and the expression pattern of Hsps do not coincide (Rallu et al. 1997; Alastalo et al. 1998; Eriksson et al. 2000). Similarly, although proteasome inhibitors activate HSF2 DNA binding (Kawazoe et al. 1998; Mathew et al. 1998), experiments with HSF1- and HSF2-deficient MEFs have demonstrated that the heat shock response induced by proteasome inhibition is regulated in an HSF1-dependent manner (Pirkkala et al. 2000; McMillan et al. 2002). Recently, however, HSF2 was shown to interact with HSF1 (Alastalo et al. 2003), and in reporter assays, HSF2-α was able to stimulate HSF1-induced transcription upon exposure to stress (He et al. 2003). The hypothesis of cooperation between HSF1 and HSF2 is supported by recent findings in our laboratory suggesting that HSF2 is recruited to the *hsp70* promoter only in the presence of HSF1 (Pia Roos-Mattjus and Päivi Östling, unpublished results). In addition to the results obtained with cultured cells, further evidence for the cooperative function of HSF1 and HSF2 is provided by an *in vivo* study, showing that disruption of both HSF1 and HSF2 leads to more dramatic defects in mouse spermatogenesis and male fertility than disruption of either HSF1 or HSF2 alone (Wang et al. 2004).

Since HSF2 *per se* is refractory to heat stress (Sarge et al. 1991; Sistonen et al. 1992), its regulatory role *in vivo* has remained obscure. However, studies on HSF2-deficient mice have started to shed light into the physiological functions of HSF2. *hsf2*$^{-/-}$ mice are viable, but they display abnormalities in brain development and gametogenesis (Kallio et al. 2002). HSF2 is abundantly expressed in the ventricular zone of embryonic and adult mouse brain, and in HSF2-deficient mice, the lateral and third ventricles are clearly enlarged. In *hsf2*$^{-/-}$ males, the testes are smaller, they contain disrupted semineferous tubules, and meiosis of the spermatocytes is disturbed, leading to increased apoptotic cell death. *hsf2*$^{-/-}$ females are hypofertile, produce abnormal eggs, and display hemorrhagic cystic follicles as well as a reduction in the total number of follicles. The penetrance of the effects reported by Kallio and coworkers (2002) might depend on the genetic background of mice; another group reported similar effects (Wang et al. 2003), whereas a third independent study showed no obvious phenotype (McMillan et al. 2002). The primary characterization of HSF2-deficient mice provides a good platform for more detailed analyses of the physiological roles for HSF2, and tissue-specific microarray analyses are likely to gain valuable information about the *in vivo* targets of HSF2-regulated transcription. Another key question is to decipher how the regulatory mechanisms, such as regulation of mRNA stability and SUMO-1 modification that have been established in cell cultures, function *in vivo* to control HSF2 expression and activity.

7 HSF3 – an avian-specific regulator of heat shock genes

Although the function of HSF1 as the main regulator of the stress-induced heat shock gene transcription is well conserved among vertebrates, avian species make an interesting exception. In birds, the heat shock response is controlled by HSF1 and HSF3, the latter being the main regulator of heat shock genes. Soon after the cloning of chicken HSF3 (Nakai and Morimoto 1993), it was shown to be responsive to heat stress (Nakai et al. 1995). Disruption of *hsf3* in chicken DT40 cells caused remarkable inhibition in stress-induced heat shock gene expression, providing compelling evidence for a dominant role of HSF3 in avian heat shock-induced transcription. Recently, Inouye and coworkers (2003) demonstrated that chicken HSF1 has little potential to activate transcription of heat shock genes, because of a unique alanine-rich stretch present in the N-terminus of cHSF1. Nevertheless, cHSF1 is not fully dispensable in the heat shock response, since it prominently protects cells against stress-induced cell death. Therefore, it appears that the two well-established functions of mammalian HSF1 are taken care of by two different members of the avian HSF family, raising an important question about the HSF-regulated anti-apoptotic target genes involved in suppression of heat stress-induced apoptosis.

8 HSF4 – regulator of eye development

The newest member of the HSF family is HSF4, which was originally shown to lack transactivating potential (Nakai et al. 1997). Subsequently, two alternatively spliced isoforms of HSF4 were discovered; HSF4a functions as a repressor of basal heat shock gene expression, whereas HSF4b is capable of activating transcription (Tanabe et al. 1999). HSF4 is mainly expressed in brain and lung, where HSF4b is found as a predominant isoform (Tanabe et al. 1999). Although the functions of HSF4 are poorly understood, the repressive HSF4a isoform has been shown to interact with TFIIF of the basal transcriptional machinery, which provides a potential mechanistic explanation to the repressive activity of HSF4a (Frejtag et al. 2001). Interestingly, mutations in the DNA-binding domain of HSF4 have been reported to be associated with autosomal dominant lamellar and Marner cataract (Bu et al. 2002). The role for HSF4 in regulating eye development is further supported by the results from HSF4-deficient mice, which display a cataract-like phenotype with abnormal lens fiber cells (Fujimoto et al. 2004). HSF4 seems to have a capacity to activate and repress transcription in the lens; while HSF4 is needed for expression of gamma-crystallins in lens fiber cells, it suppresses the expression of fibroblast growth factors in lens epithelial cells (Fujimoto et al. 2004).

9 Novel functions for HSFs in transcription

Despite the fact that many of the basic principles underlying the transcriptional regulation of the heat shock response have already been revealed, active research in the field has recently led to novel discoveries that have significantly broadened the initial concept of the heat shock response. In addition to providing a well-conserved cellular defense mechanism against stress and cytotoxic insults, the biology of heat shock transcription has many more ramifications than originally anticipated.

9.1 Broad repertoire of HSF target genes

The role of HSF1 as a transcriptional activator of classical heat shock genes has been expanded by observations that HSF1 would act also as a repressor, as has been suggested in the case of prointerleukin-1β and TNFα (Cahill et al. 1996; Singh et al. 2000, 2002; Xie et al. 2002). Regarding the versatile role of HSF1 in transcriptional regulation, recent advances in genome-wide gene expression analyses are of great interest. Hahn and coworkers (2004) identified almost 3% of the genomic loci in *S. cerevisiae* to be occupied by HSF upon heat stress. As expected, many of the target genes code for molecular chaperones known to function in the heat shock response, but this study indicates dozens of novel direct targets for HSF, including proteins involved in proteolysis, vesicular and small molecular transport, cell wall and cytoskeleton maintenance as well as carbohydrate and energy metabolism. Another surprising finding was that the binding of HSF to a large subset of its target promoters was dramatically enhanced by heat stress, raising a question of the proportion of budding yeast HSF molecules that exist as constitutively DNA-bound trimers. In addition to the promoters containing a consensus HSE, Hahn and coworkers (2004) observed binding of HSF in a heat shock-dependent manner to promoters that lack canonical HSEs, suggesting that the HSEs are not the sole determinant of whether a promoter can be occupied and activated by HSF. However, it remains to be shown how HSF recognizes and binds to the non-consensus sites.

Genome-wide analysis performed by Murray and coworkers (2004) revealed that in contrast to yeast cells (Gasch et a. 2000; Causton et al. 2001; Chen et al. 2003), mammalian cells have surprisingly specific responses to different stresses, such as heat shock, oxidative stress, and ER stress. In addition to gene products involved in protein folding, heat shock-induced genes include genes involved in post-translational modifications of proteins, protein degradation, signal transduction, membrane transport and metabolic processes, suggesting that in mammals, as in yeast, the heat shock response regulates genes involved in numerous other processes besides protein folding. As expected, many gene products important for basal metabolism or cell growth were repressed during heat shock (Murray et al. 2004). Interestingly, cells from different origins seem to respond to heat stress differently, as G2/M-specific genes were shown to be upregulated in primary fibroblasts but not in HeLa or K562 cells, indicating that perhaps primary cells, unlike cancer cells, are arrested at G2/M upon heat stress (Murray et al. 2004).

A more focused effort towards understanding the role of HSF1 as a global regulator of the heat shock response was undertaken by Trinklein and coworkers (2004). HSE-containing promoter microarrays were generated based on promoter databases, and the presence of HSF1 in specific promoters was examined by chromatin immunoprecipitation and compared with microarray expression analyses. Based on this study, HSF1 appears to bind to many, but not all, promoters containing potential HSEs, and the most conserved position within the nGAAn motif of HSE is the 'G', confirming the previous *in vivo* footprinting analyses of human *hsp70* promoter (Abravaya et al. 1991a; Sistonen et al. 1992). Furthermore, binding of HSF1 to a promoter did not always lead to transcriptional activation of the target gene, and in some cases even transcriptional repression was detected (Trinklein et al. 2004). Microarray expression analyses with *hsf1*$^{-/-}$ MEFs revealed that although heat-induced expression of most transcripts is regulated by HSF1, certain gene products are upregulated in the absence of HSF1. However, these genes are likely to be subjected to post-transcriptional regulation, since the promoter activities of HSF1-independent genes were not heat inducible (Trinklein et al. 2004).

9.2 Regulation of longevity by HSF1

The understanding of genes involved in the regulation of longevity is of great biological and clinical interest. Many genes shown to be involved in aging are pleiotropic and appear to be involved in maintaining resistance to environmental stresses and oxidation (for review see Finkel and Holbrook 2000; Söti and Csermely 2003). Furthermore, members of the insulin signaling pathway have emerged from genetic studies in *Caenorhabditis elegans* as critical regulators of life span (for review see Nelson and Padgett 2003).

Given the role of HSF1 as a major regulator of cellular stress resistance and the fact that heat shock gene expression is induced poorly in aged cells and animals (Choi et al. 1990; Fargnoli et al. 1990; Fawcett et al. 1994), it comes as no surprise that HSF1 regulates longevity in *C. elegans* (Garigan et al. 2002; Hsu et al. 2003; Morley and Morimoto 2004). A more surprising finding is that the function of HSF1 in maintaining longevity appears to be intertwined with the insulin signaling pathway (Hsu et al. 2003; Morley and Morimoto 2004). Thorough genetic analyses have revealed that HSF1 mutants display a reduction in life span similar to mutants of DAF-16, a FOXO transcription factor involved in the insulin signaling pathway of *C. elegans* (Hsu et al. 2003; Morley and Morimoto 2004). Although HSF1 and DAF-16 are known to function independently of each other, they appear to share a subset of target genes. One group of these common targets includes small heat shock proteins (sHsps), whose expression is impaired upon inactivation of DAF-16 or HSF1 (Hsu et al. 2003). Moreover, downregulation of sHsps by RNA interference reduced the life span of the worms, indicating that sHsps are likely to belong to the signaling cascade that mediates the effects of DAF-16 and HSF1 on aging and longevity. It is, however, worth noticing that downregulation of any single molecular chaperone reduced longevity considerably less than did downregulation of HSF1, emphasizing the complex genetic networks involved in the biology of aging (Morley and Morimoto 2004).

The actions of HSF1, DAF-16 and molecular chaperones influencing aging and longevity provide an interesting link to the development of numerous neurodegenerative disorders characterized by accumulation of protein aggregates, such as the polyglutamine (polyQ) expansions formed in Huntington's disease. Morley and coworkers (2002) have demonstrated that the reduced rate of aging in mutant *C. elegans* with extended life span is associated with delayed aggregation and toxicity of polyQ expansions. Furthermore, inactivation of DAF-16, HSF1 or sHsps accelerated the aggregation of polyQ-containing proteins (Hsu et al. 2003). These findings suggest that the regulation of aging as well as the detection and prevention of misfolded and aggregation-prone proteins likely share common determinants, the characterization of which will be important in order to understand the molecular mechanisms underlying age-dependent health problems, such as neurodegenerative diseases.

9.3 HSF1-dependent transcription of satellite III repeats previously considered as heterochromatin

Microscopic analyses have revealed that in primate but not rodent cells exposed to various stress stimuli, HSF1 is localized into irregular subnuclear compartments called nuclear stress granules or nuclear stress bodies (nSBs) (Fig. 7; Sarge et al. 1993; Cotto et al. 1997; Jolly et al. 1997; Holmberg et al. 2000; Denegri et al. 2002; for review see Biamonti 2004). The nSBs are structures distinct from other known subnuclear compartments such as PML bodies and Cajal bodies (Cotto et al. 1997). Although nSBs do not coincide with the sites of *hsp70* and *hsp90* transcription (Jolly et al. 1997), the observations that nSBs relocalize to the same sites upon repeated heat shocks and that the number of nSBs correlates with ploidy of the cell have indicated a physical association with chromosomal structures (Jolly et al. 1997, 1999). Interestingly, studies by Denegri and coworkers (2002) and Jolly and coworkers (2002) located the nucleation sites of nSBs to human chromosomes 9, 12, and 15, with the 9q12 region comprising the main site. This chromosomal region contains tandem repeats of nGAAn sequences, which are similar to but differently organized than the repeats within HSEs. Furthermore, *in vitro* binding experiments have shown that HSF1 can directly interact with 9q12 DNA (Jolly et al. 2002).

Although the mechanism by which the nSBs are formed is beginning to emerge, the biological functions of these structures have remained enigmatic. Recently, two laboratories demonstrated that upon formation of nSBs, active transcription of large and stable RNAs is initiated (Jolly et al. 2004; Rizzi et al. 2004; for review see Sandqvist and Sistonen 2004). This was surprising, as the 9q12 locus had previously been described as constitutive heterochromatin and transcriptionally silent. According to the recent studies, however, the chromatin region associated with nSBs has properties of euchromatin; firstly, the chromatin contains acetylated histones, secondly, Pol II and the histone acetyltransferase CREB-binding protein CBP are recruited to nSBs, and thirdly, the transcription appears to be driven by stress-activated HSF1 (Jolly et al. 2004; Rizzi et al. 2004). The transcripts are not exported to the cytoplasm but remain anchored to the chromosomal locus after recovery from heat shock and even throughout mitosis. They are therefore considered as non-coding RNAs, possibly involved in organization of the nucleus (for review see Biamonti 2004).

The HSF1-dependent transcription in nSBs most likely explains why RNA-processing factors, such as heterogeneous nuclear ribonucleoprotein HAP, RNA-binding protein Sam68, and splicing factors SRp30c, 9G8, and SF2/ASF, accumulate into nSBs upon heat shock (Weighardt et al. 1999; Denegri et al. 2001). Since the assembly and disassembly of RNA-binding proteins kinetically follow those of HSF1 and depend on RNA (Weighardt et al. 1999; Chiodi et al. 2000), it is plausible that HSF1 is indeed the master regulator of satellite III transcripts, to which RNA-processing factors are recruited. Further work will be needed to gain insight into the identity and functional relevance of these newly discovered transcripts. Moreover, it will be interesting to know whether the same regulatory mechanisms that are important for HSF1-mediated transcription of *hsp70* and other target

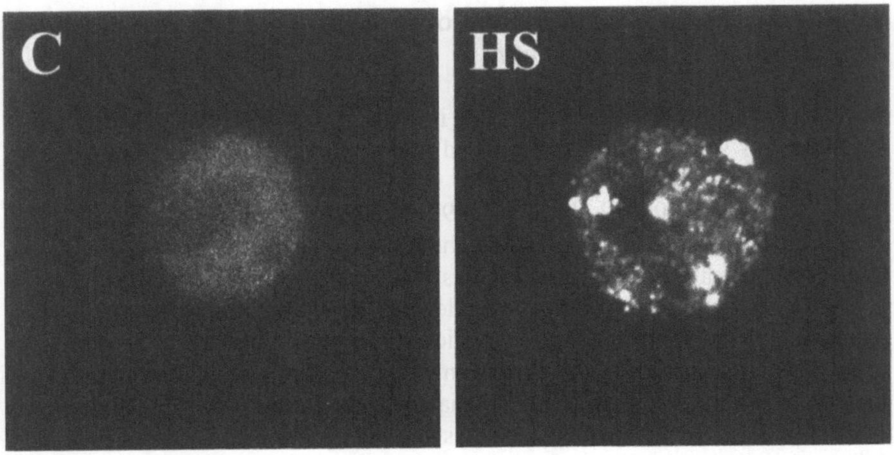

Fig. 7. Stress-induced localization of HSF1 to the nuclear stress bodies. Under normal conditions (C), HSF1 is dispersed throughout the cell, mainly in the nucleus. Upon exposure to heat shock (HS), HSF1 translocates to the punctate nuclear stress bodies (nSBs). Immunofluorescence images of HSF1 localization in HeLa cells were obtained using monoclonal anti-HSF1 antibodies (Ab-4, NeoMarkers; images kindly provided by Anton Sandqvist).

genes also apply to the transcription of satellite III repeats, and whether the localization of HSF1 to nSBs has any impact on the transcription of the classical heat shock genes.

10 Conclusions and perspectives

Transcriptional regulation of the heat shock response has been investigated for more than forty years, and despite the major discoveries discussed in this review, many fundamental questions remain open. The recently established genetic models, especially in *C. elegans* and mouse, have remarkably expanded our knowledge of HSF functions beyond the classical heat shock response, and the various physiological roles of HSFs are being unraveled at an accelerated pace. At the molecular level, post-translational modifications of HSFs, including phosphorylation and sumoylation, have proved to be complex but important regulatory mechanisms. Yet, many of the pathways mediating these modifications and regulating the activity of HSFs need to be explored. The interplay between different HSFs has yielded insights into cooperative functions among the members of the HSF family, directing future work towards more mechanistic analyses both at the molecular level and in the physiological context, where the various genetic model systems can be employed.

In contrast to the reductionistic view of the regulation of a few classical heat shock genes, genome-wide analyses have linked the heat shock response to completely unexpected cellular processes, raising questions of gene- or gene cluster-specific transcriptional regulation provided by HSFs. It will be of great interest to learn to what extent and by which mechanisms the HSFs interact with other transcriptional activators and repressors to mediate these multi-faceted regulatory programs. Localization of HSFs into the nuclear stress bodies has underlined the importance of subnuclear localization in the regulation of transcription. Furthermore, the surprising findings of HSF1-mediated transcription of satellite III repeats within the 9q12 locus and the recruitment of proteins involved in mRNA metabolism have raised possibilities of novel roles for HSF1 as an indirect regulator of mRNA splicing and/or heterochromatin assembly (for review see Biamonti 2004).

In addition to regulating longevity, Hsps and other components of the heat shock response pathway have been shown to prevent and dissociate protein aggregates that have a central role in pathogenesis of neurodegenerative diseases (for review see Slavotinek and Biesecker 2001; Sakahira et al. 2002). On the other hand, constantly elevated levels of Hsps have been reported in various cancers, which may be due to their well-established anti-apoptotic activity in promoting survival of malignant cells (for review see Jäättelä 1999; Beere 2004; Mosser and Morimoto 2004). Given the finely tuned balance in the regulation of cell survival and cell death, pharmacological interventions for generation of small molecular inhibitors and activators, targeted to either Hsps (for review see Sreedhar and Csermely 2004) or HSF-mediated transcription (for review see Westerheide and Morimoto 2005), are of increasing therapeutic interest.

Acknowledgements

We apologize to those whose original work could only be cited indirectly due to space limitations. Konstantin Denessiouk and Anton Sandqvist are acknowledged for their contributions to the figures, and the members of our laboratory for generously providing results before publication. We thank Rick Morimoto, Sandy Westerheide, Dave Denslow, Carina Holmberg, John Eriksson, Minna Poukkula, Johanna Ahlskog, Julius Anckar, Henri Blomster, and Anton Sandqvist for valuable discussions and comments on the manuscript. The work in our laboratory is supported by the Academy of Finland, the Finnish Cancer Organizations, The Sigrid Juselius Foundation, the Finnish Life and Pension Insurance Companies, and the Borg Foundation (Åbo Akademi University).

References

Abravaya K, Phillips B, Morimoto RI (1991a) Heat shock-induced interactions of heat shock transcription factor and the human hsp70 promoter examined by *in vivo* footprinting. Mol Cell Biol 11:586-592

Abravaya K, Phillips B, Morimoto RI (1991b) Attenuation of the heat shock response in HeLa cells is mediated by the release of bound heat shock transcription factor and is modulated by changes in growth and in heat shock temperatures. Genes Dev 5:2117-2127

Abravaya K, Myers MP, Murphy SP, Morimoto RI (1992) The human heat shock protein hsp70 interacts with HSF, the transcription factor that regulates heat shock gene expression. Genes Dev 6:1153-1164

Ahn SG, Thiele DJ (2003) Redox regulation of mammalian heat shock factor 1 is essential for Hsp gene activation and protection from stress. Genes Dev 17:516-528

Ahn SG, Liu PC, Klyachko K, Morimoto RI, Thiele DJ (2001) The loop domain of heat shock transcription factor 1 dictates DNA-binding specificity and responses to heat stress. Genes Dev 15:2134-2145

Alastalo TP, Lönnström M, Leppä S, Kaarniranta K, Pelto-Huikko M, Sistonen L, Parvinen M (1998) Stage-specific expression and cellular localization of the heat shock factor 2 isoforms in the rat *Seminiferous epithelium.* Exp Cell Res 240:16-27

Alastalo TP, Hellesuo M, Sandqvist A, Hietakangas V, Kallio M, Sistonen L (2003) Formation of nuclear stress granules involves HSF2 and coincides with the nucleolar localization of Hsp70. J Cell Sci 116:3557-3570

Ali A, Bharadwaj S, O'Carroll R, Ovsenek N (1998) HSP90 interacts with and regulates the activity of heat shock factor 1 in *Xenopus* oocytes. Mol Cell Biol 18:4949-4960

Amin J, Ananthan J, Voellmy R (1988) Key features of heat shock regulatory elements. Mol Cell Biol 8:3761-3769

Andrulis ED, Werner J, Nazarian A, Erdjument-Bromage H, Tempst P, Lis JT (2002) The RNA processing exosome is linked to elongating RNA polymerase II in *Drosophila.* Nature 420:837-841

Arsene F, Tomoyasu T, Bukau B (2000) The heat shock response of *Escherichia coli.* Int J Food Microbiol 55:3-9

Baler R, Welch WJ, Voellmy R (1992) Heat shock gene regulation by nascent polypeptides and denatured proteins: hsp70 as a potential autoregulatory factor. J Cell Biol 117: 1151-1159

Beere HM (2004) 'The stress of dying': the role of heat shock proteins in the regulation of apoptosis. J Cell Sci 117:2641-2651

Bevilacqua A, Fiorenza MT, Mangia F (1997) Developmental activation of an episomic hsp70 gene promoter in two-cell mouse embryos by transcription factor Sp1. Nucleic Acids Res 25:1333-1338

Bharadwaj S, Ali A, Ovsenek N (1999) Multiple components of the HSP90 chaperone complex function in regulation of heat shock factor 1 *in vivo*. Mol Cell Biol 19:8033-8041

Bharti K, Von Koskull-Doring P, Bharti S, Kumar P, Tintschl-Korbitzer A, Treuter E, Nover L (2004) Tomato heat stress transcription factor HsfB1 represents a novel type of general transcription coactivator with a histone-like motif interacting with the plant CREB binding protein ortholog HAC1. Plant Cell 16:1521-1535

Biamonti G (2004) Nuclear stress bodies: a heterochromatin affair? Nat Rev Mol Cell Biol 5:493-498

Boehm AK, Saunders A, Werner J, Lis JT (2003) Transcription factor and polymerase recruitment, modification, and movement on dhsp70 *in vivo* in the minutes following heat shock. Mol Cell Biol 23:7628-7637

Boellmann F, Guettouche T, Guo Y, Fenna M, Mnayer L, Voellmy R (2004) DAXX interacts with heat shock factor 1 during stress activation and enhances its transcriptional activity. Proc Natl Acad Sci USA 101:4100-4105

Brown SA, Imbalzano AN, Kingston RE (1996) Activator-dependent regulation of transcriptional pausing on nucleosomal templates. Genes Dev 10:1479-1490

Bu L, Jin Y, Shi Y, Chu R, Ban A, Eiberg H, Andres L, Jiang H, Zheng G, Qian M, Cui B, Xia Y, Liu J, Hu L, Zhao G, Hayden MR, Kong X (2002) Mutant DNA-binding domain of HSF4 is associated with autosomal dominant lamellar and Marner cataract. Nat Genet 31:276-278

Cahill CM, Waterman WR, Xie Y, Auron PE, Calderwood SK (1996) Transcriptional repression of the prointerleukin 1beta gene by heat shock factor 1. J Biol Chem 271:24874-24879

Causton HC, Ren B, Koh SS, Harbison CT, Kanin E, Jennings EG, Lee TI, True HL, Lander ES, Young RA (2001) Remodeling of yeast genome expression in response to environmental changes. Mol Biol Cell 12:323-337

Chen D, Toone WM, Mata J, Lyne R, Burns G, Kivinen K, Brazma A, Jones N, Bahler J (2003) Global transcriptional responses of fission yeast to environmental stress. Mol Biol Cell 14:214-229

Chen Y, Barlev NA, Westergaard O, Jakobsen BK (1993) Identification of the C-terminal activator domain in yeast heat shock factor: independent control of transient and sustained transcriptional activity. EMBO J 12:5007-5018

Chiodi I, Biggiogera M, Denegri M, Corioni M, Weighardt F, Cobianchi F, Riva S, Biamonti G (2000) Structure and dynamics of hnRNP-labelled nuclear bodies induced by stress treatments. J Cell Sci 113:4043-4053

Choi H-S, Lin Z, Li B, Liu AY-C (1990) Age-dependent decrease in the heat-inducible DNA sequence-specific binding activity of human diploid fibroblasts. J Biol Chem 265:18005-18011

Christians E, Davis AA, Thomas SD, Benjamin IJ (2000) Maternal effect of Hsf1 on reproductive success. Nature 407:693-694

Chu B, Soncin F, Price BD, Stevenson MA, Calderwood SK (1996) Sequential phosphorylation by mitogen-activated protein kinase and glycogen synthase kinase 3 represses transcriptional activation by heat shock factor-1. J Biol Chem 271:30847-30857

Chu B, Zhong R, Soncin F, Stevenson MA, Calderwood SK (1998) Transcriptional activity of heat shock factor 1 at 37 degrees C is repressed through phosphorylation on two distinct serine residues by glycogen synthase kinase 3 and protein kinases Calpha and Czeta. J Biol Chem 273:18640-18646

Clos J, Westwood JT, Becker PB, Wilson S, Lambert K, Wu C (1990) Molecular cloning and expression of a hexameric *Drosophila* heat shock factor subject to negative regulation. Cell 63:1085-1097

Corey LL, Weirich CS, Benjamin IJ, Kingston RE (2003) Localized recruitment of a chromatin-remodeling activity by an activator *in vivo* drives transcriptional elongation. Genes Dev 17:1392-1401

Cotto JJ, Kline M, Morimoto RI (1996) Activation of heat shock factor 1 DNA binding precedes stress-induced serine phosphorylation. Evidence for a multistep pathway of regulation. J Biol Chem 271:3355-3358

Cotto J, Fox S, Morimoto R (1997) HSF1 granules: a novel stress-induced nuclear compartment of human cells. J Cell Sci 110:2925-2934

Dai R, Frejtag W, He B, Zhang Y, Mivechi NF (2000) c-Jun NH2-terminal kinase targeting and phosphorylation of heat shock factor-1 suppress its transcriptional activity. J Biol Chem 275:18210-18218

Dai Q, Zhang C, Wu Y, McDonough H, Whaley RA, Godfrey V, Li HH, Madamanchi N, Xu W, Neckers L, Cyr D, Patterson C (2003) CHIP activates HSF1 and confers protection against apoptosis and cellular stress. EMBO J 22:5446-5458

Damberger FF, Pelton JG, Harrison CJ, Nelson HC, Wemmer DE (1994) Solution structure of the DNA-binding domain of the heat shock transcription factor determined by multidimensional heteronuclear magnetic resonance spectroscopy. Protein Sci 3:1806-1821

Denegri M, Chiodi I, Corioni M, Cobianchi F, Riva S, Biamonti G (2001) Stress-induced nuclear bodies are sites of accumulation of pre-mRNA processing factors. Mol Biol Cell 12:3502-3514

Denegri M, Moralli D, Rocchi M, Biggiogera M, Raimondi E, Cobianchi F, De Carli L, Riva S, Biamonti G (2002) Human chromosomes 9, 12, and 15 contain the nucleation sites of stress-induced nuclear bodies. Mol Biol Cell 13:2069-2079

Eriksson M, Jokinen E, Sistonen L, Leppä S (2000) Heat shock factor 2 is activated during mouse heart development. Int J Dev Biol 44:471-477

Fargnoli J, Kunisada T, Fornace AJ Jr, Schneider EL, Holbrook NJ (1990) Decreased expression of heat shock protein 70 mRNA and protein after heat treatment in cells of aged rats. Proc Natl Acad Sci USA 87:846-850

Fawcett TW, Sylvester SL, Sarge KD, Morimoto RI, Holbrook NJ (1994) Effects of neurohormonal stress and aging on the activation of mammalian heat shock factor 1. J Biol Chem 269:32272-32278

Finkel T, Holbrook NJ (2000) Oxidants, oxidative stress and the biology of ageing. Nature 408:239-247

Fiorenza MT, Farkas T, Dissing M, Kolding D, Zimarino V (1995) Complex expression of murine heat shock transcription factors. Nucleic Acids Res 11:467-474

Frejtag W, Zhang Y, Dai R, Anderson MG, Mivechi NF (2001) Heat shock factor-4 (HSF-4a) represses basal transcription through interaction with TFIIF. J Biol Chem 276:14685-14694

Fujimoto M, Izu H, Seki K, Fukuda K, Nishida T, Yamada S, Kato K, Yonemura S, Inouye S, Nakai A (2004) HSF4 is required for normal cell growth and differentiation during mouse lens development. EMBO J 23:4297-4306

Gallo GJ, Schuetz TJ, Kingston RE (1991) Regulation of heat shock factor in *Schizosaccharomyces pombe* more closely resembles regulation in mammals than in *Saccharomyces cerevisiae*. Mol Cell Biol 11:281-288

Gallo GJ, Prentice H, Kingston RE (1993) Heat shock factor is required for growth at normal temperatures in the fission yeast *Schizosaccharomyces pombe*. Mol Cell Biol 13:749-761

Garigan D, Hsu AL, Fraser AG, Kamath RS, Ahringer J, Kenyon C (2002) Genetic analysis of tissue aging in *Caenorhabditis elegans*: a role for heat-shock factor and bacterial proliferation. Genetics 161:1101-1112

Gasch AP, Spellman PT, Kao CM, Carmel-Harel O, Eisen MB, Storz G, Botstein D, Brown PO (2000) Genomic expression programs in the response of yeast cells to environmental changes. Mol Biol Cell 11:4241-4257

Gilmour DS, Lis JT (1985) *In vivo* interactions of RNA polymerase II with genes of *Drosophila melanogaster*. Mol Cell Biol 5:2009-2018

Goodson ML, Sarge KD (1995) Heat-inducible DNA binding of purified heat shock transcription factor 1. J Biol Chem 270:2447-2450

Goodson ML, Park-Sarge OK, Sarge KD (1995) Tissue-dependent expression of heat shock factor 2 isoforms with distinct transcriptional activities. Mol Cell Biol 15:5288-5293

Goodson ML, Hong Y, Rogers R, Matunis MJ, Park-Sarge OK, Sarge KD (2001) Sumo-1 modification regulates the DNA binding activity of heat shock transcription factor 2, a promyelocytic leukemia nuclear body associated transcription factor. J Biol Chem 276:18513-18518

Green M, Schuetz TJ, Sullivan EK, Kingston RE (1995) A heat shock-responsive domain of human HSF1 that regulates transcription activation domain function. Mol Cell Biol 15:3354-3362

Greene JM, Larin Z, Taylor IC, Prentice H, Gwinn KA, Kingston RE (1987) Multiple basal elements of a human hsp70 promoter function differently in human and rodent cell lines. Mol Cell Biol 7:3646-3655

Grossman AD, Erickson JW, Gross CA (1984) The htpR gene product of *E. coli* is a sigma factor for heat-shock promoters. Cell 38:383-390

Guo Y, Guettouche T, Fenna M, Boellmann F, Pratt WB, Toft DO, Smith DF, Voellmy R (2001) Evidence for a mechanism of repression of heat shock factor 1 transcriptional activity by a multichaperone complex. J Biol Chem 276:45791-45799

Hahn JS, Thiele DJ (2004) Activation of the *Saccharomyces cerevisiae* heat shock transcription factor under glucose starvation conditions by Snf1 protein kinase. J Biol Chem 279:5169-5176

Hahn JS, Hu Z, Thiele DJ, Iyer VR (2004) Genome-wide analysis of the biology of stress responses through heat shock transcription factor. Mol Cell Biol 24:5249-5256

Harrison CJ, Bohm AA, Nelson HC (1994) Crystal structure of the DNA binding domain of the heat shock transcription factor. Science 263:224-227

Hashikawa N, Sakurai H (2004) Phosphorylation of the yeast heat shock transcription factor is implicated in gene-specific activation dependent on the architecture of the heat shock element. Mol Cell Biol 24:3648-3659

He H, Soncin F, Grammatikakis N, Li Y, Siganou A, Gong J, Brown SA, Kingston RE, Calderwood SK (2003) Elevated expression of heat shock factor (HSF) 2A stimulates HSF1-induced transcription during stress. J Biol Chem 278:35465-35475

Hietakangas V, Ahlskog JK, Jakobsson AM, Hellesuo M, Sahlberg NM, Holmberg CI, Mikhailov A, Palvimo JJ, Pirkkala L, Sistonen L (2003) Phosphorylation of serine 303 is a prerequisite for the stress-inducible SUMO modification of heat shock factor 1. Mol Cell Biol 23:2953-2968

Holmberg CI, Leppä S, Eriksson JE, Sistonen L (1997) The phorbol ester 12-*O*-tetradecanoylphorbol 13-acetate enhances the heat-induced stress response. J Biol Chem 272:6792-6798

Holmberg CI, Illman SA, Kallio M, Mikhailov A, Sistonen L (2000) Formation of nuclear HSF1 granules varies depending on stress stimuli. Cell Stress Chap 5:219-228

Holmberg CI, Hietakangas V, Mikhailov A, Rantanen JO, Kallio M, Meinander A, Hellman J, Morrice N, MacKintosh C, Morimoto RI, Eriksson JE, Sistonen L (2001) Phosphorylation of serine 230 promotes inducible transcriptional activity of heat shock factor 1. EMBO J 20:3800-3810

Holmberg CI, Tran SEF, Eriksson JE, Sistonen L (2002) Multisite phosphorylation provides sophisticated regulation of transcription factors. Trends Biochem Sci 27:619-627

Hong S, Kim SH, Heo MA, Choi YH, Park MJ, Yoo MA, Kim HD, Kang HS, Cheong J (2004) Coactivator ASC-2 mediates heat shock factor 1-mediated transactivation dependent on heat shock. FEBS Lett 559:165-170

Hong Y, Rogers R, Matunis MJ, Mayhew CN, Goodson ML, Park-Sarge OK, Sarge KD, Goodson M (2001) Regulation of heat shock transcription factor 1 by stress-induced SUMO-1 modification. J Biol Chem 276:40263-40267

Hsu AL, Murphy CT, Kenyon C (2003) Regulation of aging and age-related disease by DAF-16 and heat-shock factor. Science 300:1142-1145

Imbriano C, Bolognese F, Gurtner A, Piaggio G, Mantovani R (2001) HSP-CBF is an NF-Y-dependent coactivator of the heat shock promoters CCAAT boxes. J Biol Chem 276:26332-26339

Inouye S, Katsuki K, Izu H, Fujimoto M, Sugahara K, Yamada S, Shinkai Y, Oka Y, Katoh Y, Nakai A (2003) Activation of heat shock genes is not necessary for protection by heat shock transcription factor 1 against cell death due to a single exposure to high temperatures. Mol Cell Biol 23:5882-5895

Izu H, Inouye S, Fujimoto M, Shiraishi K, Naito K, Nakai A (2004) Heat-shock transcription factor 1 is involved in quality control mechanisms in male germ cells. Biol Reprod 70:18-24

Jakobsen BK, Pelham HR (1988) Constitutive binding of yeast heat shock factor to DNA *in vivo*. Mol Cell Biol 8:5040-5042

Jedlicka P, Mortin MA, Wu C (1997) Multiple functions of *Drosophila* heat shock transcription factor *in vivo*. EMBO J 16:2452-2462

Jolly C, Morimoto R, Robert-Nicoud M, Vourc'h C (1997) HSF1 transcription factor concentrates in nuclear foci during heat shock: relationship with transcription sites. J Cell Sci 110:2935-2941

Jolly C, Usson Y, Morimoto RI (1999) Rapid and reversible relocalization of heat shock factor 1 within seconds to nuclear stress granules. Proc Natl Acad Sci USA 96:6769-6774

Jolly C, Konecny L, Grady DL, Kutskova YA, Cotto JJ, Morimoto RI, Vourc'h C (2002) *In vivo* binding of active heat shock transcription factor 1 to human chromosome 9 heterochromatin during stress. J Cell Biol 156:775-781

Jolly C, Metz A, Govin J, Vigneron M, Turner BM, Khochbin S, Vourc'h C (2004) Stress-induced transcription of satellite III repeats. J Cell Biol 164:25-33

Jurivich DA, Sistonen L, Kroes RA, Morimoto RI (1992) Effect of sodium salicylate on the human heat shock response. Science 255:1243-1245

Jurivich DA, Pachetti C, Qiu L, Welk JF (1995) Salicylate triggers heat shock factor differently than heat. J Biol Chem 270:24489-24495

Jäättelä M (1999) Escaping cell death: survival proteins in cancer. Exp Cell Res 248:30-43

Kallio M, Chang Y, Manuel M, Alastalo TP, Rallu M, Gitton Y, Pirkkala L, Loones MT, Paslaru L, Larney S, Hiard S, Morange M, Sistonen L, Mezger V (2002) Brain abnormalities, defective meiotic chromosome synapsis and female subfertility in HSF2 null mice. EMBO J 21:2591-2601

Kawazoe Y, Nakai A, Tanabe M, Nagata K (1998) Proteasome inhibition leads to the activation of all members of the heat-shock-factor family. Eur J Biochem 255:356-362

Kline MP, Morimoto RI (1997) Repression of the heat shock factor 1 transcriptional activation domain is modulated by constitutive phosphorylation. Mol Cell Biol 17:2107-2115

Knauf U, Newton EM, Kyriakis J, Kingston RE (1996) Repression of human heat shock factor 1 activity at control temperature by phosphorylation. Genes Dev 10:2782-2793

Kroeger PE, Morimoto RI (1994) Selection of new HSF1 and HSF2 DNA-binding sites reveals difference in trimer cooperativity. Mol Cell Biol 14:7592-7603

Kroeger PE, Sarge KD, Morimoto RI (1993) Mouse heat shock transcription factors 1 and 2 prefer a trimeric binding site but interact differently with the HSP70 heat shock element. Mol Cell Biol 13:3370-3383

Landsberger N, Wolffe AP (1995) Role of chromatin and *Xenopus laevis* heat shock transcription factor in regulation of transcription from the *X. laevis* hsp70 promoter *in vivo*. Mol Cell Biol 15:6013-6024

Larson JS, Schuetz TJ, Kingston RE (1988) Activation *in vitro* of sequence-specific DNA binding by a human regulatory factor. Nature 335:372-375

Leppä S, Pirkkala L, Saarento H, Sarge KD, Sistonen L (1997) Overexpression of HSF2-beta inhibits hemin-induced heat shock gene expression and erythroid differentiation in K562 cells. J Biol Chem 272:15293-15298

Lis J (1998) Promoter-associated pausing in promoter architecture and postinitiation transcriptional regulation. Cold Spring Harbor Symp Quant Biol 63:347-356

Lis JT, Mason P, Peng J, Price DH, Werner J (2000) P-TEFb kinase recruitment and function at heat shock loci. Genes Dev 14:792-803

Littlefield O, Nelson HC (1999) A new use for the 'wing' of the 'winged' helix-turn-helix motif in the HSF-DNA cocrystal. Nat Struct Biol 6:464-470

Mathew A, Mathur SK, Morimoto RI (1998) Heat shock response and protein degradation: regulation of HSF2 by the ubiquitin-proteasome pathway. Mol Cell Biol 18:5091-5098

McMillan DR, Xiao X, Shao L, Graves K, Benjamin IJ (1998) Targeted disruption of heat shock transcription factor 1 abolishes thermotolerance and protection against heat-inducible apoptosis. J Biol Chem 273:7523-7528

McMillan DR, Christians E, Forster M, Xiao X, Connell P, Plumier JC, Zuo X, Richardson J, Morgan S, Benjamin IJ (2002) Heat shock transcription factor 2 is not essential for embryonic development, fertility, or adult cognitive and psychomotor function in mice. Mol Cell Biol 22:8005-8014

Mezger V, Rallu M, Morimoto RI, Morange M, Renard JP (1994) Heat shock factor 2-like activity in mouse blastocysts. Dev Biol 166:819-822

Mieusset R, Bujan L (1995) Testicular heating and its possible contributions to male infertility: a review. Int J Androl 18:169-184

Mishra SK, Tripp J, Winkelhaus S, Tschiersch B, Theres K, Nover L, Scharf KD (2002) In the complex family of heat stress transcription factors, HsfA1 has a unique role as master regulator of thermotolerance in tomato. Genes Dev 16:1555-1567

Mogk A, Homuth G, Scholz C, Kim L, Schmid FX, Schumann W (1997) The GroE chaperonin machine is a major modulator of the CIRCE heat shock regulon of *Bacillus subtilis*. EMBO J 16:4579-4590

Morimoto RI (1998) Regulation of the heat shock transcriptional response: cross talk between a family of heat shock factors, molecular chaperones, and negative regulators. Genes Dev 12:3788-3796

Morimoto RI (2002) Dynamic remodeling of transcription complexes by molecular chaperones. Cell 110:281-284

Morley JF, Morimoto RI (2004) Regulation of longevity in *Caenorhabditis elegans* by heat shock factor and molecular chaperones. Mol Biol Cell 15:657-664

Morley JF, Brignull HR, Weyers JJ, Morimoto RI (2002) The threshold for polyglutamine-expansion protein aggregation and cellular toxicity is dynamic and influenced by aging in *Caenorhabditis elegans*. Proc Natl Acad Sci USA 99:10417-10422

Mosser DD, Morimoto RI (2004) Molecular chaperones and the stress of oncogenesis. Oncogene 23:2907-2918

Mosser DD, Theodorakis NG, Morimoto RI (1988) Coordinate changes in heat shock element-binding activity and HSP70 gene transcription rates in human cells. Mol Cell Biol 8:4736-4744

Murphy SP, Gorzowski JJ, Sarge KD, Phillips B (1994) Characterization of constitutive HSF2 DNA-binding activity in mouse embryonal carcinoma cells. Mol Cell Biol 14:5309-5317

Murray JI, Whitfield ML, Trinklein ND, Myers RM, Brown PO, Botstein D (2004) Diverse and specific gene expression responses to stresses in cultured human cells. Mol Biol Cell 15:2361-2374

Nakai A, Morimoto RI (1993) Characterization of a novel chicken heat shock transcription factor, heat shock factor 3, suggests a new regulatory pathway. Mol Cell Biol 13:1983-1997

Nakai A, Kawazoe Y, Tanabe M, Nagata K, Morimoto RI (1995) The DNA-binding properties of two heat shock factors, HSF1 and HSF3, are induced in the avian erythroblast cell line HD6. Mol Cell Biol 15:5268-5278

Nakai A, Tanabe M, Kawazoe Y, Inazawa J, Morimoto RI, Nagata K (1997) HSF4, a new member of the human heat shock factor family, which lacks properties of a transcriptional activator. Mol Cell Biol 17:469-481

Nakai A, Suzuki M, Tanabe M (2000) Arrest of spermatogenesis in mice expressing an active heat shock transcription factor 1. EMBO J 19:1545-1554

Nelson DW, Padgett RW (2003) Insulin worms its way into the spotlight. Genes Dev 17:813-818

Newton EM, Knauf U, Green M, Kingston RE (1996) The regulatory domain of human heat shock factor 1 is sufficient to sense heat stress. Mol Cell Biol 16:839-846

Ni Z, Schwartz BE, Werner J, Suarez J-R, Lis JT (2004) Coordination of transcription, RNA processing, and surveillance by P-TEFb kinase on heat shock genes. Mol Cell 13:55-65

Nover L, Bharti K, Doring P, Mishra SK, Ganguli A, Scharf KD (2001) *Arabidopsis* and the heat stress transcription factor world: how many heat stress transcription factors do we need? Cell Stress Chap 6:177-189

Nykänen P, Alastalo TP, Ahlskog J, Horelli-Kuitunen N, Pirkkala L, Sistonen L (2001) Genomic organization and promoter analysis of the human heat shock factor 2 gene. Cell Stress Chap 6:377-385

Park JM, Werner J, Kim JM, Lis JT, Kim YJ (2001) Mediator, not holoenzyme, is directly recruited to the heat shock promoter by HSF upon heat shock. Mol Cell 8:9-19

Pelham HR (1982) A regulatory upstream promoter element in the *Drosophila* hsp 70 heat-shock gene. Cell 30:517-528

Pirkkala L, Alastalo TP, Nykänen P, Seppä L, Sistonen L (1999) Differentiation lineage-specific expression of human heat shock transcription factor 2. FASEB J 13:1089-1098

Pirkkala L, Alastalo TP, Zuo X, Benjamin IJ, Sistonen L (2000) Disruption of heat shock factor 1 reveals an essential role in the ubiquitin proteolytic pathway. Mol Cell Biol 20:2670-2675

Pirkkala L, Nykänen P, Sistonen L (2001) Roles of the heat shock transcription factors in regulation of the heat shock response and beyond. FASEB J 15:1118-1131

Proudfoot NJ, Furger A, Dye MJ (2002) Integrating mRNA processing with transcription. Cell 108:501-512

Rabindran SK, Giorgi G, Clos J, Wu C (1991) Molecular cloning and expression of human heat shock factor, HSF1. Proc Natl Acad Sci USA 88:6906-6910

Rabindran SK, Haroun RI, Clos J, Wisniewski J, Wu C (1993) Regulation of heat shock factor trimer formation: role of a conserved leucine zipper. Science 259:230-234

Rallu M, Loones M, Lallemand Y, Morimoto R, Morange M, Mezger V (1997) Function and regulation of heat shock factor 2 during mouse embryogenesis. Proc Natl Acad Sci USA 94:2392-2397

Rasmussen EB, Lis JT (1993) *In vivo* transcriptional pausing and cap formation on three *Drosophila* heat shock genes. Proc Natl Acad Sci USA 90:7923-7927

Ritossa F (1962) A new puffing pattern induced by temperature shock and DNP in *Drosophila*. Experientia 18:571-573

Rizzi N, Denegri M, Chiodi I, Corioni M, Valgardsdottir R, Cobianchi F, Riva S, Biamonti G (2004) Transcriptional activation of a constitutive heterochromatic domain of the human genome in response to heat shock. Mol Biol Cell 15:543-551

Rougvie AE, Lis JT (1988) The RNA polymerase II molecule at the 5' end of the uninduced hsp70 gene of *D. melanogaster* is transcriptionally engaged. Cell 54:795-804

Sakahira H, Breuer P, Hayer-Hartl MK, Hartl FU (2002) Molecular chaperones as modulators of polyglutamine protein aggregation and toxicity. Proc Natl Acad Sci USA 99 Suppl 4:16412-16418

Sandqvist A, Sistonen L (2004) Nuclear stress granules: the awakening of a sleeping beauty? J Cell Biol 164:15-17

Sarge KD, Zimarino V, Holm K, Wu C, Morimoto RI (1991) Cloning and characterization of two mouse heat shock factors with distinct inducible and constitutive DNA-binding ability. Genes Dev 5:1902-1911

Sarge KD, Murphy SP, Morimoto RI (1993) Activation of heat shock gene transcription by heat shock factor 1 involves oligomerization, acquisition of DNA-binding activity, and nuclear localization and can occur in the absence of stress. Mol Cell Biol 13:1392-1407

Sarge KD, Park-Sarge OK, Kirby JD, Mayo KE, Morimoto RI (1994) Expression of heat shock factor 2 in mouse testis: potential role as a regulator of heat-shock protein gene expression during spermatogenesis. Biol Reprod 50:1334-1343

Sarge KD, Bray AE, Goodson ML (1995) Altered stress response in testis. Nature 374:126

Schuetz TJ, Gallo GJ, Sheldon L, Tempst P, Kingston RE (1991) Isolation of a cDNA for HSF2: evidence for two heat shock factor genes in humans. Proc Natl Acad Sci USA 88:6911-6915

Shi Y, Kroeger PE, Morimoto RI (1995) The carboxyl-terminal transactivation domain of heat shock factor 1 is negatively regulated and stress responsive. Mol Cell Biol 15:4309-4318

Shi Y, Mosser DD, Morimoto RI (1998) Molecular chaperones as HSF1-specific transcriptional repressors. Genes Dev 12:654-666

Singh IS, Viscardi RM, Kalvakolanu I, Calderwood S, Hasday JD (2000) Inhibition of tumor necrosis factor-alpha transcription in macrophages exposed to febrile range temperature. A possible role for heat shock factor-1 as a negative transcriptional regulator. J Biol Chem 275:9841-9848

Singh IS, He JR, Calderwood S, Hasday JD (2002) A high affinity HSF-1 binding site in the 5'-untranslated region of the murine tumor necrosis factor-alpha gene is a transcriptional repressor. J Biol Chem 277:4981-4988

Sistonen L, Sarge KD, Phillips B, Abravaya K, Morimoto RI (1992) Activation of heat shock factor 2 during hemin-induced differentiation of human erythroleukemia cells. Mol Cell Biol 12:4104-4111

Sistonen L, Sarge KD, Morimoto RI (1994) Human heat shock factors 1 and 2 are differentially activated and can synergistically induce hsp70 gene transcription. Mol Cell Biol 14:2087-2099

Slavotinek AM, Biesecker LG (2001) Unfolding the role of chaperones and chaperonins in human disease. Trends Genet 17:528-535

Sorger PK (1990) Yeast heat shock factor contains separable transient and sustained response transcriptional activators. Cell 62: 793-805

Sorger PK, Nelson HC (1989) Trimerization of a yeast transcriptional activator via a coiled-coil motif. Cell 59:807-813

Sorger PK, Pelham HR (1988) Yeast heat shock factor is an essential DNA-binding protein that exhibits temperature-dependent phosphorylation. Cell 54:855-864

Sorger PK, Lewis MJ, Pelham HR (1987) Heat shock factor is regulated differently in yeast and HeLa cells. Nature 329:81-84

Söti C, Csermely P (2003) Aging and molecular chaperones. Exp Gerontol 38:1037-1040

Sreedhar AS, Csermely P (2004) Heat shock protein in the regulation of apoptosis: new strategies in tumor therapy. A comprehensive review. Pharmacol Ther 101:227-257

Straus DB, Walter WA, Gross CA (1987) The heat shock response of *E. coli* is regulated by changes in the concentration of sigma 32. Nature 329:348-351

Sullivan EK, Weirich CS, Guyon JR, Sif S, Kingston RE (2001) Transcriptional activation domains of human heat shock factor 1 recruit human SWI/SNF. Mol Cell Biol 21:5826-5837

Svejstrup JQ (2004) The RNA polymerase II transcription cycle: cycling through chromatin. Biochim Biophys Acta 1677:64-73

Takagaki Y, Manley JL (2000) Complex protein interactions within the human polyadenylation machinery identify a novel component. Mol Cell Biol 20:1515-1525

Tanabe M, Kawazoe Y, Takeda S, Morimoto RI, Nagata K, Nakai A (1998) Disruption of the HSF3 gene results in the severe reduction of heat shock gene expression and loss of thermotolerance. EMBO J 17:1750-1758

Tanabe M, Sasai N, Nagata K, Liu XD, Liu PC, Thiele DJ, Nakai A (1999) The mammalian HSF4 gene generates both an activator and a repressor of heat shock genes by alternative splicing. J Biol Chem 274:27845-27856

Tatsuta T, Tomoyasu T, Bukau B, Kitagawa M, Mori H, Karata K, Ogura T (1998) Heat shock regulation in the ftsH null mutant of *Escherichia coli*: dissection of stability and activity control mechanisms of sigma32 *in vivo*. Mol Microbiol 30:583-593

Tilly K, McKittrick N, Zylicz M, Georgopoulos C (1983) The dnaK protein modulates the heat-shock response of *Escherichia coli*. Cell 34:641-646

Tissieres A, Mitchell HK, Tracy UM (1974) Protein synthesis in salivary glands of *Drosophila melanogaster*: relation to chromosome puffs. J Mol Biol 84:389-398

Tomoyasu T, Ogura T, Tatsuta T, Bukau B (1998) Levels of DnaK and DnaJ provide tight control of heat shock gene expression and protein repair in *Escherichia coli*. Mol Microbiol 30:567-581

Trinklein ND, Murray JI, Hartman SJ, Botstein D, Myers RM (2004) The role of heat shock transcription factor 1 in the genome-wide regulation of the mammalian heat shock response. Mol Biol Cell 15:1254-1261

Vuister GW, Kim SJ, Orosz A, Marquardt J, Wu C, Bax A (1994a) Solution structure of the DNA-binding domain of *Drosophila* heat shock transcription factor. Nat Struct Biol 1:605-614

Vuister GW, Kim SJ, Wu C, Bax A (1994b) NMR evidence for similarities between the DNA-binding regions of *Drosophila melanogaster* heat shock factor and the helix-turn-helix and HNF-3/forkhead families of transcription factors. Biochemistry 33:10-16

Wang G, Zhang J, Moskophidis D, Mivechi NF (2003) Targeted disruption of the heat shock transcription factor (hsf)-2 gene results in increased embryonic lethality, neuronal defects, and reduced spermatogenesis. Genesis 36:48-61

Wang G, Ying Z, Jin X, Tu N, Zhang Y, Phillips M, Moskophidis D, Mivechi NF (2004) Essential requirement for both hsf1 and hsf2 transcriptional activity in spermatogenesis and male fertility. Genesis 38:66-80

Wang X, Grammatikakis N, Siganou A, Calderwood SK (2003) Regulation of molecular chaperone gene transcription involves the serine phosphorylation, 14-3-3 epsilon binding, and cytoplasmic sequestration of heat shock factor 1. Mol Cell Biol 23:6013-6026

Weighardt F, Cobianchi F, Cartegni L, Chiodi I, Villa A, Riva S, Biamonti G (1999) A novel hnRNP protein (HAP/SAF-B) enters a subset of hnRNP complexes and relocates in nuclear granules in response to heat shock. J Cell Sci 112:1465-1476

Westerheide SD, Morimoto RI (2005) Heat shock response modulators as therapeutic tools for diseases of protein conformation. J Biol Chem 280:33097-33100

Westwood JT, Wu C (1993) Activation of *Drosophila* heat shock factor: conformational change associated with a monomer-to-trimer transition. Mol Cell Biol 13:3481-3486

Wiederrecht G, Seto D, Parker CS (1988) Isolation of the gene encoding the *S. cerevisiae* heat shock transcription factor. Cell 54:841-853

Williams GT, Morimoto RI (1990) Maximal stress-induced transcription from the human HSP70 promoter requires interactions with the basal promoter elements independent of rotational alignment. Mol Cell Biol 10:3125-3136

Williams GT, McClanahan TK, Morimoto RI (1989) E1a transactivation of the human HSP70 promoter is mediated through the basal transcriptional complex. Mol Cell Biol 9:2574-2587

Wu C (1984) Two protein-binding sites in chromatin implicated in the activation of heat-shock genes. Nature 309:229-234

Wu C (1995) Heat shock transcription factors: structure and regulation. Annu Rev Cell Dev Biol 11:441-469

Wu C, Wilson S, Walker B, Dawid I, Paisley T, Zimarino V, Ueda H (1987) Purification and properties of *Drosophila* heat shock activator protein. Science 238:1247-1253

Xia W, Voellmy R (1997) Hyperphosphorylation of heat shock transcription factor 1 is correlated with transcriptional competence and slow dissociation of active factor trimers. J Biol Chem 272:4094-4102

Xia W, Guo Y, Vilaboa N, Zuo J, Voellmy R (1998) Transcriptional activation of heat shock factor HSF1 probed by phosphopeptide analysis of factor ^{32}P-labeled *in vivo*. J Biol Chem 273:8749-8755

Xiao H, Lis JT (1988) Germline transformation used to define key features of heat-shock response elements. Science 239:1139-1142

Xiao X, Zuo X, Davis AA, McMillan DR, Curry BB, Richardson JA, Benjamin IJ (1999) HSF1 is required for extra-embryonic development, postnatal growth and protection during inflammatory responses in mice. EMBO J 18:5943-5952

Xie Y, Chen C, Stevenson MA, Auron PE, Calderwood SK (2002) Heat shock factor 1 represses transcription of the IL-1beta gene through physical interaction with the nuclear factor of interleukin 6. J Biol Chem 277:11802-11810

Xing H, Mayhew CN, Cullen KE, Park-Sarge OK, Sarge KD (2004) HSF1 modulation of Hsp70 mRNA polyadenylation via interaction with symplekin. J Biol Chem 279:10551-10555

Yan LJ, Christians ES, Liu L, Xiao X, Sohal RS, Benjamin IJ (2002) Mouse heat shock transcription factor 1 deficiency alters cardiac redox homeostasis and increases mitochondrial oxidative damage. EMBO J 21:5164-5172

Yoshima T, Yura T, Yanagi H (1998) Function of the C-terminal transactivation domain of human heat shock factor 2 is modulated by the adjacent negative regulatory segment. Nucleic Acids Res 26:2580-2585

Yura T, Nakahigashi K (1999) Regulation of the heat-shock response. Curr Opin Microbiol 2:153-158

Zhong M, Orosz A, Wu C (1998) Direct sensing of heat and oxidation by *Drosophila* heat shock transcription factor. Mol Cell 2:101-108

Zou J, Guo Y, Guettouche T, Smith DF, Voellmy R (1998) Repression of heat shock transcription factor HSF1 activation by HSP90 (HSP90 complex) that forms a stress-sensitive complex with HSF1. Cell 94:471-480

Zuo J, Baler R, Dahl G, Voellmy R (1994) Activation of the DNA-binding ability of human heat shock transcription factor 1 may involve the transition from an intramolecular to an intermolecular triple-stranded coiled-coil structure. Mol Cell Biol 14:7557-7568

Hietakangas, Ville

Turku Centre for Biotechnology, P.O. Box 123, FIN-20521 Turku, Finland. Current address: EMBL Heidelberg, Meyerhofstrasse 1, D-69117 Heidelberg, Germany

Sistonen, Lea

Turku Centre for Biotechnology, P.O. Box 123, FIN-20521 Turku, Finland
lea.sistonen@btk.fi

The unfolded protein response unfolds

Maho Niwa

Abstract

As a key organelle of protein targeting and secretion, the endoplasmic reticulum (ER) plays host to a wide variety of protein maturation steps including folding, modification, and complex formation. Homeostasis of ER function is therefore critical to cell function. The unfolded protein response (UPR), a conserved eukaryotic signal transduction pathway, regulates the ER's capacity to perform protein folding according to cellular demand. UPR signaling is initiated by ER transmembrane components that sense unfolded protein levels within the ER. In yeast, the only known UPR initiator is IRE1, a transmembrane serine/threonine kinase/ endoribonuclease. In higher eukaryotes, the UPR also comprises signals initiated by the ER-transmembrane kinase PERK and the ER-transmembrane transcription factor ATF6. A major consequence of UPR initiator activation is transcription induction of a wide variety of genes for ER-resident chaperons and protein folding enzymes, in order to increase ER protein folding capacity. Ultimately, UPR activation leads to remodeling the entire secretory pathway in order to meet cellular demand. The identification of these initiators and recent studies of their behaviors is revealing fascinating aspects of the overall UPR. This review discusses highlights of these discoveries and relationships between the UPR signaling branches initiated by each ER component.

1 Endoplasmic reticulum (ER): the journey to secretion

Virtually all proteins found throughout the secretory pathway – ER, Golgi, plasma membrane – are initially targeted to the protein translocation channel on the ER membrane and subsequently translocated through the pore into the lumen of the ER (Gilmore 1993; Brown et al. 1995; Matlack et al. 1998; Voeltz et al. 2002). Upon entering the ER lumen, these nascent proteins must undergo different folding and modification processes to produce fully functional proteins before leaving the ER to their final destinations. ER resident chaperons such as Grp78 (BiP) and Grp98 associate with nascent proteins in the lumen as they emerge from the pores of the ER. Some chaperones assist in folding nascent proteins – ER molecular chaperons, such as calnexin and calreticulum, bind to folding intermediates to facilitate the completion of folding – while others keep proteins from aggregating. Nascent proteins also associate with various modification enzymes that either glycosylate or induce disulfide bond formation according to the presence of appropriate modification sites. In some cases, multiple protein subunits may have to come together to form a functional complex within the lumen of the ER, such as the case with the T cell receptor complex. Many proteins undergo multiple and sequential maturation steps through many folding intermediates. As the initial and critical site for protein maturation of secreted proteins and proteins that reside within the secretory pathway, the ER is constantly challenged to both measure protein traffic levels and provide appropriate levels of protein folding capacity (Gething 1999; Kaufman 1999; Mori 2000; Fewell et al. 2001; Patil and Walter 2001; Harding et al. 2002; Ellgaard and Helenius 2003).

The ER ensures sufficient protein folding capacity at any given time while performing quality control of protein folding states in order to ensure that only properly folded proteins are allowed to leave for their final destinations. This quality control or clearance mechanism of the ER, which identifies misfolded proteins for retrograde translocation to the cytoplasm and subsequent degradation by the ubiquitin proteosome system, is termed ERAD, for ER-associated protein degradation. Proteins that bear mutations disrupting proper folding, or irreversibly trapping them in low energy intermediate states are prohibited from entering the secretory pathway and accumulate in the ER. Proteins that are transported from the ER without appropriate maturation carry severe consequences in the form of many human diseases: type I diabetes, cystic fibrosis, and emphysema (discussed in other reviews; Kim and Arvan 1998; Kuznetsov and Nigam 1998; Harding and Ron 2002; Kaufman 2002; Shi et al. 2003).

The ER is constantly challenged to distinguish permanently misfolded or unfolded proteins from folding intermediates. The molecular basis behind this critical decision is yet not well understood. For N-glycosylated proteins, the current model suggests that the amount of time it takes to trim the glycans added to the proteins plays a role in earmarking folding intermediates from misfolded proteins (Abeijon and Hirschberg 1992; Hammond et al. 1994; Hebert et al. 1995; Holkeri and Makarow 1998; Sitia et al. 2001; Helenius and Aebi 2004; Kleizen and Braakman 2004). N-linked carbohydrates are synthesized as lipid precursors in the ER membrane and contain three extra glucoses on one of the mannose branches. After the sugar is transferred to a nascent protein, the three glucoses are trimmed from the carbohydrates; this step is tightly coupled to the folding states of the protein. Trimming of the glucoses only takes place upon completion of folding. If the protein is not finished folding, re-glycosylation takes place to mark and prevent the folding intermediate from being prematurely transported out to the Golgi.

The ER must upregulate its protein folding capacity in cells of specific developmental stages or in certain tissues that secrete high levels of proteins. For example, high levels of insulin secreting β-cells in the pancreas require higher ER protein folding capacity. Thus, the ER must continually evaluate constantly fluctuating protein folding demands and respond to changes in capacity need in a timely fashion. How does the ER measure its protein folding needs at any given time and keep up with protein folding demands? In eukaryotic cells, one mechanism that mediates this regulation is a pathway called the unfolded protein response (UPR) pathway.

2 The unfolded protein response (UPR) pathway

The UPR signaling pathway, an inter-organelle signaling pathway that connects the ER to nuclear functions, responds to changes in the ER's protein handling capacities. UPR activation induces gene transcription of ER resident molecular chaperones and other protein folding enzymes as well as genes involved in phospholipid biosynthesis to increase both folding capacity and physical size of the ER. The UPR also regulates the transcription of genes coding for ERAD-associated functions to facilitate the removal of permanently unfolded proteins, and several post-ER secretory pathway components to maintain increased overall capacity for protein traffic (Harding et al. 2002; Kaufman 1999; Mori 1999; Patil and Walter 2001). The identification of proteins that mediate and regulate the UPR has revealed fascinating molecular signaling cascades connecting the ER to the nucleus that help us understand how a cell regulates the dynamic secretory pathway.

2.1 Yeast UPR

The UPR is conserved in all eukaryotic cells. One remarkable aspect of the UPR involves a non-conventional splicing regulatory step. In *S. cerevisiae*, high protein folding needs trigger and activate the N-terminal luminal domain of Ire1p, a type I ER-transmembrane kinase/endoribonuclease, which initiates the UPR pathway (Cox et al. 1993; Mori et al. 1993; Shamu and Walter 1996). This signal is transmitted across the ER membrane to Ire1 cytosolic domains causing Ire1p oligomerization and autophosphorylation in a manner similar to hormone receptor kinases at the plasma membrane. Unlike most receptor kinases, however, Ire1p contains an additional enzymatic activity and functions as an endoribonuclease to cleave a single 252 nucleotide intron from the mRNA encoding the UPR-specific transcription factor Hac1p (Cox and Walter 1996; Mori et al. 1996, 1997; Sidrauski and Walter 1997). The resulting exons of Ire1p-cleaved *HAC1* mRNA are then joined by tRNA ligase to produce the spliced, mature form of *HAC1* mRNA (Sidrauski et al. 1996). The Ire1p-dependent splicing of *HAC1* mRNA is an obligate step in the UPR signal transduction pathway. The un-spliced form of *HAC1* mRNA associates with polyribosomes but the presence of the intron prevents Hac1p translation (Ruegsegger et al. 2001). Ire1p mediated splicing seems to occur while *HAC1* mRNA is on polyribosomes, allowing for quick production of Hac1p and subsequent UPR target gene transcription.

2.2 Non-conventional HAC1 mRNA splicing

Switching on a signal transduction pathway by mRNA splicing is one unique aspect of the UPR. *HAC1* mRNA splicing is a novel RNA processing pathway mediated by Ire1p and tRNA ligase, not by splicesomal components (Sidrauski et al. 1996, 1998; Mori et al. 1997; Sidrauski and Walter 1997; Shamu 1998; Gonzalez et al. 1999). The chemistry of *HAC1* splicing resembles that of pre-tRNA splicing rather than mRNA splicing; there is no obligate order for cleavage at either the 5' or 3' splice junctions. Ire1p cleavage generates 2'3'-cyclic phopshodiester termini, and the joining of the two exons is mediated by tRNA ligase in a second step (Abelson 1994; Hopper and Phizicky 2003). One major difference between tRNA splicing and *HAC1* splicing is the direct involvement of Ire1p as the endoribonuclease for *HAC1*, in contrast to tRNA endoribonuclease for pre-tRNA introns. Curiously, Ire1p shares no sequence similarity with the multiple subunit-containing tRNA endoribonuclease (Trotta et al. 1997). Differences in the enzymes involved in the cleavage step are reflected in cis RNA sequences required for the cleavage. Ire1 recognizes 7 nucleotide loop closed by a stem (Gonzalez et al. 1999). Furthermore, Ire1p mediated cleavage is sensitive to the Guanosine residue next to the cleavage site. On the other hand, tRNA endonuclease recognizes the tertiary structure of pre-tRNA rather than sequence of specific nucleotides (Trotta et al. 1997). tRNA ligase mediated exon ligation steps leaves a 2' phosphate group on the spliced tRNA junction, subsequently removed by a NAD-dependent 2'-phosphotransferase (Spinelli et al. 1997). Presence of the similar chemistry for the HAC1 exon junction and its potential contribution to the UPR has not been investigated. Finally, in addition to IRE1 and tRNA ligase, ADA5 has been suggested to play a role in the *HAC1* splicing, as it is impaired in the *ada5* knockout cells (Welihinda et al. 2000). ADA5 is a component of a large 1.8 mDa SAGA complex, responsible for histone acetylation during transcriptional activation, although its functional role for the *HAC1* splicing is not well understood.

Structurally, the protein most closely related to Ire1p is the mammalian endoribonuclease, RNase L, a non-sequence specific endoribonuclease activated in cells infected with double stranded RNA viruses; activated RNase L functions to degrade cellular RNA during viral infection (Bork and Sander 1993; Zhou et al. 1993; Dong and Silverman 1997). RNase L has a riboendonuclease domain followed by a kinase domain, similar to the architecture of the cytosolic portion of Ire1p. RNase L is a soluble enzyme found in the cytosol, but instead of having similar transmembrane and ER luminal domains to Ire1p, the amino terminus of RNase L contains an ankiryn repeat domain. Both Ire1 and RNaseL dimerize (or oligomerize) as an obligate step to activating their endoribonuclease activities. RNase L and IRE1, however, differ in their enzymatic activities (Tirasophon et al. 2000; Dong et al. 2001). No apparent kinase activity has been detected with RNase L, while the kinase domain function is required for IRE1 function during the UPR. Furthermore, IRE1 is a sequence specific endoribonuclease while RNase L does not have sequence specificity. A recent study has suggested that nucleotide binding to the kinase domain drives oligomerization of Ire1p, which is critical for the nuclease activation (Papa et al. 2003). The sequence specificity of IRE1 may derive from how it binds its substrate, *HAC1* mRNA. There is still much unknown about the mechanisms of Ire1 function.

2.3 Are there additional RNA substrates for yeast Ire1p?

Sequence analysis of the 5' and 3' cleavage sites of *HAC1* mRNA reveal that both share similar stem-loop structures (Gonzalez et al. 1999). Cleavage occurs after a conserved guanosine at the third position within a seven nucleotide loop. Change in this guanosine is critical for cleavage. Mutational studies also show that the stem loop secondary structure flanking the cleavage sites contains information necessary and sufficient for Ire1p cleavage. Computational analysis of the yeast genome searching for additional RNAs containing *HAC1* mRNA consensus stem loops has identified multiple candidate genes, although none of them share the same spacing of two such stem loops with that seen in *HAC1*. So far, all available data points to *HAC1* as the only RNA substrate cleaved by Ire1p during the UPR (Niwa et al. 2004). Thus, in *S. cerevisiae*, IRE1-*HAC1* may be an exclusive signaling pathway between the ER and the nucleus, insulated from other signaling events in the cell.

2.4 UPR specific transcription factors in yeast

Hac1p belongs to the leucine zipper family of transcription factors, and activates transcription by binding directly to the yeast unfolded protein response element (UPRE) present within the 5' UTR of target genes that code for ER resident molecular chaperones (BiP, Calnexin, Calreticulum), protein disulfide isomerase (PDI), and other protein folding enzymes (Mori et al. 1992, 1997; Cox et al. 1997). Microarray profiling of transcription induced in yeast under the UPR activated conditions has shown that genes involved in ERAD and those involved in protein trafficking beyond the ER are upregulated during the UPR (Casagrande et al. 2000; Friedlander et al. 2000; Travers et al. 2000). Not only does the UPR increase ER protein folding capacity, but it also increases the ability to transport proteins throughout the secretory pathway, and the capacity to eliminate permanently unfolded proteins from the ER. Ultimately, the UPR activation affects the transcription of ~400 genes in yeast and leads to re-programming of the entire secretory pathway. Hac1p induces transcription of a subset of these target genes by binding to the 5' UTR UPRE (Mori et al. 1992). Recently two additional transcription enhancer elements were identified within UPR target genes and a subsequent genetic screen identified Gcn4p as an additional UPR responsive transcriptional activator (Patil et al. 2004).

Gcn4p also belongs to the basic leucine zipper family of transcription factors, and is known to induce expression of certain stress response genes during amino acid starvation and glucose limitation. The activation of Gcn4p under these general stress conditions requires activation of Gcn2 kinase, which phosphoryates a translation initiation factor, eIF2α (Hinnebusch 1985; Hinnebusch et al. 2000). eIF2α phosphorylation prevents efficient formation of 80S ribosomes, and thus inhibits translation initiation, resulting in translational repression of most mRNA. Translation of GCN4, however, is preferentially increased because of the presence of four short ORFs within the 5' UTR. Curiously, eIF2α phosphorylation does not occur in yeast during the UPR although GCN2 is required for the activation of Gcn4p; the gcn2 knockout strain is deficient for UPR transcriptional activation (Patil et al. 2004). The discovery that Gcn4p is involved in the UPR has opened many interesting directions for research, including identifying GCN2 kinase targets during the UPR, and elucidating the molecular mechanism of Gcn4p induction.

3 Mammalian UPR

The UPR is conserved in all eukaryotic cells, although the scope of the response is much greater in mammalian cells; it affects not only target gene transcription but also translational repression, cell cycle arrest, and cell fate decisions. Metazoan cells have at least three ER transmembrane components - IRE1, PERK, ATF6 - involved in initiating the UPR. Each signaling branch initiated from the ER by these three UPR components induces different sets of downstream events, including the activation of three distinct transcription factors (Fig. 1). Below, we discuss the three UPR initiating components and their respective downstream events.

3.1 IRE1

The non-conventional splicing critical for the yeast UPR also plays an important role in mammalian cells. There are at least two isoforms of IRE1 (IRE1α & IRE1β) in human and mouse cells (Tirasophon et al. 1998; Wang et al. 1998). Currently, the functional significance of having two isoforms is not well understood. In both human and mouse, IRE1α is expressed ubiquitously, while IRE1β is reported to be restricted to epithelial cells of the intestinal epithelial cells (Bertolotti et al. 2001). As in yeast, the cytosolic portion of mammalian IRE1 contains both kinase and endoribonuclease activities that are required for the non-conventional splicing of XBP1 mRNA, a mammalian homologue of the yeast HAC1 substrate (Niwa et al. 1999; Tirasophon et al. 2000; Yoshida et al. 2001a; Calfon et al. 2002; Shen et al. 2002b). XBP1 encodes a CREB/ATF family transcription factor, originally identified as important for B-cell differentiation into immunoglobulin secreting plasma cells (Reimold et al. 1999, 2001). The XBP1 gene contains five "standard" exons separated by "standard" introns that are spliced by conventional splicesome mediated steps in the nucleus. In addition, a single non-standard 26 nucleotide intron located within exon 4 is spliced by IRE1 during the UPR; we will refer to this as the "UPR specific intron". In contrast to yeast HAC1, the protein coded by the un-spliced form of mammalian XBP1 mRNA is present at detectable levels prior to the UPR. Splicing of the UPR specific intron in XBP1 mRNA results in a translational frame shift that produces a second form of XBP1 protein with a unique C-terminus; this second form of XBP1 protein is a far more potent transcriptional activator than the un-spliced form. The IRE1-dependent splicing of an UPR-specific transcription factor mRNA

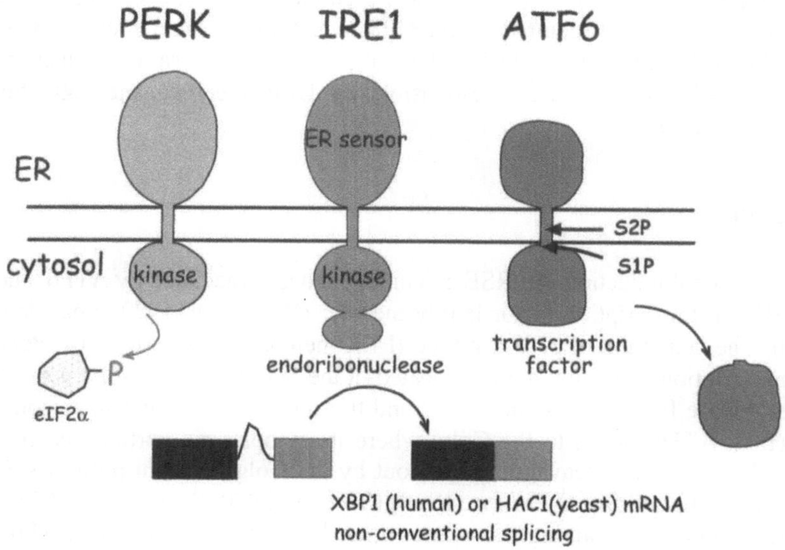

Fig. 1. In mammalian cells, there are, at least, three ER transmembrane components, IRE1, PERK, and ATF6, initiating the UPR. IRE1 is a bifunctional kinase/endoribonuclease ER-transmembrane protein involved in non-conventional splicing of either HAC1 or XBP1 mRNA. Since the spliced form of XBP1 is a more potent transcription factor, Ire1p-dependent splicing is an obligate step in UPR pathway. PERK is another transmembrane kinase, that phosphorylates a translation initiation factor, eIF2α, leading to the repression of overall translation to alleviate secretory burden on the ER. ATF6 a transcription factor, initially synthesized as an ER transmembrane protein. Upon UPR induction, cytosolic portion of ATF6 containing the transcription activation domain is released by proteases from the membrane and participates in trascription program of the UPR. Together, broad scope of responses induced by IRE1, PERK, and ATF6 functions to increase ER protein folding capacity.

is a key regulatory step both in yeast and mammalian UPR signaling pathways (Yoshida et al. 2001a; Calfon et al. 2002; Shen et al. 2002b). XBP1 binds directly to the mammalian UPRE to activate transcription during the UPR. Genes containing the mammalian UPRE include EDEM (ER degradation enhancing α-mannosidase like protein), which is involved in substrate recognition for ER mediated protein degradation (Hosokawa et al. 2003; Oda et al. 2003; Yoshida et al. 2003). EDEM is a type II ER transmembrane protein and its luminal domain shares homology with α-1,2-mannosidase (Hosokawa et al. 2001). Overexpression of EDEM enhances degradation of misfolded glycoproteins carrying/containing the Man8 structure/motif, suggesting an involvement in ERAD.

Curiously, transcriptional induction of classical UPR targets, such as GRP78 (BiP) occurs normally in mouse fibroblasts derived from hIRE1α knockout embryos, while transcription of EDEM is greatly repressed (Urano et al. 2000b; Lee et al. 2002; Yoshida et al. 2003). Analyses of promoter sequences among the genes regulated during the UPR also identified the ERSE, an additional transcription enhancer element required for the UPR induction (Yoshida et al. 1998; Wang et al. 2000; Kokame et al. 2001). Thus, at least two transcription enhancer elements, UPRE and ERSE, are responsible for UPR specific gene regulation in mammalian cells.

3.2 ATF6

Transcriptional induction of ERSE containing genes is mediated by ATF6, another CREB/ATF transcription factor family member (Haze et al. 1999; Yoshida et al. 2000). The activation of ATF6, a type-II ER-membrane transcription factor with the transcription activation domain located in the N-terminal cytosolic portion, requires release from the ER membrane and transport into the nucleus. Upon UPR induction, ATF6 moves to the Golgi where its cytoplasmic portion becomes released by sequential cleavages carried out by the Golgi resident proteases Site 1 Protease (S1P) and Site 2 Protease (S2P) (Ye et al. 2000; Shen et al. 2002a). The cytosolic portion containing the transcriptional activation domain of ATF6 contains a nuclear localization signal (NLS) that mediates nuclear transport. Curiously, both S1P and S2P are also involved in the cleavage of sterol response element binding protein (SREBP) (Wang et al. 1994; Hua et al. 1996; Brown and Goldstein 1997; Sakai et al. 1998), a transcription factor for the cholesterol biosynthesis pathway, in response to changes in cellular cholesterol levels. SREBP is a basic helix-loop-helix transcription factor and induces transcription of cholesterol biosynthetic genes during cholesterol deprivation. Under non-stressed conditions, both ATF6 and SREBP are kept in inactive forms in the ER and sequestered away from S1P and S2P. The key activation step is the transport of either ATF6 or SREBP from the ER to the Golgi and becoming accessible to the proteases.

ATF6 is sequestered from activation by binding to an ER chaperone, BiP, through its C-terminal domain in the lumen of the ER. Accumulation of unfolded proteins in the ER causes BiP to release ATF6, which then moves freely into the Golgi compartment (Shen et al. 2002a). In the meantime, dissociated BiP binds to unfolded proteins in order to facilitate their folding and prevent aggregation. In contrast, SREBP does not bind BiP in the ER. Instead, it associates with an SREBP cleavage activating protein called SCAP during cholesterol deprivation (Nohturfft et al. 1999, 2000). Through its association with SCAP, SREBP becomes packaged into vesicles for transport from the ER. Once in the Golgi, S1P cleaves within the transmembrane domain of both ATF6 and SREBP and S2P cleaves at a site adjacent to the cytosolic portion of these proteins.

In the nucleus, ATF6 activates transcription of genes by directly binding to the ER stress response element, ERSE, defined by the sequence CCAAT(N_9)CCACG. Within the ERSE is constitutively bound by the CCAAT binding factor (NF-Y), while activated ATF6 binds to CCACG, the 3' half of the ERSE, which is required for transcriptional activation (Haze et al. 1999; Roy and Lee 1999; Li et al. 2000; Yoshida et al. 2000, 2001b). ERSE containing genes include genes coding for the ER resident chaperons BiP (GRP78), and GRP94. SREBP binds to the sterol response element (SRE) found within the 5' UTR of genes coding for enzymes involved in cholesterol biosynthesis.

The involvement of S1P and S2P proteases in both UPR and cholesterol biosynthetic pathways is interesting, particularly because the ER is the site of protein folding and cholesterol biosynthesis. However, experimental data available thus far suggests that cleavage of ATF6 occurs independent of SREBP cleavage. ATF6 is processed by S1P and S2P only during the UPR, and is not induced in response to a change in the concentration of cellular cholesterol levels; likewise, SREBP is not cleaved in response to the UPR. The involvement of the same two proteases suggests potential cross-regulation between the UPR and cholesterol biosynthetic pathway, though evidence of such does not currently exist. However, the disease homocysteinemia may present a condition where the UPR and the cholesterol biosynthetic pathways are intimately linked (Werstuck et al. 2001). Homocysteinemia is caused by a mutation in the cystathionine β-synthetase gene that prevents the conversion of homocysteine to cysteine, which results in high homocysteine accumulation (Outinen et al. 1999; Huang et al. 2001; Zhang et al. 2001). Elevated levels of homocysteine disrupt disulfide bond formation in the ER and lead to the accumulation of unfolded proteins. Homocysteinemia also activates SREBP, resulting in the accumulation of intracellular cholesterol levels. It is possible that simultaneous activation of both UPR and SREBP pathways is deleterious to cells and that a mechanism exists to prevent both from happening at the same time.

It is interesting that ectopically expressed transcription activation domains of both SREBP and ATF6 can be co-immunoprecipitated with histondeacetylase (HDAC) bound to the sterol response element (Zeng et al. 2004). Under this experimental condition, ATF6 target genes – but not SREBP target genes – are activated, suggesting that ATF6 suppresses transcription of the cholesterol pathway, while activating transcription of the UPR target genes. Negative regulation of the cholesterol pathway by ATF6 might serve to ensure an immediate switch in transcription program in response to the UPR.

Finally, there are at least two isoforms of ATF6 in mammals, ATF6α and ATF6β, encoded by two independent genes (Yoshida et al. 2001b; Thuerauf et al. 2004). While their overall amino acid sequences are very similar, ATF6β is missing some amino acids from the serine cluster found within the luminal portion. While both forms ATF6 contain similar transcriptional activation domains, functional differences between the two isoforms are not well understood (Lee et al. 2003).

3.3 PERK

Among the three UPR inducers, the most functions are assigned to the PERK mediated signaling branch, including transcriptional responses by activation of the PERK induced transcription factor, ATF4, and translational repression (Shi et al. 1998; Harding et al. 1999, 2000a; Scheuner et al. 2001). PERK also induces subsets of both apoptotic and survival pathways possibly to prepare for critical cell fate decisions (Harding et al. 2000). Finally, PERK activation leads to cell cycle arrest at G1/S phase (Brewer et al. 1999; Brewer and Diehl 2000). During this arrest, a decision between the cell death and resumption of growth is made depending upon the extent of ER stress and the degree the UPR can induce cell recovery. Below, we will discuss each of these events in more detail.

One cellular response that takes place during the UPR in mammalian cells is a reduction in overall translation levels, which is initiated by PERK (or PEK). Similar to IRE1, PERK is a type-I ER-transmembrane serine/threonine kinase and measures ER protein folding needs with its N-terminus located within the ER lumen. Upon activation, PERK phosphorylates the α subunit of eukaryotic translation initiation factor 2 (eIF2α) on Serine 51. Phosphorylation of eIF2α traps it in an inactive form and thus interferes with the formation of the 43S translation initiation complex (Hinnebusch et al. 2000); this leads to an overall translational repression and ultimately alleviates protein folding stress by reducing the influx of newly synthesized proteins into the ER.

PERK-mediated eIF2α phosphorylation also inhibits the production of cyclin D1, leading to G1/S phase cell cycle arrest. Presumably, this G1/S phase arrest provides an opportunity for cells to evaluate both the extent of ER stress and the level of UPR response to make a critical cell fate decision between survival and death.

In addition to translational repression and cell cycle arrest at G1/S phase, PERK activation causes the production of a transcription factor, ATF4 (Harding et al. 2000a, 2002, 2003; Scheuner et al. 2001). ATF4 is a basic leucine zipper transcriptional activator that belongs to the CREB family of transcription factors. Activation of ATF4 during the UPR is unique; ATF4 transcript exists in cells all the time even prior to ER stress, though under normal conditions, its translation is repressed. Translation of the ATF4 mRNA during the UPR induces ATF4. The 5' UTR region of both human and mouse ATF4 mRNAs contain two small uORFs coding for short stretches of amino acids (Vattem and Wek 2004). The uORFs allow re-initiation complexes to form efficiently at uORF2 after scanning downstream of uORF1. uORF2 overlaps with the first three nucleotides of the ATF4 coding region; translation of uORF2 results in no ATF4 production. However, when eIF2α is phosphorylated, the re-initiation complex takes longer to form. This causes re-initiation at the ATF4 ORF to dominate over uORF2, resulting in the production of ATF4 protein.

Transcriptional target genes for ATF4 include CHOP/GADD153 (Fawcett et al. 1999; Harding et al. 2000a). In mouse embryonic fibroblasts deleted for the PERK gene, transcription of CHOP during the UPR is diminished, demonstrating that CHOP is a direct target of PERK (Harding et al. 2000a; Scheuner et al. 2001; Ma et al. 2002b; Novoa et al. 2003). CHOP/GADD153 codes for a transcription factor which heterodimerizes with the C/EBP family of transcription factors to inhibit their functions (Ron and Habener 1992). Experimental data show a strong correlation between CHOP expression and apoptotic cell death. Overexpression of CHOP causes cell cycle arrest and apoptosis, while cell death levels decrease in CHOP knockout cells. Thus, PERK plays a role in activating apoptotic genes (Barone et al. 1994; Amundson et al. 1998; Zinszner et al. 1998).

PERK is also responsible for the proteolytic cleavage and activation of procaspase-12, an ER associated caspase (Nakagawa et al. 2000). While the exact mechanism by which PERK induces procaspase-12 is not understood, caspase-12 is not activated in PERK -/- MEFs (Harding et al. 2000). Therefore, the PERK signaling branch appears to function to prepare the cell for the worst case scenario of being overwhelmed with protein folding stress. However, PERK activation does not always result in cell death, rather only selected events leading towards cell death. Paradoxically, PERK-/- cells are more susceptible to cell death during the UPR (Harding et al. 2000). These observations suggest that PERK activation leads to events to protect cells against cell death, while inducing some apoptotic events at the same time. In other words, PERK mediates downstream events promoting both cell survival and cell death simultaneously, without commitment to either one. What are the specific protective events activated by PERK during the UPR? How do cells achieve activation of two antagonistic events without favoring one over another? Furthermore, cells must make the ultimate decision at some point during ER stress. When and how do cells make this decision? How does PERK mediate partial activation of both pro- and anti-apoptotic signaling pathways? For cells undergoing the ER stress, PERK may facilitate or shorten the time required to commit to a specific outcome once the cell fate decision has been made. For example, if upregulation of UPR target genes is sufficient to alleviate ER stress during G1/S phase arrest, cells will resume a normal cell cycle. On the other hand, if the ER stress is beyond repair, the remaining apoptotic events will be activated to carry out cell death. How ER stress levels are measured, which molecules measure the stress and what molecular signals trigger the life or death decision, are all outstanding questions that remain to be answered.

Curiously, PERK activation leads to a signaling event shared by other types of cellular conditions and stress. Phosphorylation of Serine 51 on eIF2α can be carried out by at least three other kinases including double strand-activated protein kinase (PKR), heme-regulated kinase (HRI) and amino acid biosynthesis kinase (GCN2) (Chen et al. 1994; Samuel et al. 1997; Hinnebusch et al. 2000; Han et al. 2001). These kinases become activated in response to different cellular signals. PKR is activated by double stranded RNA during viral infection and HRI phosphorylates eIF2α in response to changes in concentration of heme. In contrast to PERK, however, eIF2α phosphorylation by PKR induces apoptotic cell death rather than activation of both pro- and anti-apoptotic events. Presumably apoptosis mediated by PKR is to eliminate virus infected cells. Thus, it appears that multiple cellular stresses (or events) other than the UPR share the eIF2α phosphorylation mediated signal transduction pathway, although resulting in different outcomes. How does this eIF2α phosphorylation signaling pathway distinguish among the upstream kinases in order to provide appropriate downstream responses? Genes regulated by ATF4 contain the binding site, amino acid responsive element (AARE) within the 5' UTR (Fawcett et al. 1999; Ron 2002; Rutkowski and Kaufman 2004). Analysis of promoter sequences has defined the AARE consensus sequence to be RTTKCATCA, which is also found in genes activated during the UPR. Curiously, overexpression of ATF4 does not induce AARE containing UPR target genes including CHOP even though it is sufficient for induction of other AARE containing general stress genes (Harding et al. 2000). These observations suggest that UPR specific modification of ATF4 or an additional protein(s) induced by PERK may required for activating AARE-containing target genes during the UPR.

Alternatively, PERK and other eIF2α kinases have unique phosphorylation substrates beyond eIF2α, which contribute towards different transcriptional responses. Recently, an additional protein phosphorylated by PERK has been reported, called Nrf2, which will be discussed later in this review.

3.4 Transcriptional output of three UPR signaling branches

Activation of the three UPR inducers, IRE1, ATF6, PERK, ultimately leads to transcriptional reprogramming by inducing three transcription factors: XBP1, ATF6, and ATF4. Each of these transcription factors is activated by a unique mechanism: XBP1 is produced by the IRE1 mediated splicing of *XBP1* mRNA; ATF6 is activated by proteolytic cleavage of its full-length precursor; ATF4 translation is enhanced through its preferential association with polyribosomes during the UPR. The identification of at least two UPR specific transcriptional enhancer elements, UPRE and ERSE, within the promoter regions of some of the target genes further evidences the complexity of the mammalian UPR. Specific outcomes can be assigned to each UPR inducer based on the transcription output. Genome wide transcription profiles of fibroblasts derived from XBP1 or ATF4 knockout mouse embryos or tissue culture cells lacking ATF6 have revealed that each transcription factor activates subsets of the UPR target genes (Harding et al. 2000a, 2003; Scheuner et al. 2001; Yoshida et al. 2001a; Calfon et al. 2002; Okada et al. 2002; Lee et al. 2003). For example, transcription of EDEM, DnaJ/Hsp40, p58^{IPK}, ErdJ4, HEDJ, PDI-P5, PAMP4 are regulated by XBP1; UPR transcriptional induction of these genes is greatly reduced in XBP1 knockout cells. Many proteins encoded by this set of genes function towards ERAD. For example, EDEM codes for an ER degradation enhancing α-mannosidase like protein, that might be involved in recognizing substrate proteins targeted for degradation (Wang and Hebert 2003). Thus, a major role of the IRE1-XBP1 signaling arm during the UPR is the clearance of permanently unfolded proteins (Yoshida et al. 2003). In contrast, the lack of ATF6 has no effect on the UPR induction of these genes. The PERK-ATF4 signaling branch might be responsible for cell fate decisions during the UPR, since target genes induced include CHOP, MGP, GADD45, HERP. The three UPR signaling arms responsible for activating genes involved in different and specific aspects of the UPR illustrate a complex regulatory mechanism through a multi-faceted pathway.

However, the story grows more complicated as some UPR target gene transcription appears to depend upon multiple UPR transcription factors. cDNAs coding for the N-terminal fragment of ATF6 were originally identified as a transcription factor which activates an ERSE containing reporter gene, suggesting that ERSE is sufficient for transcriptional activation by ATF6 (Haze et al. 1999; Li et al. 2000; Yoshida et al. 2000, 2001b). Consistently, CHO cells deficient in either

S1P or S2P exhibit minimal transcriptional activation of GRP78 (BiP) and GRP98 during the UPR, suggesting that activating ATF6 is necessary for inducing ERSE containing genes during the UPR (Urano et al. 2000a, 2000b; Ye et al. 2000; Lee et al. 2002). However, transcriptional induction of GRP78 is also depressed in perk-/- cells, suggesting that transcriptional induction of the ER chaperone genes GRP78 and GRP98 may also depend upon PERK activated ATF4 (Harding et al. 2000a; Novoa et al. 2001; Scheuner et al. 2001). Thus, transcription of some of the UPR target genes may be regulated by multiple transcriptional elements or, novel transcriptional enhancer elements that bind more than one UPR transcription factor exist. To date, genes that are activated exclusively by ATF6 are yet to be defined.

4 Phospholipids and the UPR

In addition to protein folding/maturation, phospholipid synthesis also takes place in the ER. ER membrane proliferation concomitant with increased immunoglobin secretion during plasma cell differentiation demonstrates a tight link between the expansion of protein folding capacity and phospholipid biosynthesis (Nikawa and Yamashita 1992; Reimold et al. 2001; Calfon et al. 2002). Thus, the UPR may also play a role in phospholipid biosynthesis in order to coordinate the increase of both protein handling capacity and physical space to accommodate increased levels of proteins in the ER. Recently, it was reported that the ectopic expression of the spliced form of XBP1 results in ER membrane expansion (Sriburi et al. 2004); this supports a connection between the two. The effect of XBP1 is specific to the ER lipids since no significant effect on cholesterol synthesis has been observed.

In addition to protein folding capacity, the ER is also sensitive to changes in lipid levels. Changes in ER lipid content also activate the UPR. Free cholesterol accumulation in the ER membrane has been found to cause UPR mediated apoptosis through ER calcium release in macrophages (Feng et al. 2003). Furthermore, abnormal accumulation of ganglioside, a major sialoglycolipid of neuronal cell membranes, also causes UPR activation (Tessitore et al. 2004). Buildup of ganglioside in the ER caused by a diminished degradative capacity of β-gal knockout lysosomes induces ER calcium release and leads to the induction of all three UPR signaling branches.

A link between the UPR and phospholipid levels in yeast has also been proposed. INO1, a major phosphoinositol regulator, is a transcriptional target of the UPR. Thus, the UPR activation induces transcription of genes to increase not only protein folding capacity but also phospholipids synthesis, expanding the physical space. Ability of not only protein folding but also lipid synthesis are required for cell growth as yeast cells deficient in UPR activation cannot sustain cell growth in the presence of high levels of unfolded proteins or in the absence of inositol. Transcriptional induction of INO1 during the UPR takes place by the heterodimeric transcription factor Ino2/4 upon inactivation of the transcriptional repressor, Opi1(Graves and Henry 2000; Chang et al. 2002). Under the normal conditions,

Opi1 binds to the hetrodimeric transcription factors Ino2/4, which are constitutively bound at the 5' UTR; Opi1 binding inhibits efficient transcription of INO1 gene. During the UPR, the Opi1 negative regulator dissociates from Ino2/4 to allow transcription of INO1 (Brickner and Walter 2004). The dissociated Opi1 binds to Scs2, a transmembrane protein; this interaction seems to be sensitive to levels of phosphoethanolamine, a major ER lipid (Loewen et al. 2004).

In mammalian cells, a similar link has been seen in antibody secreting B-lymphocytes. During the differentiation of resting mature B-cells to the antibody secreting plasma cells, IRE1 is activated to produce the spliced form of XBP1. During this differentiation, plasma cells acquire both higher secretory capacity for immunoglobulin molecules and proliferated ER. B cells deficient in XBP1 have normal numbers of activated B-lymphocytes but cannot differentiate into antibody secreting plasma cells, underscoring the importance of XBP1 for protein folding capacity and membrane proliferation. .

While these observations are consistent with a link between the UPR and ER lipid biosynthesis, ER expansion can also take place independently of the UPR in yeast. Overexpression of the multi-span ER transmembrane protein HMG-CoA reductase causes massive proliferation of the ER membrane, resulting in the formation of a structure called karmellae, which is highly proliferated ER (Masuda et al. 1983). Formation of karmellae can take place even in the absence of IRE1 or HAC1, demonstrating that the UPR is not required for membrane expansion. Thus, there may exist UPR independent mechanisms underlying ER proliferation. In addition, the shapes of proliferated membrane can come in multiple flavors. Based on EM studies, the membranes of karmellae appear to be rough ER associated with polyribosomes. Overexpression of the peroxisome protein Pex5 in yeast results in proliferation of smooth ER (Elgersma et al. 1997). What determines UPR dependent or independent proliferation of the ER? The mechanisms by which overexpression of ER proteins cause proliferated ER with different morphologies remain unclear.

5 UPR signaling arm specific components identified to date

An emerging idea based on the transcription outcomes is that individual signals mediated separately by IRE1, PERK, and ATF6 might be optimized towards specific roles within the UPR response. The mammalian UPR includes events other than target gene transcriptional responses such as translation repression, cell cycle arrest, and cell fate decision. In order to gain a better understanding of the functional roles of each signaling arm of the UPR, it will be critical to identify components and events associated with individual UPR initiators beyond their transcriptional responses. Some of these components may include phosphorylation targets since both IRE1 and PERK are kinases. A brief summary of each UPR downstream component is summarized here.

5.1 IRE1 signaling branch

A yeast two-hybrid screen with the kinase domain of IRE1β identified TRAF2, an adaptor protein for c-Jun NH2-terminal kinases (JNKs), as an IRE1 interacting protein (Urano et al. 2000b). During the UPR, TRAF2 dissociates from IRE1β leading to JNK kinase activation, which in turn activates JNK. JNK is activated by multiple environmental stresses and in response to cytokines. Activation of JNK results in activating the AP1 (Activating Protein-1) transcription factor, which is involved in embryonic morphogenesis, cellular proliferation and apoptosis. This implies that the embryonic lethal phenotype of IRE1α knockout mice may be caused in part by the lack of JNK kinase activation.

Other protein reported to interact with IRE1α include the heat shock protein Hsp90 (Marcu et al. 2001) and Jun activation domain binding protein 1 (JAB1) (Oono et al. 2004), although the functional significance of these interactions is not yet clear. Additional IRE-interacting proteins probably exist, including potential kinase substrates. So far, the only substrate known in both yeast and mammalian systems is IRE1 itself.

IRE1 endoribonuclease may also have more RNA substrates. The only RNA substrate for mammalian IRE1 identified so far is XBP1. IRE1α and IRE1β may also have different substrates and substrate specificity. While both isoforms can cleave the UPR intron from XBP1 mRNA *in vitro*, they exhibit different cleavage efficiencies suggesting biochemical differences between the two isoforms of IRE1. Previous studies have suggested that in metazoans IRE1 mRNA itself (Tirasophon et al. 2000) or 28S rRNA is cleaved in cells overexpressing IRE1β (Iwawaki et al. 2001). Although the IRE1 cleavage consensus stem/loop structure has not been found in either RNA, IRE1 dependent cleavage might downregulate either IRE1α mRNA or 28S rRNA, playing a role in feedback regulation or translational repression of the UPR. Identification of additional bona fide RNA substrates, and understanding the consequences of IRE1 dependent splicing of such RNAs should provide a better understanding of the range of functions of the IRE1 signaling branch during the UPR.

5.2 PERK signaling branch

Until recently, the best characterized substrate for the PERK kinase was eIF2α. The transcription factor Nrf2 also undergoes PERK dependent phosphorylation in NIH 3T3 cells (Cullinan et al. 2003; Cullinan and Diehl 2004). Nrf2 regulates the transcription of phase II detoxifying enzymes upon exposure to reactive oxygen species. Normally, Nrf2 is kept in the cytosol in association with Keap1; PERK mediated phosphorylation causes Nrf2 to dissociate from Keap1 and translocate to the nucleus. Cells lacking Nrf2 exhibit increased cell death upon UPR induction, suggesting one functional role of Nrf2 is cell survival. Thus, Nrf2 activation is one protective event activated by PERK during the UPR.

The E2 Hepatitis C virus envelope glycoprotein has also been shown to bind to the C-terminus of PERK (Pavio et al. 2003). This association represses the kinase activity of PERK and causes release from translational repression normally induced by PERK during the UPR. In contrast, BDVD (bovine viral diarrhea virus) activates PERK (Jordan et al. 2002). BDVD is an ER-tropic virus that belongs to the *Flaviviridae* family. For BDVD and other viruses in the *Flaviviridae* family, polyprotein processing, envelope glycoprotein biogenesis and virion formation occur in the ER. BDVD may modulate ER protein handling capacity for viral particle proliferation upon infection. Detailed mechanisms of how these viral proteins modulate PERK activity are not well understood.

5.3 ATF6 signaling branch

To date, there are no known protein(s) that associate with the N-terminus of ATF6 and contain transcription activation domains. However, proteins that regulate ATF6, including those like SCAP or SREBP that facilitate transport from the ER to the Golgi, or other factors that ensure that S1P and S2P protease cleaved ATF6 properly transports to the nucleus, might exist.

6 BiP associates with the luminal domains of IRE1, PERK, and ATF6

Signaling components that associate with the luminal portions of IRE1, PERK, and ATF6 also play a role in the UPR. Both IRE1 and PERK are transmembrane receptor kinases with their N-terminus located within the lumen of the ER; ligands specific to either IRE1 or PERK might exist. The overall architecture of these proteins is similar to that of hormone receptor kinases at the plasma membrane. However, amino acid sequences between the luminal domains of IRE1 and PERK share minimal similarity. The only similarity is contained within one hundred amino acids beyond the signal sequence at the N-terminal end. This region contains no identifiable protein motifs. Thus, the presence of specific ligands or proteins that associate with either IRE1 or PERK may be difficult to predict; novel motifs may exist on a structural level.

BiP has been the only protein shown to associate with the luminal domains of IRE1, PERK, and ATF6 (Bertolotti et al. 2000; Liu et al. 2002; Ma et al. 2002a; Shen et al. 2002a). BiP associates with all UPR inducers under normal conditions and its association becomes significantly diminished upon activation of the UPR. BiP is one of the major chaperones in the ER and is known to assist folding of nascent proteins as they are imported through Sec 61 protein channels (Hendershot 2004; Ma and Hendershot 2004). A current model proposes that BiP dissociation upon the accumulation of unfolded proteins during the UPR activates UPR components by allowing their oligomerization. During the UPR, BiP dissociates to bind to unfolded proteins, allowing either IRE1 or PERK to oligomerize and become activated (Bertolotti et al. 2000; Liu et al. 2000, 2002; Shen et al. 2002a; Kimata et al. 2003). Dissociation of BiP from ATF6 unmasks the Golgi localization signal within ATF6 allowing its transport to the Golgi for proteolytic activation (Shen et al. 2002a; Hong et al. 2004). BiP dissociation as the activation step for the UPR components relieves the necessity for the individual ligands for each UPR component. Homologous regions within the luminal domains of each respective protein may serve as the binding site for BiP, although the exact binding sites have not been identified (Liu et al. 2003)[154]. The stoichiometric relationship between BiP and the UPR components remains to be determined. A biochemical study has revealed multiple BiP binding sites within ATF6, suggesting that more than one BiP may associate with ATF6 under normal conditions. This suggests a gradation of response by releasing one BiP molecule at one time; how each or all molecules of BiP dissociate is not understood.

This model predicts that the affinity of BiP to unfolded proteins is intrinsically stronger than that to the UPR components. Therefore, BiP could easily dissociate from IRE1, PERK, or ATF6 at low concentrations of unfolded proteins; it is not clear what assures the relatively week association of BiP to the UPR components under non-stressed conditions. BiP dissociation by itself could be sufficient for activation of the UPR components, or alternatively additional steps and/or protein components could be required for facilitating dimerization. This might be the case for ATF6 as BiP dissociation is not sufficient for ATF6 activation. These observations suggest an additional component is required for transporting of ATF6 from the ER lumen to the Golgi. Another interesting question is why IRE1 and PERK preferentially homodimerize rather than forming an IRE1/PERK heterodimer upon BiP dissociation; so far, no examples of IRE1-PERK heterodimers have been reported. Additional component(s) might exist to ensure either homodimerization or inhibit heterodimerization of IRE1 and PERK.

7 Time dependent shift of the UPR response

The presence of three ER proximal components provides unique opportunities for regulating the UPR response according to cellular needs. Certain situations might exist where signaling events induced by only subsets of the UPR responses are sufficient. IRE1, PERK, and ATF6 are expressed ubiquitously, although the relative concentration of these proteins in specific tissues or at certain developmental stages has not been carefully examined. Preferential activation of parts of the UPR signaling branches could be achieved by modulating the concentration of the UPR components. Activation of the three UPR signaling branches may also change to provide best-suited response depending upon the type of ER stress. Pharmacological agents used to induce the UPR are reported to uniformly activate all three ER membrane UPR components as well as their respective downstream signaling branches. The specificities of the sensor domains of each of the three ER proximal components towards various types of the ER stress conditions have not been investigated. It remains to be seen if any physiological conditions exist where only one or two signaling branches become activated.

The current view of the kinetic behaviors of the three UPR signaling branches holds that translational repression caused by phosphorylation of eIF2α takes place immediately after induction of the UPR (Harding et al. 2000; Yoshida et al. 2003). Among all three ER proximal components become activated, phosphorylation of eIF2α is thought to require the least amount of time. This step provides the first line of defense towards alleviating the unfolded protein stress of the ER by reducing the overall translation of cells. The production of ATF4 transcription factor is thought to occur next in the timeline. ATF4 mRNA is present in the cell prior to the UPR induction and its translation takes place on this pre-existing mRNA. The proteolytic activation step of the precursor form of ATF6 is thought to be similarly fast (Yoshida et al. 2001a, 2003). Consequently, transcription of genes induced by either ATF4 or ATF6 is thought to occur relatively early in the UPR. On the other hand, producing the more potent spliced form of XBP1 protein presumably takes more time as it requires the IRE1 mediated splicing step. ATF6 binds tightly to the ERSE, whereas XBP1 binds to the UPRE. Genes containing the ERSE include those coding for ER resident chaperons and protein folding enzymes. Therefore, the machinery required to fold proteins should become available first. In contrast, the UPRE containing genes include those involved in ERAD (ER mediated degradation) that clear permanently misfolded proteins, and are not required until after folding efforts are made.

In summary, these events occur in a specific temporal sequence; translation levels decrease to minimize further burden to the ER, followed by an increase in protein folding capacity. Together, these events work to enhance the protein folding capacity of the ER (Harding et al. 2000a, 2003; Scheuner et al. 2001; Okada et al. 2002; Lee et al. 2003). In the meantime, ERAD components are produced to eliminate permanently unfolded proteins. Finally, components required for cell fate decisions become available to make critical cell fate decisions according to the condition of the ER.

8 Physiological roles of the UPR

Understanding the behaviors of the UPR in biologically relevant conditions is critical for evaluating the physiological roles of the response. Experimentally, pharmacological drugs are often used to study the UPR. For example, DTT induces the UPR by disrupting disufide bonds, while tunicamycin causes the accumulation of unfolded proteins by inhibiting gycosylation. Using these drugs has facilitated the identification of UPR components and the dissection of their functions. However, further study of UPR component behaviors under physiologically activating conditions are required to uncover their regulatory functions in the context of the UPR.

Multiple physiological conditions where the UPR becomes induced have been described. IRE1α is important for embryonic development as evidenced by the

IRE1 mouse knockout phenotype; mice deficient for IRE1α die between days 9.5 and 11.5 of gestation (Urano et al. 2000b). The molecular basis for this phenotype is not well understood but it demonstrates a critical role for IRE1 and thus suggests that IRE1 mediated UPR signaling events play a role during early mammalian development. It is also possible that importance of IRE1 during the development is mediated by its function beyond UPR.

The UPR has been shown to play critical roles in cells specialized for secretion, such as antibody secreting B-lymphocytes (Calfon et al. 2002; Iwakoshi et al. 2003). XBP1 was originally identified as a transcription factor critical for plasma cell development (Reimold et al. 2001). In the absence of XBP1, plasma cell differentiation from mature B-lymphocytes is severely diminished, even though earlier B-cell maturation steps occur normally. IRE1 dependent splicing of XBP1 mRNA takes place in resting mature B-cells as they differentiate into plasma cells, inducing the UPR to prepare cells for increased protein handling and secretion in the ER (Iwakoshi et al. 2003; van Anken et al. 2003).

Another example demonstrating the importance of the UPR in cells specialized for secretion is the insulin secreting βcell in pancreas. Juvenile diabetic patients have no pancreatic β-cells (Harding and Ron 2002; Kaufman 2002; Ron 2002). The cause of this phenotype was recently mapped to the gene coding for PERK. This particular mutation truncates full-length PERK causing patients with this mutation to have no PERK activity. While both IRE1 and ATF6 are presumably functional in these β-cells, increased demand of the ER protein folding capacity must require all three UPR components in order to produce insulin.

Roles of the UPR in cells other than secretory cells have not yet been described. Evidence supporting a physiological importance of the UPR beyond secretory function lies in the fact that each signaling arm of the UPR may be optimized for certain functions, and that differential expression levels of each UPR component may exist in individual tissues. These proteins may function in other signal transduction pathways not directly related to the UPR.

9 Conclusions

The UPR is a signal transduction pathway filled with exciting findings and surprises. Ever since the first UPR component, IRE1 was identified in yeast, discoveries of other UPR components have happened at incredibly fast rate. IRE1 is an ER transmembrane receptor kinase with sequence specific endoribonuclease activity. IRE1 mediated non-conventional splicing of mRNA coding for the UPR specific transcription factor is the major switch to turn on the UPR pathway in yeast. Extending the understanding of the yeast UPR, studies in higher eukaryotic cells started to provide full scope of the signaling pathway and to uncover physiological role(s) of the UPR. IRE1 and IRE1 mediated splicing is conserved in mammalian cells and is one of the critical regulatory steps of activating the UPR. In addition, two ER transmembrane components, PERK and ATF6, function as the UPR initiators in mammalian cells. PERK is an ER transmembrane receptor kinase, which phosphorylates a translation initiator eIF2α, while ATF6 is an ER transmembrane transcription factor, activated by proteolytic processing within the transmembrane domain. Activation of these proteins leads to variety of events beyond transcription.

Functional consequences of activating three UPR initiators are far more complex in mammalian cells and extend beyond induction of both ER protein handling and secretory capacity. Furthermore, while the UPR mediates signaling between the ER and the nucleus, additional communication between the ER and other organelles may participate in establishment of ER homeostasis. Ultimately, dissecting complexity of the mammalian UPR and providing deeper understanding will guide our efforts towards designing strategies to combat human diseases caused by failure of ER protein folding.

Acknowledgements

We apologize to those whose work could only be cited by reference to review articles.

References

Abeijon C, Hirschberg CB (1992) Topography of glycosylation reactions in the endoplasmic reticulum. Trends Biochem Sci 17:32-36

Abelson J (1994) Transfer-Rna Splicing. FASEB J 8:A1259-A1259

Amundson SA, Zhan Q, Penn LZ, Fornace AJ (1998) Myc suppresses induction of the growth arrest genes gadd34, gadd45, and gadd153 by DNA-damaging agents. Oncogene 17:2149-2154

Barone MV, Crozat A, Tabaee A, Philipson L, Ron D (1994) Chop (Gadd153) and its oncogenic variant, Tls-Chop, have opposing effects on the induction of G(1)/S arrest. Genes Dev 8:453-464

Bertolotti A, Wang XZ, Novoa I, Jungreis R, Schlessinger K, Cho JH, West AB, Ron D (2001) Increased sensitivity to dextran sodium sulfate colitis in IRE1 beta-deficient mice. J Clin Invest 107:585-593

Bertolotti A, Zhang Y, Hendershot LM, Harding HP, Ron D (2000) Dynamic interaction of BiP and ER stress transducers in the unfolded-protein response. Nat Cell Biol 2:326-332

Bork P, Sander C (1993) A hybrid protein kinase-Rnase in an interferon-induced pathway. FEBS Lett 334:149-152

Brewer JW, Diehl JA (2000) PERK mediates cell-cycle exit during the mammalian unfolded protein response. Proc Natl Acad Sci USA 97:12625-12630

Brewer JW, Hendershot LM, Sherr CJ, Diehl JA (1999) Mammalian unfolded protein response inhibits cyclin D1 translation and cell-cycle progression. Proc Natl Acad Sci USA 96:8505-8510

Brickner JH, Walter P (2004) Gene recruitment of the activated INO1 locus to the nuclear membrane. PLOS Biology 2:1843-1853

Brown JD, Ng DTW, Ogg SC, Walter P (1995) Targeting pathways to the endoplasmic reticulum membrane. Cold Spring Harbor Symp Quant Biol 60:23-30

Brown MS, Goldstein JL (1997) The SREBP pathway: Regulation of cholesterol metabolism by proteolysis of a membrane-bound transcription factor. Cell 89:331-340

Calfon M, Zeng H, Urano F, Till JH, Hubbard SR, Harding HP, Clark SG, Ron D (2002) IRE1 couples endoplasmic reticulum load to secretory capacity by processing the XBP-1 mRNA. Nature 415:92-96

Casagrande R, Stern P, Diehn M, Shamu C, Osario M, Zuniga M, Brown PO, Ploegh H (2000) Degradation of proteins from the ER of *S. cerevisiae* requires an intact unfolded protein response pathway. Mol Cell 5:729-735

Chang HJ, Jones EW, Henry SA (2002) Role of the unfolded protein response pathway in regulation of INO1 and in the sec14 bypass mechanism in *Saccharomyces cerevisiae.* Genetics 162:29-43

Chen JJ, Crosby JS, London IM (1994) Regulation of heme-regulated Eif-2-alpha kinase and its expression in erythroid-cells. Biochimie 76:761-769

Cox JS, Chapman RE, Walter P (1997) The unfolded protein response coordinates the production of endoplasmic reticulum protein and endoplasmic reticulum membrane. Mol Biol Cell 8:1805-1814

Cox JS, Shamu CE, Walter P (1993) Transcriptional induction of genes encoding endoplasmic reticulum resident proteins requires a transmembrane protein kinase. Cell 73:1197-1206

Cox JS, Walter P (1996) A novel mechanism for regulating activity of a transcription factor that controls the unfolded protein response. Cell 87:391-404

Cullinan SB, Diehl JA (2004) PERK-dependent activation of Nrf2 contributes to redox homeostasis and cell survival following endoplasmic reticulum stress. J Biol Chem 279:20108-20117

Cullinan SB, Zhang D, Hannink M, Arvisais E, Kaufman RJ, Diehl JA (2003) Nrf2 is a direct PERK substrate and effector of PERK-dependent cell survival. Mol Cell Biol 23:7198-7209

Dong BH, Niwa M, Walter P, Silverman RH (2001) Basis for regulated RNA cleavage by functional analysis of RNase L and Ire1p. RNA 7:361-373

Dong BH, Silverman RH (1997) A bipartite model of 2-5A-dependent RNase L. J Biol Chem 272:22236-22242

Elgersma Y, Kwast L, van den Berg M, Snyder WB, Distel B, Subramani S, Tabak HF (1997) Overexpression of Pex15p, a phosphorylated peroxisomal integral membrane protein required for peroxisome assembly in *S. cerevisiae*, causes proliferation of the endoplasmic reticulum membrane. EMBO J 16:7326-7341

Ellgaard L, Helenius A (2003) Quality control in the endoplasmic reticulum. Nat Rev Mol Cell Biol 4:181-191

Fawcett TW, Martindale JL, Guyton KZ, Hai T, Holbrook NJ (1999) Complexes containing activating transcription factor (ATF)/cAMP-responsive-element-binding protein (CREB) interact with the CCAAT enhancer-binding protein (C/EBP)-ATF composite site to regulate Gadd153 expression during the stress response. Biochem J 339:135-141

Feng B, Yao PM, Li YK, Devlin CM, Zhang DJ, Harding HP, Sweeney M, Rong JX, Kuriakose G, Fisher EA, Marks AR, Ron D, Tabas I (2003) The endoplasmic reticulum is the site of cholesterol-induced cytotoxicity in macrophages. Nat Cell Biol 5:781-792

Fewell SW, Travers KJ, Weissman JS, Brodsky JL (2001) The action of molecular chaperones in the early secretory pathway. Ann Rev Genet 35:149-191

Friedlander R, Jarosch E, Urban J, Volkwein C, Sommer T (2000) A regulatory link between ER-associated protein degradation and the unfolded-protein response. Nat Cell Biol 2:379-384

Gething MJ (1999) Role and regulation of the ER chaperone BiP. Sem Cell Dev Biol 10:465-472

Gilmore R (1993) Protein translocation across the endoplasmic-reticulum - a tunnel with toll booths at entry and exit. Cell 75:589-592

Gonzalez TN, Sidrauski C, Dorfler S, Walter P (1999) Mechanism of non-spliceosomal mRNA splicing in the unfolded protein response pathway. EMBO J 18:3119-3132

Graves JA, Henry SA (2000) Regulation of the yeast INO1 gene: The products of the IN02, IN04 and OPI1 regulatory genes are not required for repression in response to inositol. Genetics 154:1485-1495

Hammond C, Braakman I, Helenius A (1994) Role of N-linked oligosaccharide recognition, glucose trimming, and calnexin in glycoprotein folding and quality-control. Proc Natl Acad Sci USA 91:913-917

Han AP, Yu C, Lu LR, Fujiwara Y, Browne C, Chin G, Fleming M, Leboulch P, Orkin SH, Chen JJ (2001) Heme-regulated eIF2 alpha kinase (HRI) is required for translational regulation and survival of erythroid precursors in iron deficiency. EMBO J 20:6909-6918

Harding HP, Calfon M, Urano F, Novoa I, Ron D (2002) Transcriptional and translational control in the mammalian unfolded protein response. Ann Rev Cell Mol Biol 18:575-599

Harding HP, Novoa I, Zhang YH, Zeng HQ, Wek R, Schapira M, Ron D (2000a) Regulated translation initiation controls stress-induced gene expression in mammalian cells. Mol Cell 6:1099-1108

Harding HP, Ron D (2002) Endoplasmic reticulum stress and the development of diabetes - a review. Diabetes 51:S455-S461

Harding HP, Zhang Y, Ron D (1999) Protein translation and folding are coupled by an endoplasmic-reticulum-resident kinase. Nature 397:271-274

Harding HP, Zhang YH, Bertolotti A, Zeng HQ, Ron D (2000) PERK is essential for translational regulation and cell survival during the unfolded protein response. Mol Cell 5:897-904

Harding HP, Zhang YH, Zeng HQ, Novoa I, Lu PD, Calfon M, Sadri N, Yun C, Popko B, Paules R, Stojdl DF, Bell JC, Hettmann T, Leiden JM, Ron D (2003) An integrated stress response regulates amino acid metabolism and resistance to oxidative stress. Mol Cell 11:619-633

Haze K, Yoshida H, Yanagi H, Yura T, Mori K (1999) Mammalian transcription factor ATF6 is synthesized as a transmembrane protein and activated by proteolysis in response to endoplasmic reticulum stress. Mol Biol Cell 10:3787-3799

Hebert DN, Foellmer B, Helenius A (1995) Glucose trimming and reglucosylation determine glycoprotein association with calnexin in the endoplasmic-reticulum. Cell 81:425-433

Helenius A, Aebi M (2004) Roles of N-linked glycans in the endoplasmic reticulum. Ann Rev Biochem 73:1019-1049

Hendershot LM (2004) The ER chaperone BiP is a master regulator of ER function. Mt Sinai J Med 71:289-297

Hinnebusch AG (1985) A hierarchy of trans-acting factors modulates translation of an activator of amino acid biosynthetic genes in *Saccharomyces cerevisiae*. Mol Cell Biol 5:2349-2360

Hinnebusch AG (1997) Translational regulation of yeast GCN4. A window on factors that control initiator-trna binding to the ribosome. J Biol Chem 272:21661-21664

Holkeri H, Makarow M (1998) Different degradation pathways for heterologous glycoproteins in yeast. FEBS Lett 429:162-166

Hong M, Luo SZ, Baumeister P, Huang JM, Gogia RK, Li MQ, Lee AS (2004) Underglycosylation of ATF6 as a novel sensing mechanism for activation of the unfolded protein response. J Biol Chem 279:11354-11363

Hopper AK, Phizicky EM (2003) tRNA transfers to the limelight. Genes Dev 17:162-180

Hosokawa N, Tremblay LO, You ZP, Herscovics A, Wada I, Nagata K (2003) Enhancement of endoplasmic reticulum (ER) degradation of misfolded null Hong Kong alpha(1)-antitrypsin by human ER mannosidase I. J Biol Chem 278:26287-26294

Hosokawa N, Wada I, Hasegawa K, Yorihuzi T, Tremblay LO, Herscovics A, Nagata K (2001) A novel ER alpha-mannosidase-like protein accelerates ER-associated degradation. EMBO Rep 2:415-422

Hua XX, Nohturfft A, Goldstein JL, Brown MS (1996) Sterol resistance in CHO cells traced to point mutation in SREBP cleavage-activating protein. Cell 87:415-426

Huang RFS, Huang SM, Lin BS, Wei JS, Liu TZ (2001) Homocysteine thiolactone induces apoptotic DNA damage mediated by increased intracellular hydrogen peroxide and caspase 3 activation in HL-60 cells. Life Sci 68:2799-2811

Iwakoshi NN, Lee AH, Vallabhajosyula P, Otipoby KL, Rajewsky K, Glimcher LH (2003) Plasma cell differentiation and the unfolded protein response intersect at the transcription factor XBP-1. Nat Immunol 4:321-329

Iwawaki T, Hosoda A, Okuda T, Kamigori Y, Nomura-Furuwatari C, Kimata Y, Tsuru A, Kohno K (2001) Translational control by the ER transmembrane kinase/ribonuclease IRE1 under ER stress. Nat Cell Biol 3:158-164

Jordan R, Wang LJ, Graczyk TM, Block TM, Romano PR (2002) Replication of a cytopathic strain of bovine viral diarrhea virus activates PERK and induces endoplasmic reticulum stress- mediated apoptosis of MDBK cells. J Virol 76:9588-9599

Kaufman RJ (1999) Stress signaling from the lumen of the endoplasmic reticulum: coordination of gene transcriptional and translational controls. Genes Dev 13:1211-1233

Kaufman RJ (2002) Orchestrating the unfolded protein response in health and disease. J Clin Invest 110:1389-1398

Kim PS, Arvan P (1998) Endocrinopathies in the family of endoplasmic reticulum (ER) storage diseases: disorders of protein trafficking and the role of ER molecular chaperones. Endocrine Revi 19:173-202

Kimata YL, Shimizu Y, Abe H, Farcasanu RC, Takeuchi M, Rose MD, Kohno K (2003) Genetic evidence for a role of BiP/Kar2 that regulates Ire1 in response to accumulation of unfolded proteins. Mol Biol Cell 14:2559-2569

Kimata Y, Oikawa D, Shimizu Y, Ishiwata-Kimata Y, Kohno KA (2004) A role for BiP as an adjustor for the endoplasmic reticulum stress-sensing protein Ire1. J Cell Biol 167:445-456

Kleizen B, Braakman I (2004) Protein folding and quality control in the endoplasmic reticulum. Curr Opin Cell Biol 16:597-597

Kokame K, Kato H, Miyata T (2001) Identification of ERSE-II, a new cis-actin element responsible for the ATF6-dependent mammalian unfolded protein response. J Biol Chem 276:9199-9205

Kuznetsov G, Nigam SK (1998) Mechanisms of disease - folding of secretory and membrane proteins. New Eng J Med 339:1688-1695

Lee AH, Iwakoshi NN, Glimcher LH (2003) XBP-1 regulates a subset of endoplasmic reticulum resident chaperone genes in the unfolded protein response. Mol Cell Biol 23:7448-7459

Lee K, Tirasophon W, Shen XH, Michalak M, Prywes R, Okada T, Yoshida H, Mori K, Kaufman RJ (2002) IREI-mediated unconventional mRNA splicing and S2P-mediated ATF6 cleavage merge to regulate XBP1 in signaling the unfolded protein response. Genes Dev 16:452-466

Li MQ, Baumeister P, Roy B, Phan T, Foti D, Luo SZ, Lee AS (2000) ATF6 as a transcription activator of the endoplasmic reticulum stress element: Thapsigargin stress-induced changes and synergistic interactions with NF-Y and YY1. Mol Cell Biol 20:5096-5106

Liu CY, Schroder M, Kaufman RJ (2000) Ligand-independent dimerization activates the stress response kinases IRE1 and PERK in the lumen of the endoplasmic reticulum. J Biol Chem 275:24881-24885

Liu CY, Wong HN, Schauerte JA, Kaufman RJ (2002) The protein kinase/endoribonuclease IRE1alpha that signals the unfolded protein response has a luminal N-terminal ligand-independent dimerization domain. J Biol Chem 277:18346-18356

Liu CY, Xu ZH, Kaufman RJ (2003) Structure and intermolecular interactions of the luminal dimerization domain of human IRE1 alpha. J Biol Chem 278:17680-17687

Loewen CJR, Gaspar ML, Jesch SA, Delon C, Ktistakis NT, Henry SA, Levine TP (2004) Phospholipid metabolism regulated by a transcription factor sensing phosphatidic acid. Science 304:1644-1647

Ma K, Vattem KM, Wek RC (2002a) Dimerization and release of molecular chaperone inhibition facilitate activation of eukaryotic initiation factor-2 kinase in response to endoplasmic reticulum stress. J Biol Chem 277:18728-18735

Ma Y, Brewer JW, Diehl JA, Hendershot LM (2002b) Two distinct stress signaling pathways converge upon the CHOP promoter during the mammalian unfolded protein response. J Mol Biol 318:1351-1365

Ma YJ, Hendershot LM (2004) ER chaperone functions during normal and stress conditions. J Chem Neuroanat 28:51-65

Marcu MG, Hendershot L, Bertolotti A, Ron D, Neckers L (2001) Heat shock protein 90 (Hsp90) stabilizes IRE1 and PERK, transmembrane kinases of the endoplasmic reticulum that mediate the unfolded protein response. Clin Cancer Res 7:620

Masuda A, Kuwano M, Shimada T (1983) Ultrastructural-changes during the enhancement of cellular 3-hydroxy-3-methylglutaryl-coenzyem-a reductase in a Chinese-hamster cell mutant resistant to compactin (Ml236b). Cell Struct Func 8:309-312

Matlack KES, Mothes W, Rapoport TA (1998) Protein translocation: tunnel vision. Cell 92:381-390

Mori K (1999) Cellular response to endoplasmic reticulum stress mediated by unfolded protein response pathway. Tanpakushitsu Kakusan Koso 44:2442-2448

Mori K (2000) Tripartite management of unfolded proteins in the endoplasmic reticulum. Cell 101:451-454

Mori K, Kawahara T, Yanagi H, Yura T (1997) ER stress-induced mRNA splicing permits synthesis of transcription factor Hac1p/Ern4p that activates the unfolded protein response. Mol Biol Cell 8:2056-2056

Mori K, Kawahara T, Yoshida H, Yanagi H, Yura T (1996) Molecular cloning and characterization of a major component of the unfolded protein-response transcription factor. Mol Biol Cell 7:770-770

Mori K, Ma WZ, Gething MJ, Sambrook J (1993) A Transmembrane protein with a Cdc2+/Cdc28-related kinase- activity is required for signaling from the ER to the nucleus. Cell 74:743-756

Mori K, Sant A, Kohno K, Normington K, Gething MJ, Sambrook JF (1992) A 22 bp cis-acting element is necessary and sufficient for the induction of the yeast KAR2 (BiP) gene by unfolded proteins. EMBO J 11:2583-2593

Nakagawa T, Zhu H, Morishima N, Li E, Xu J, Yankner BA, Yuan JY (2000) Caspase-12 mediates endoplasmic-reticulum-specific apoptosis and cytotoxicity by amyloid-beta. Nature 403:98-103

Nikawa JI, Yamashita S (1992) Ire1 encodes a putative protein kinase containing a membrane-spanning domain and is required for inositol phototrophy in *Saccharomyces-cerevisiae*. Mol Microbiol 6:1441-1446

Niwa M, Sidrauski C, Kaufman RJ, Walter P (1999) A role for presenilin-1 in nuclear accumulation of Ire1 fragments and induction of the mammalian unfolded protein response. Cell 99:691-702

Niwa M, Patil CK, DeRisi J, Walter P (2004) Genome-scale approaches for discovering novel non-conventional splicing substrates of the Ire1 nuclease. Genome Biol 6:R3.1-10

Nohturfft A, DeBose-Boyd RA, Scheek S, Goldstein JL, Brown MS (1999) Sterols regulate cycling of SREBP cleavage-activating protein (SCAP) between endoplasmic reticulum and Golgi. Proc Natl Acad Sci USA 96:11235-11240

Nohturfft A, Yabe D, Goldstein JL, Brown MS, Espenshade PJ (2000) Regulated step in cholesterol feedback localized to budding of SCAP from ER membranes. Cell 102:315-323

Novoa I, Zeng HQ, Harding HP, Ron D (2001) Feedback inhibition of the unfolded protein response by GADD34- mediated dephosphorylation of eIF2 alpha. J Cell Biol 153:1011-1021

Novoa I, Zhang YH, Zeng HQ, Jungreis R, Harding HP, Ron D (2003) Stress-induced gene expression requires programmed recovery from translational repression. EMBO J 22:1180-1187

Oda Y, Hosokawa N, Wada I, Nagata K (2003) EDEM as an acceptor of terminally misfolded glycoproteins released from calnexin. Science 299:1394-1397

Okada T, Yoshida H, Akazawa R, Negishi M, Mori K (2002) Distinct roles of activating transcription factor 6 (ATF6) and double-stranded RNA-activated protein kinase-like endoplasmic reticulum kinase (PERK) in transcription during the mammalian unfolded protein response. Biochem J 366:585-594

Oono K, Yoneda T, Manabe T, Yamagishi S, Matsuda S, Hitomi J, Miyata S, Mizuno T, Imaizumi K, Katayama T, Tohyama M (2004) JAB1 participates in unfolded protein responses by association and dissociation with IRE1. Neurochem Int 45:765-772

Outinen PA, Sood SK, Pfeifer SI, Pamidi S, Podor TJ, Li J, Weitz JI, Austin RC (1999) Homocysteine-induced endoplasmic reticulum stress and growth arrest leads to specific changes in gene expression in human vascular endothelial cells. Blood 94:959-967

Papa FR, Zhang C, Shokat K, Walter P (2003) Bypassing a kinase activity with an ATP-competitive drug. Science 302:1533-1537

Patil C, Walter P (2001) Intracellular signaling from the endoplasmic reticulum to the nucleus: the unfolded protein response in yeast and mammals. Curr Opin Cell Biol 13:349-355

Patil CK, Li H, Walter P (2004) Gcn4p and novel upstream activating sequences regulate targets of the unfolded protein response. Plos Biology 2:1208-1223

Pavio N, Romano PR, Graczyk TM, Feinstone SM, Taylor DR (2003) Protein synthesis and endoplasmic reticulum stress can be modulated by the hepatitis C virus envelope protein E2 through the eukaryotic initiation factor 2 alpha kinase PERK. J Virol 77:3578-3585

Reimold A, Friend D, Alt F, Glimcher L (1999) Control of terminal B cell differentiation by transcription factor XBP-1. Arthritis Rheum 42:S177-S177

Reimold AM, Iwakoshi NN, Manis J, Vallabhajosyula P, Szomolanyi-Tsuda E,

Gravallese EM, Friend D, Grusby MJ, Alt F, Glimcher LH (2001) Plasma cell differentiation requires the transcription factor XBP-1. Nature 412:300-307

Ron D (2002) Translational control in the endoplasmic reticulum stress response. J Clin Invest 110:1383-1388

Ron D, Habener JF (1992) Chop, a novel developmentally regulated nuclear-protein that dimerizes with transcription factors C/Ebp and Lap and functions as a dominant-negative inhibitor of gene-transcription. Genes Dev 6:439-453

Roy B, Lee AS (1999) The mammalian endoplasmic reticulum stress response element consists of an evolutionarily conserved tripartite structure and interacts with a novel stress-inducible complex. Nucleic Acids Res 27:1437-1443

Ruegsegger U, Leber JH, Walter P (2001) Block of HAC1 mRNA translation by long-range base pairing is released by cytoplasmic splicing upon induction of the unfolded protein response. Cell 107:103-114

Rutkowski DT, Kaufman RJ (2004) A trip to the ER: coping with stress. Trends Cell Biol 14:20-28

Sakai J, Rawson RB, Espenshade PJ, Cheng D, Seegmiller AC, Goldstein JL,

Brown MS (1998) Molecular identification of the sterol-regulated luminal protease that cleaves SREBPs and controls lipid composition of animal cells. Mol Cell 2:505-514

Samuel CE, Kuhen KL, George CX, Ortega LG, RendeFournier R, Tanaka H (1997) The PKR protein kinase - An interferon-inducible regulator of cell growth and differentiation. Int J Hematol 65:227-237

Scheuner D, Song B, McEwen E, Liu C, Laybutt R, Gillespie P, Saunders T, Bonner-Weir S, Kaufman RJ (2001) Translational control is required for the unfolded protein response and in vivo glucose homeostasis. Mol Cell 7:1165-1176

Shamu CE (1998) Splicing: HACking into the unfolded-protein response. Curr Biol 8:R121-R123

Shamu CE, Walter P (1996) Oligomerization and phosphorylation of the Ire1p kinase during intracellular signaling from the endoplasmic reticulum to the nucleus. EMBO J 15:3028-3039

Shen JS, Chen X, Hendershot L, Prywes R (2002a) ER stress regulation of ATF6 localization by dissociation of BiP/GRP78 binding and unmasking of Golgi localization signals. Dev Cell 3:99-111

Shen XJ, Ellis RE, Lee K, Liu CY, Yang K, Solomon A, Yoshida H, Morimoto R, Kurnit DM, Mori K, Kaufman RJ (2002b) Complementary signaling pathways regulate the unfolded protein response and are required for *C. elegans* development. FASEB J 16:A891-A891

Shi YG, Taylor SI, Tan SL, Sonenberg N (2003) When translation meets metabolism: Multiple links to diabetes. Endocr Rev 24:91-101

Shi YG, Vattem KM, Sood R, An J, Liang JD, Stramm L, Wek RC (1998) Identification and characterization of pancreatic eukaryotic initiation factor 2 alpha-subunit kinase, PEK, involved in translational control. Mol Cell Biol 18:7499-7509

Sidrauski C, Chapman R, Walter P (1998) The unfolded protein response: an intracellular signalling pathway with many surprising features. Trends Cell Biol 8:245-249

Sidrauski C, Cox JS, Walter P (1996) tRNA ligase is required for regulated mRNA splicing in the unfolded protein response. Cell 87:405-413

Sidrauski C, Walter P (1997) The transmembrane kinase Ire1p is a site-specific endonuclease that initiates mRNA splicing in the unfolded protein response. Cell 90:1031-1039

Sitia R, Helenius A, Swoboda BEP (2001) Quality control in the secretory assembly line - discussion. Philos Trans R Soc Lond B Biol Sci 356:150-150

Spinelli SL, Consaul SA, Phizicky EM (1997) A conditional lethal yeast phosphotransferase (tpt1) mutant accumulates tRNAs with a 2 '-phosphate and an undermodified base at the splice junction. RNA 3:1388-1400

Sriburi R, Jackowski S, Mori K, Brewer JW (2004) XBP1: a link between the unfolded protein response, lipid biosynthesis, and biogenesis of the endoplasmic reticulum. J Cell Biol 167:35-41

Tessitore A, del P Martin M, Sano R, Ma YJ, Mann L, Ingrassia A, Laywell ED, Steindler DA, Hendershot LM, d'Azzo A (2004) G(M1)-ganglioside-mediated activation of the unfolded protein response causes neuronal death in a neurodegenerative gangliosidosis. Mol Cell 15:753-766

Thuerauf DJ, Morrison L, Glembotski CC (2004) Opposing roles for ATF6 alpha and ATF6 beta in endoplasmic reticulum stress response gene induction. J Biol Chem 279:21078-21084

Tirasophon W, Lee K, Callaghan B, Welihinda A, Kaufman RJ (2000) The endoribonuclease activity of mammalian IRE1 autoregulates its mRNA and is required for the unfolded protein response. Genes Dev 14:2725-2736

Tirasophon W, Welihinda AA, Kaufman RJ (1998) A stress response pathway from the endoplasmic reticulum to the nucleus requires a novel bifunctional protein kinase/endoribonuclease (Ire1p) in mammalian cells. Genes Dev 12:1812-1824

Travers KJ, Patil CK, Wodicka L, Lockhart DJ, Weissman JS, Walter P (2000) Functional and genomic analyses reveal an essential coordination between the unfolded protein response and ER-associated degradation. Cell 101:249-258

Trotta CR, Miao F, Arn EA, Stevens SW, Ho CK, Rauhut R, Abelson, JN (1997) The yeast tRNA splicing endonuclease: A tetrameric enzyme with two active site subunits homologous to the archaeal tRNA endonucleases. Cell 89:849-858

Urano F, Bertolotti A, Ron D (2000a) IRE1 and efferent signaling from the endoplasmic reticulum. J Cell Sci 113:3697-3702

Urano F, Wang X, Bertolotti A, Zhang Y, Chung P, Harding HP, Ron D (2000b) Coupling of stress in the ER to activation of JNK protein kinases by transmembrane protein kinase IRE1. Science 287:664-666 van Anken E, Romijn EP, Maggioni C, Mezghrani A, Sitia R, Braakman I, Heck AJR (2003) Sequential waves of functionally related proteins are expressed when B cells prepare for antibody secretion. Immunity 18:243-253

Vattem KM, Wek RC (2004) Re-initiation involving upstream ORFs regulates ATF4 mRNA translation in mammalian cells. Proc Natl Acad Sci USA 101:11269-11274

Voeltz GK, Rolls MM, Rapoport TA (2002) Structural organization of the endoplasmic reticulum. EMBO Rep 3:944-950

Wang T, Hebert DN (2003) EDEM an ER quality control receptor. Nat Struct Biol 10:319-321

Wang XD, Sato R, Brown MS, Hua XX, Goldstein JL (1994) Srebp-1, a Membrane-bound transcription factor released by sterol-regulated proteolysis. Cell 77:53-62

Wang XZ, Harding HP, Zhang YH, Jolicoeur EM, Kuroda M, Ron D (1998) Cloning of mammalian Ire1 reveals diversity in the ER stress responses. EMBO J 17:5708-5717

Wang Y, Shen J, Arenzana N, Tirasophon W, Kaufman RJ, Prywes R (2000) Activation of ATF6 and an ATF6 DNA binding site by the endoplasmic reticulum stress response. J Biol Chem 275:27013-27020

Welihinda AK, Tirasophon W, Kaufman RJ (2000) The transcriptional co-activator ADA5 is required for HAC1 mRNA processing *in vivo*. J Biol Chem 275:3377-3381

Werstuck GH, Lentz SR, Dayal S, Hossain GS, Sood SK, Shi YY, Zhou J, Maeda N, Krisans SK, Malinow MR, Austin RC (2001) Homocysteine-induced endoplasmic reticulum stress causes dysregulation of the cholesterol and triglyceride biosynthetic pathways. J Clin Invest 107:1263-1273

Ye J, Rawson RB, Komuro R, Chen X, Dave UP, Prywes R, Brown MS, Goldstein JL (2000) ER stress induces cleavage of membrane-bound ATF6 by the same proteases that process SREBPs. Mol Cell 6:1355-1364

Yoshida H, Haze K, Yanagi H, Yura T, Mori K (1998) Identification of the cis-acting endoplasmic reticulum stress response element responsible for transcriptional induction of mammalian glucose-regulated proteins - Involvement of basic leucine zipper transcription factors. J Biol Chem 273:33741-33749

Yoshida H, Matsui T, Hosokawa N, Kaufman RJ, Nagata K, Mori K (2003) A time-dependent phase shift in the mammalian unfolded protein response. Dev Cell 4:265-271

Yoshida H, Matsui T, Yamamoto A, Okada T, Mori K (2001a) XBP1 mRNA is induced by ATF6 and spliced by IRE1 in response to ER stress to produce a highly active transcription factor. Cell 107:881-891

Yoshida H, Okada T, Haze K, Yanagi H, Yura T, Negishi M, Mori K (2000) ATF6 activated by proteolysis binds in the presence of NF-Y (CBF) directly to the cis-acting element responsible for the mammalian unfolded protein response. Mol Cell Biol 20:6755-6767

Yoshida H, Okada T, Haze K, Yanagi H, Yura T, Negishi M, Mori K (2001b) Endoplasmic reticulum stress-induced formation of transcription factor complex ERSF including NF-Y (CBF) and activating transcription factors 6alpha and 6beta that activates the mammalian unfolded protein response. Mol Cell Biol 21:1239-1248

Zeng LF, Lu M, Mori K, Luo SZ, Lee AS, Zhu Y, Shyy JYJ (2004) ATF6 modulates SREBP2-mediated lipogenesis. EMBO J 23:950-958

Zhang C, Cai Y, Adachi MT, Oshiro S, Aso T, Kaufman RJ, Kitajima S (2001) Homocysteine induces programmed cell death in human vascular endothelial cells through activation of the unfolded protein response. J Biol Chem 276:35867-35874

Zhou AM, Hassel BA, Silverman RH (1993) Expression cloning of 2-5a-dependent Rnase - a uniquely regulated mediator of interferon action. Cell 72:753-765
Zinszner H, Kuroda M, Wang X, Batchvarova N, Lightfoot RT, Remotti H, Stevens JL, Ron D (1998) CHOP is implicated in programmed cell death in response to impaired function of the endoplasmic reticulum. Genes Dev 12:982-995

Niwa, Maho
Division of Biological Sciences, Section of Molecular Biology, University of California, San Diego, 9500 Gilman Drive, La Jolla, California, 92093-0377
niwa@ucsd.edu

Hsp104p: a protein disaggregase

Johnny M. Tkach and John R. Glover

Abstract

All newly synthesized proteins must fold to their correct native conformation in order to function. That protein folding in the crowded macromolecular environment of the cell is as efficient as it appears to be is remarkable in itself. However, physical or chemical stresses or the accumulation of aberrant proteins encoded by mutated genes can easily perturb protein folding homeostasis in cells. In the protein biochemistry laboratory the aggregation of proteins can be frustrating and even aggravating, but when partially folded or misfolded proteins aggregate in the cell or even, in the case of systemic amyloidoses or Alzheimer's Disease, outside the cell, the consequences can be devastating. Molecular chaperones typically play a key role in preventing protein aggregation. However, this monograph describes the biology and biochemistry of yeast Hsp104p, an unconventional molecular chaperone that specializes, not in preventing, but in reversing protein aggregation.

1 Hsp104p and thermotolerance in yeast

Thermotolerance is a universal characteristic of all cellular life and is defined as a transient state of resistance to lethal heat stress in organisms that have been preconditioned by exposure to a mild, non-lethal heat shock. One of a number of events elicited by the preconditioning heat shock is the coordinated expression of genes encoding proteins collectively known as the heat shock proteins or HSPs. Most of the heat shock proteins are molecular chaperones (Parsell and Lindquist 1994). Thus, perhaps in anticipation of a life-threatening thermal stress, organisms accumulate substantially higher levels of many molecular chaperones as a strategy for dealing with a larger number of thermally unfolded, aggregation-prone proteins.

In yeast, as in other organisms, the expression of heat shock proteins is not the only transient adaptation elicited by mild heat shock. Indeed, yeast accumulate very high concentrations of trehalose that acts as a chemical chaperone to stabilize folded proteins and suppresses the aggregation of non-native proteins (Singer and Lindquist 1998). By knocking out the genes involved in trehalose biosynthesis, it can be shown that trehalose accumulation makes a substantial contribution to thermotolerance (Attfield et al. 1994; Elliott et al. 1996). In addition, the lipid composition of the membrane is modified and yeast mutants that are unable to respond in this manner have also been shown to be less thermotolerant (Swan and

Watson 1999). However, it is among the HSPs of yeast that one of the most potent thermotolerance factors is found.

Metabolic labeling of newly synthesized proteins with radioactive amino acids shows that among the proteins that are synthesized in response to heat shock in yeast are those typically seen in many prokaryotic and eukaryotic organisms including HSPs in the 90 kDa, 70 kDa, 40-45 kDa molecular weight range as well as small heat shock proteins (Parsell et al. 1993). One prominent HSP that is highly induced by heat shock in yeast is a protein of 104 kDa (Sanchez and Lindquist 1990). Disruption of the gene encoding Hsp104p has no effect on the growth of yeast under non-stress conditions and does not cause a compensatory induction of other heat shock proteins. Instead, cells lacking Hsp104p exhibit a profound defect in the acquisition of thermotolerance (Fig. 1). In fact, *hsp104* mutant yeast die almost as rapidly as wild type yeast that have not been preconditioned by mild heat shock and are not, by definition, thermotolerant.

These experiments showed that Hsp104p was necessary for thermotolerance. But because a number of changes in cell physiology and gene expression are coincident with Hsp104p accumulation during the pretreatment, it is impossible to tell from experiments with *hsp104* mutant cells whether high Hsp104p expression alone is sufficient for thermotolerance. To test this directly, Hsp104p was expressed in *hsp104* mutant cells from various heat shock-independent promoters (Lindquist and Kim 1996). These experiments demonstrated that the expression of Hsp104p, in the absence of a preconditioning heat treatment that would induce the whole spectrum of heat shock-mediated events, is alone sufficient to confer the thermotolerant state.

To examine the function of Hsp104p as a chaperone *in vivo*, a fusion protein consisting of the two subunits of bacterial luciferase was expressed in both wild type and *hsp104* mutant yeast (Parsell et al. 1994b). The activity of the bacterial luciferase is readily detected as light emission from intact cells after addition of luciferase substrate to the medium. Furthermore, the enzyme is exquisitely temperature sensitive and inactivates rapidly during a non-lethal heat treatment (Escher et al. 1989). The recovery of luciferase activity after heat inactivation was measured in presence of cycloheximide, a translation inhibitor, to prevent the synthesis of new luciferase. In wild type cells, the activity of preexisting, inactive luciferase was restored to almost preheat shock levels but remained inactive in *hsp104* mutant cells (Parsell et al. 1994b) indicating that Hsp104p has an important role in protein refolding *in vivo*.

An even more intriguing aspect of these experiments was the observation that bacterial luciferase became largely insoluble in both wild type and *hsp104* mutant yeast during the heat shock but that solubility was restored only in wild type cells (Parsell et al. 1994b). This raised the possibility that Hsp104p might actually resolubilize inactive proteins that are aggregated, a property that is distinct from other molecular chaperones that act primarily to prevent aggregation. To determine if the dispersal of protein aggregates was a general feature of Hsp104p function, heat shocked yeast were examined by transmission electron microscopy (Parsell et al. 1994b). Heat shocked cells contained electron-dense anomalies,

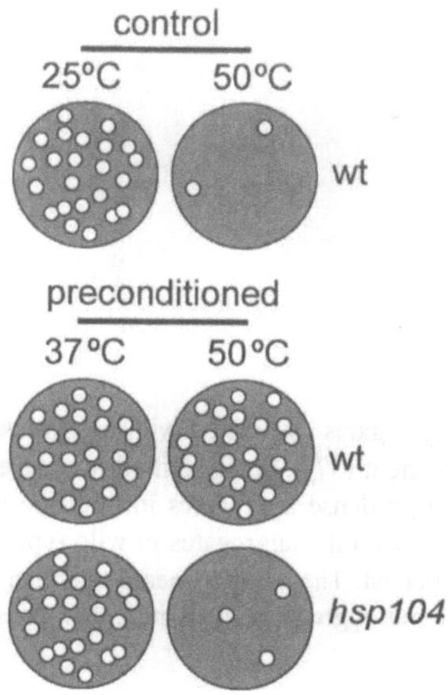

Fig. 1. Hsp104p is a determinant of thermotolerance in yeast. The viability of yeast is measured by the ability of cells to divide and form colonies. In the illustration, all culture plates on the left of each pair demonstrate the viability of cells prior to the severe heat challenge. The culture plates on the right represent the same cells plated after exposure to severe heat treatment. Control cells shifted directly from normal temperature (25°C) to the severe temperature (50°C) die rapidly. The survival of wild type cells (wt) that have been made thermotolerant by a mild non-lethal heat shock (37°C) is enhanced. Disruption of the gene encoding Hsp104p (*hsp104* mutant) largely abolishes this survival advantage indicating its pivotal role in the establishment of thermotolerance.

probably protein aggregates, in their cytoplasm and nucleoplasm. During recovery at normal temperature these anomalies disappeared in wild type cells but persisted in the *hsp104* mutant cells (Fig. 2) suggesting that Hsp104p plays a general and pivotal role in the dispersion of protein aggregates.

Little is known about the identities of proteins that might be targeted for refolding by Hsp104p in severely heat shocked cells or if these can be directly linked to cell survival. mRNA splicing is one example of a heat-sensitive cellular process that is dependent on Hsp104p for rapid recovery (Vogel et al. 1995). Biochemical analysis of the splicing complexes in heat shocked and recovered cells indicate that the small ribonuclear protein complexes involved in splicing become largely disassembled after heat shock and that reassembly of these complexes is, in part, dependent on Hsp104p (Bracken and Bond 1999).

Another approach to identify targets of Hsp104p-mediated disaggregation and refolding is to compare the soluble protein fraction of wild type and *hsp104* mutant cells after recovery from a severe, but non-lethal heat shock. Proteins that appear in the soluble fraction of recovering type cells and that are absent from the *hsp104* mutant cell extract would be likely substrates for Hsp104p-mediated refolding. Although no formal report has yet been published, three potential

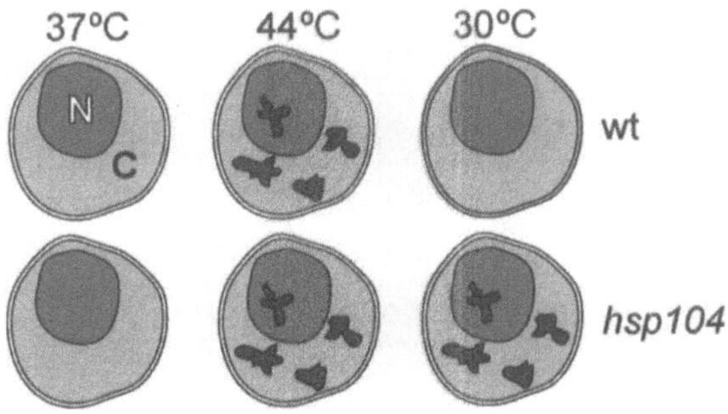

Fig. 2. Hsp104p disperses aggregates associated with heat stress in yeast. After preconditioning with a mild heat treatment (37°C) cells shifted to a severe but non-lethal temperature (44°C) accumulate electron-dense aggregates in both the cytoplasm (C) and nucleoplasm (N). During recovery (25°C) the aggregates in wild type cells disappear whereas, in *hsp104* mutants, aggregates persist. These experiments establish that Hsp104p is directly or indirectly involved in dispersing aggregates in living cells. (illustration modeled after data in Parsell 1994).

Hsp104p substrates have been identified using this technique: Tef3p, the yeast translation elongation factor 3; Rrp46p a 3'-5' exoribonuclease component of the yeast exosome; and a protein with unknown function encoded by the open reading frame ykl056c (Cashikar et al. 2002). An essential, aggregation prone protein that is a known target of Hsp104p is the yeast translation termination factor Sup35p (Chernoff et al. 1995), which is also associated with a prion-like phenomenon in yeast and will be discussed in greater detail below.

2 *In vitro* reconstitution of Hsp104p refolding activity

Typically molecular chaperones possess one or both of two hallmark properties in the test tube: the ability to suppress aggregation of non-native proteins and the ability to promote refolding of non-native proteins to the native state. To monitor the first property, model substrate proteins are either heated in the presence of the molecular chaperone under investigation or alternatively, denatured using high concentrations of chaotropic agents such as urea or guanidinium (Gdn) and then dispersed into non-denaturing buffer containing chaperones (Lee 1995). As large protein aggregates appear in suspension they scatter light. The ability of a protein to suppress light scattering is generally considered to be a good indication of its molecular chaperone properties. In such experiments, a yeast Hsp70/Hsp40 chaperone/cochaperone system consisting of Ssa1p/Ydj1p efficiently suppress aggregation (Cyr 1995; Glover and Lindquist 1998) while aggregation in presence of Hsp104p is similar to that observed in control experiments containing no chaperones at all (Glover and Lindquist 1998) (Fig. 3A).

To monitor the second property, protein refolding, the chaperone-dependent recovery of enzyme activity of a denatured model substrate protein is measured. In

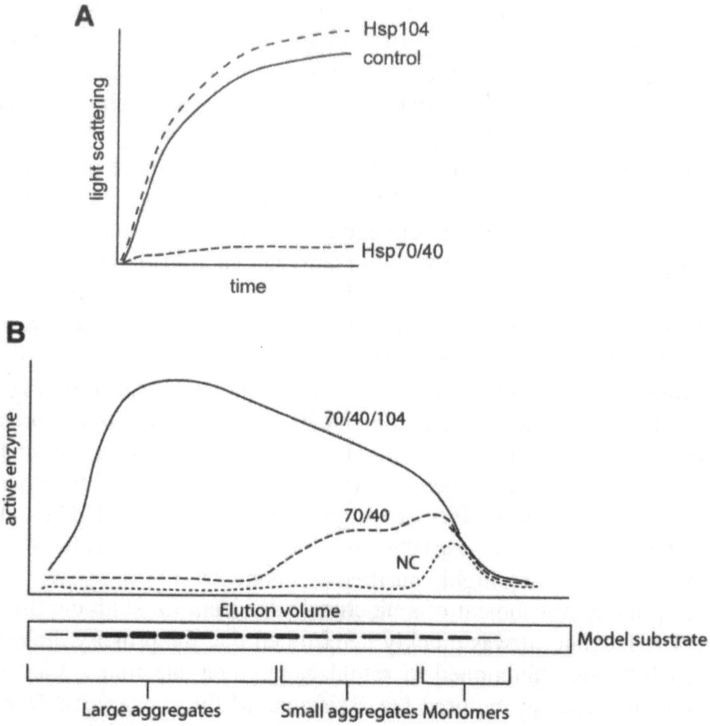

Fig. 3. Reconstitution of Hsp104p molecular chaperone activity. A. Most molecular chaperones, such as yeast Hsp70/Hsp40, can suppress the formation of large light-scattering aggregates, but this is not true of Hsp104p. B. To demonstrate the protein refolding properties of Hsp104p, aggregates of a model substrate protein can be divided into different size classes by gel filtration chromatography and fractions from the column used in refolding reactions. In reactions lacking chaperones (NC), only monomeric non-native proteins can spontaneously refold to the active state. Hsp70/40 chaperones can promote refolding of non-native proteins that are otherwise trapped in small aggregates. Hsp104p by itself does not refold aggregated substrates (not indicated in the illustration) but together with Hsp70/40 can refold protein, even from large aggregates (illustrations based on data from Glover and Lindquist 1998).

refolding assays using firefly luciferase as a substrate, the same Ssa1/Ydj1 chaperone/cochaperone system that is capable of suppressing aggregation also promotes modest refolding of non-native luciferase. In contrast, Hsp104p alone fails to display any refolding capability (Glover and Lindquist 1998). Thus, at first glance, Hsp104p displays neither of the characteristics— suppression of aggregation and protein refolding activity— associated with conventional molecular chaperones.

Remarkably, whole cell protein extracts from wild type cells efficiently promote refolding of chemically denatured firefly luciferase. In comparison, extracts from *hsp104* mutant cells are only marginally active in refolding (Glover and

Lindquist 1998). The addition of recombinant, purified Hsp104p, however, restored the refolding activity of *hsp104* mutant cell lysates demonstrating that indeed Hsp104p was active in protein refolding but that its function was dependent on other factors present in the extracts. A combination of biochemical techniques revealed that Hsp104p required the addition of the Hsp70/40 chaperone/cochaperone system to refold protein and that these three proteins alone reconstitute the observed refolding activity. The most striking feature of these experiments is the extraordinary effectiveness or synergy in protein refolding of this chaperone combination relative to either component acting alone.

The key experiment that demonstrated the role of Hsp104p in refolding was to permit chemically denatured substrate protein to aggregate in the absence of chaperones (Glover and Lindquist 1998). These aggregates were fractionated by gel filtration chromatography and subsequently used as substrates for refolding reactions (Fig. 3B). The results indicate that the Hsp70/40 system effectively promotes refolding from fractions containing non-native, monomeric protein or small aggregates. Only when the three chaperones, Hsp104/70/40, are added together are proteins trapped in very high molecular weight aggregates refolded. Thus the unique disaggregation properties of Hsp104p, first observed in living cells, can be fully recapitulated *in vitro* using highly purified components.

Following the establishment of a biochemical system for studying the refolding capability of Hsp104p, it was quickly determined that the prokaryotic ortholog of Hsp104p, ClpB, also functioned to refold aggregated substrates. Like Hsp104p, ClpB-mediated refolding required the assistance of the prokaryotic DnaK, DnaJ, GrpE chaperone/cochaperone system analogous to the Hsp70/40 system in yeast (Goloubinoff et al. 1999; Motohashi et al. 1999; Zolkiewski 1999). Bernd Bukau and colleagues coined the term "bichaperone network" to describe this functional interaction (Goloubinoff et al. 1999). Interestingly, the bacterial Hsp70 chaperone system cannot substitute for the yeast system in Hsp104p-dependent refolding. This observation suggests that there is some requirement for specificity among the components of the bichaperone network but the molecular basis of this specificity remains unknown.

Importantly, these observations lead to a satisfactory explanation for the critical importance of Hsp104p in thermotolerance. During moderate heat shock some proteins may unfold and become aggregation-prone but the abundance of molecular chaperones that prevent aggregation is sufficient to ensure survival of cells. However, during exposure to acute, severe heat stress, the overwhelming numbers of unfolded proteins lead to significant protein aggregation. Under these circumstances survival becomes critically dependent on the ability of Hsp104p to disperse the aggregates and to promote cell survival.

3 Hsp100 structure and function

Hsp104p and its bacterial ortholog ClpB are members of the Hsp100/Clp family of proteins which itself is a division within the AAA$^+$ (ATPases Associated with a

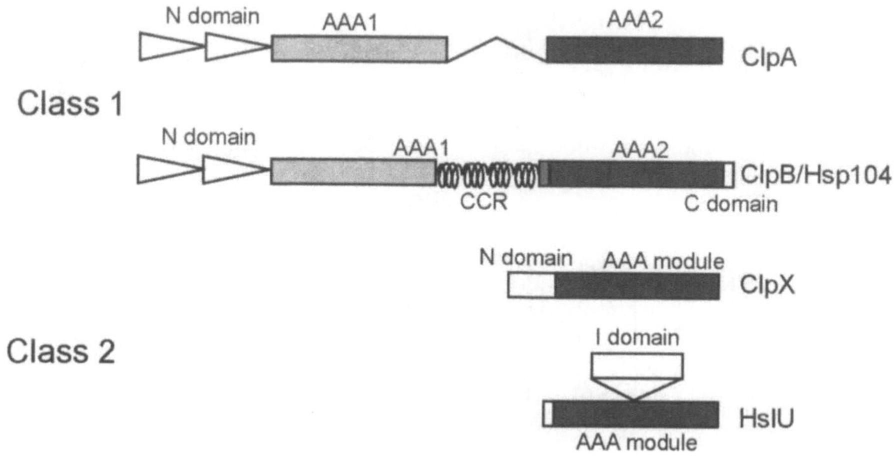

Fig. 4. The Hsp100/Clp family of ATPases. The primary sequence elements of representative Class1 (two AAA modules) and Class 2 (one AAA module) Clp/Hsp100 proteins are illustrated. For a detailed explanation see text.

variety of cellular Activities) superfamily of proteins (for reviews see Neuwald et al. 1999; Ogura and Wilkinson 2001; Lupas and Martin 2002). One of the defining features of an AAA^+ superfamily member is a conserved sequence module of approximately 250 amino acids predicted to bind and hydrolyze ATP. Based on numerous x-ray crystallographic studies, a single AAA^+ module consists of two independently folded domains. The first, or large domain (LD), adopts a Rossman fold consisting of a β-sheet region flanked by α-helices and contains highly conserved Walker A (GX_4GKT) and Walker B (h_4DE) motifs that are involved in nucleotide binding and hydrolysis. The second domain is referred to as the small domain (SD) and adopts a primarily compact α-helical structure.

In addition to its role in ATP binding and hydrolysis, the architecture of the AAA^+ module also dictates the quaternary arrangement of AAA^+ proteins. AAA^+ proteins assemble into ring-shaped complexes that are usually hexameric. In this arrangement, the SD of one AAA^+ module tightly packs against the LD of the adjacent module. Nucleotide is bound near the interface of this interaction and each subunit contributes amino acids critical to nucleotide binding and hydrolysis. Notably, this molecular arrangement results in two key characteristics. First, the physical interaction between the AAA^+ modules allows for coordination of ATP hydrolysis among the modules and second, the ring-shaped oligomers contain a channel that penetrates through the center of the ring.

The Hsp100/Clp family itself can be broadly divided into two classes based on the number of AAA^+ modules contained within the protein; Class 1 Hsp100s contain two AAA^+ modules (AAA1 and AAA2) while Class 2 Hsp100s contains a single AAA^+ module (Schirmer et al. 1996). The four most well characterized members of the family for which structural data is available, *T*ClpB (Lee et al. 2003) (ClpB from the hyperthermophile *Thermus thermophilus*), ClpA (Guo et al. 2002b), ClpX (Donaldson et al. 2003; Kim and Kim 2003), and HslU or ClpY

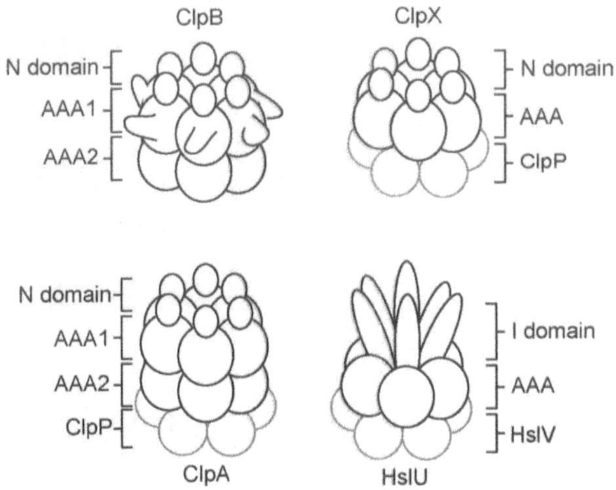

Fig. 5. Tertiary and quaternary structures of Hsp100/Clp ATPases. The cartoons illustrate the general organization of domains within assembled ATPases and are based on much more detailed models obtained from x-ray crystallographic data from papers cited in the text. For ClpX, ClpA, and HslU, one ring of the appropriate protease complex is illustrated to show the orientation of the ATPase relative to the protease.

(Sousa et al. 2000), exemplify the structural diversity within the family (Fig. 4). ClpX has a short N-terminal domain with a zinc finger motif followed by a single AAA^+ module. HslU has no substantial N-terminal domain but instead has an insertion domain or "I" domain that interrupts the primary sequence of the AAA^+ module. In models of HslU based on crystallographic data, the I domain extends upward, out of the plane of the HslU ring (Sousa et al. 2000), and may functionally substitute for the absent N domain. ClpA and *T*ClpB have similar pseudodimer N domains. Hsp104p/ClpB is distinguished from ClpA by a large insertion in the SD of AAA1 that form a coiled-coil (CC) domain that projects outward from the hexamer ring. Thus, each of the primary sequence elements within the Hsp100/Clp genes maps onto structural domains of their respective proteins in models that have been proposed based on x-ray crystallographic data (Fig. 5).

Although Hsp104p itself has not yet been crystallized, the existing structural models of Hsp100/Clp proteins, and particularly that *T*ClpB (Lee et al. 2003), serve as a reliable paradigms for Hsp104p structure. In general, Hsp104p is predicted to consist of two stacked rings of AAA^+ modules. As a result, an assembled Hsp104p molecule contains 12 ATP binding sites —six AAA1 and six AAA2 modules. The importance of ATP binding and hydrolysis in Hsp104p function is highlighted by numerous mutagenesis studies that focused on the conserved motifs and residues in the AAA^+ module. Hsp104p molecules that contain nonconservative amino acid substitutions in the Walker A (Parsell et al. 1991) and/or B motifs (R. Lum and J.R. Glover, unpublished) in either or both AAA^+ modules no longer function in thermotolerance or in protein folding *in vitro*. Interestingly, the disruption of nucleotide binding at AAA1 or AAA2 has distinct biochemical effects. Amino acid substitutions at the Walker A motif of AAA1 virtually eliminates the ATPase activity of Hsp104p while the analogous substitution in AAA2 diminishes the ATPase activity of Hsp104p somewhat but more dramatically impairs the ability of the protein to assemble into hexamers in a nucleotide-dependent manner (Parsell et al. 1994a; Schirmer et al. 1998, 2001). The functional significance of this specialization of the AAA1 and AAA2 modules is uncertain especially since it is not a conserved feature. Notably, the apparent specialization is reversed in ClpA (Seol et al. 1995) and in ClpB both domains contribute the ATPase activity of the protein (Mogk et al. 2003). In any case, the full function of Hsp100/Clp family members is dependent on their ATPase activity.

The allosteric influence among AAA^+ modules of the assembled complex has also been investigated. The kinetics of ATP hydrolysis within both of Hsp104p's AAA^+ rings demonstrate positive cooperativity implying that AAA^+ modules communicate their nucleotide status within the plane of each ring (Hattendorf and Lindquist 2002a, 2002b). Although the exact mechanism of this communication is unknown, conserved residues (especially several arginine residues) within the module undoubtedly play a role in 'sensing' the ATP status at one AAA^+ module and transmit this information to the adjacent module within the ring. A more detailed kinetic analysis of Hsp104p mutants with ATP hydrolysis defects in either AAA1 or AAA2 lead to the conclusion that when the AAA2 ring is in the ADP bound state, the rate of ATP hydrolysis in the AAA1 ring is greater than when the AAA2 ring is in the ATP bound state (Hattendorf and Lindquist 2002b). Thus, Hsp104p exhibits a complex network of allosteric control of ATP hydrolysis, both *within* the plane of each AAA+ ring, and *between* the two AAA+ rings. Although the specific structural consequences of ATP hydrolysis by Hsp104p are largely unknown, the intricate rounds of ATP binding and hydrolysis undoubtedly control the conformational changes within the molecule and are linked to its disaggregation activity.

4 Mechanism of protein disaggregation

To initiate the reactivation of aggregated proteins Hsp104p/ClpB must discriminate between assemblies of aggregated non-native proteins and native macromolecular complexes. Based on a peptide array approach, it was recently determined that ClpB preferentially interacts with peptides that are rich in basic amino acids, especially lysine, and also enriched in aromatic amino acids (Schlieker et al. 2004). Using the same approach, we determined that Hsp104p displays a similar amino acid predilection (M. Niggemann and J.R. Glover, unpublished observation). In addition, poly-L-lysine stimulates the ATPase activity of Hsp104p suggesting that it might be a substrate mimetic (Cashikar et al. 2002). Although positively charged regions will generally be displayed on the exterior of natively folded proteins, it is possible that a high density of unstructured, positively charged regions might also be a general characteristic of aggregated proteins and serve as a general recognition motif exploited by Hsp104p.

The molecular determinants involved in substrate recognition within Hsp104p itself are not well understood. The N-terminal domain of other Hsp100s, ClpA and ClpX, play a pivotal role in modulating their substrate repertoire. First, the N-domains of these proteins might directly interact with some, but not all substrates by well-defined polypeptide sequences (Singh et al. 2001; Donaldson et al. 2003; Wojtyra et al. 2003; Xia et al. 2004). Second, the N-domains interact with adaptor proteins that either directly target substrates to ClpA and ClpX, or modulate substrate specificity (Dougan et al. 2002, 2003; Guo et al. 2002a; Wah et al. 2003). This role for N-domains in substrate recognition, however, does not seem to be conserved in Hsp104p/ClpB. *In vivo*, ClpB is expressed as a full-length, 95 kDa protein and, from an internal translation site, as an 80 kDa protein lacking the N-domain (ClpBΔN) (Park et al. 1993). The expression of either translation product alone results in wild type levels of thermotolerance indicating that the N-terminal domain is not essential for ClpB activity (Beinker et al. 2002; Mogk et al. 2003). In addition, ClpBΔN is fully capable of mediating the refolding of model substrates *in vitro*. We have found that Hsp104p lacking amino acids 1 - 152 (Hsp104deltaN) provides wild type level;s of thermotolerance in yeast (Ronnie Lum and JRG, unpublished observation) further supporting the idea that these domains do not play a pivotal role in Hsp104p-mediated protein refolding.

Recently, a more direct approach to locating a substrate-binding site of ClpB was used. A strong ClpB-binding peptide derived from the peptide array experiments previously mentioned, was preferentially cross-linked to a tyrosine residue that is part of a motif, GAKYRGEFEE, present in AAA1 that is conserved among the ClpB proteins (Schlieker et al. 2004). Based on the *T*ClpB crystal structure (Lee et al. 2003), this region is flexible and is predicted to line a narrow pore near the N-terminal end of the axial channel. If this motif is involved in initial substrate binding, it is possible that only assembled Hsp104p/ClpB complexes can form a functional substrate-binding site. In support of this idea, a screen of Hsp104p molecules harboring random amino acid substitutions in the C-terminal SD revealed that only Hsp104p molecules that were completely blocked in assembly failed to associate with heat induced aggregates in living yeast (Tkach and Glover 2004).

Once Hsp104p/ClpB has detected an aggregated substrate, it must disassemble the aggregate to free non-native proteins from protein/protein contacts that prevent them from spontaneously refolding. At least two models can be envisioned for the mechanism of disaggregation (for a detailed discussion see Glover and Tkach 2001; Weibezahn et al. 2004a). In the first, Hsp104p/ClpB acts as a "molecular crowbar" to pry open the adjoining surfaces between aggregated proteins. This model was originally proposed to accommodate the inability of Hsp104p or ClpB to release spontaneously refoldable proteins from aggregates (Glover and Lindquist 1998; Goloubinoff et al. 1999). By revealing new surfaces for conventional chaperone binding, iterative action of the crowbar would progressively produce single non-native proteins or much smaller aggregates that could be assisted in refolding by the Hsp70/40 chaperone system.

It has been suggested that ATP-driven scything of the CC domains of *T*ClpB that are displayed on the exterior of the AAA1 ring might play a role in prying aggregates open (Lee et al. 2003). Indeed the introduction of amino acid substitutions that are predicted to constrain the motion of the CC domains, reduce the refolding capacity of the protein. A role for the CC domain in refolding is supported by experiments involving the complete deletion of the CC domain in ClpB (Kedzierska et al. 2003; Mogk et al. 2003). The CC domain is apparently dispensable for assembly but not for refolding activity *in vitro* or thermotolerance *in vivo*.

The second model of Hsp104p function, which we call the "molecular ratchet" model, envisions that polypeptides are extracted from aggregates and extruded from through the axial channel of Hsp104p. As segments of extended polypeptide emerge from the axial channel they can be readily bound by the Hsp70/40 components of the bichaperone network, further protecting the protein from aggregation and thereby promoting refolding. This model is parallel to the way in which Hsp100-dependent proteases are thought to function. Both ClpA and ClpX form complexes on the ends of a barrel formed by the ClpP protease while HslU forms a similar complex with the HslV protease (see Fig. 5). A number of studies indicate that substrates that are targeted for degradation by these ATP-dependent protease bind to the ends of the ATPase/protease complex where they are unfolded and translocated through the central channel of the Hsp100/Clp hexamer and into the protease where they are degraded (Singh et al. 2000; Ishikawa et al. 2001; Reid et al. 2001). Crystal structures of HslU in different nucleotide bound states provide evidence that the axial channel contains a flexible loop composed of a motif (GYVG), which is conserved among all Hsp100s, whose precise positioning within the channel is modulated by nucleotide binding (Wang et al. 2001). The bulky Tyr residue at the apex of this loop could act as a clamp that interacts transiently with translocating polypeptides and prevents them from sliding backward in the channel. In support of a critical role for the GYVG loop in Clp/Hsp100 function, different groups found that substitution of the Tyr residue in the GYVG motif with non-conservative amino acids results in HslU (Song et al. 2000) and ClpX (Siddiqui et al. 2004) molecules with normal quaternary structures and ATPase activities but impaired degradation of protein substrates.

By substituting a unique Trp residue at the Tyr position in the GYVG loop of Hsp104p, we showed that tryptophan fluorescence of the mutant Hsp104p is sensitive to nucleotide binding suggesting that the loop in Hsp104p is mobile and that its position in the axial channel is controlled by nucleotide binding to AAA2 (Lum et al. 2004). Importantly, substitution of the Tyr with a variety of non-conservative amino acid residues result in proteins that have normal assembly properties and ATPase activities but that do not function in protein refolding or thermotolerance. Very recently, support for protein unfolding/threading by ClpB was obtained by incorporating motifs derived from ClpA and ClpX that permitted binding of ClpB to the ClpP protease (Weibezahn et al. 2004b). Aggregated substrates are not refolded but degraded by the hybrid ClpB/ClpP complex. Crosslinking of denatured luciferase to the GYVG loop in ClpB strongly supports the notion that the pathway of protein extraction from aggregates is *via* the axial channel.

5 Organization of the bichaperone network

In the crowbar or ratchet models of Hsp104p disaggregation, Hsp104p cooperates with the Hsp70 chaperone system by permitting conventional molecular chaperones access to otherwise hidden hydrophobic surfaces or peptides that are sequestered within the aggregates. However, evidence points to a new role for the Hsp70 chaperone system *prior* to the action of ClpB. In the experiments with the hybrid ClpB/ClpP complex described above, the degradation of aggregated protein is accelerated in the presence of bacterial DnaK and its cochaperones (Weibezahn et al. 2004b). With access to peptides emerging from the exit pore of ClpB blocked by its association with ClpP the only opportunity for the DnaK system to act *before* ClpB. These results are consistent with another recent study that examined the initial rates of aggregated GFP refolding (Zietkiewicz et al. 2004). Incubation of the substrate with the *E. coli* Hsp70/40 chaperone machine (DnaK/DnaJ) increased the rate of initial refolding upon the addition of ClpB but not if the order of addition was reversed. These observations raise the possibility that conventional chaperones components of the bichaperone complex perhaps act on the surface of aggregates, possibly isolating loops of extended polypeptide that are accessible for entry into the ClpB/Hsp104p pore.

Although not an obligatory feature of the mechanistic models described above, we can speculate that the bichaperone network would be expected to work most efficiently if it was physically organized. A physical interaction between Hsp104p and the Hsp70/40 chaperones would efficiently channel the substrate from one chaperone system to the next, shielding the unfolded polypeptide from the cellular milieu. Recently, a direct interaction between *T*ClpB and DnaK was reported using purified proteins (Schlee et al. 2004), but the exact location of DnaK binding on *T*ClpB was not determined.

If Hsp70 or DnaK recognizes and partially remodels aggregated substrates first as has recently been suggested (Zietkiewicz et al. 2004), it might be possible that the efficiency of Hsp104p- or ClpB-mediated refolding would be enhanced if Hsp70 or DnaK acted as adaptor proteins analogous to those known to be involved in substrate binding and selection by the ClpA and ClpX. Since Clp adaptor proteins interact with the N-domains of the Clp chaperones it is plausible that an Hsp104p or ClpB adaptor would interact with the N-domains. This possibility, however, is disfavored by the observation that the N-domain of ClpB is dispensable for the function of ClpB (Beinker et al. 2002; Mogk et al. 2003) and that DnaK interacts with a *T*ClpB variant lacking its N-domain (Schlee et al. 2004).

Although the most recent evidence favors Hsp70 or DnaK acting before Hsp104p or ClpB, it is still possible that they can act on the extended polypeptide as it emerges from the exit pore of the axial channel. Interestingly, during growth on carbon sources requiring respiration, Hsp104p directly interacts with Sti1p (Abbas-Terki et al. 2001), a TPR domain protein that interacts with conserved C-terminal acidic motifs and that is involved in organizing a complex of chaperones including Hsp70 and Hsp90 in yeast (Lee et al. 2004). The interaction with Sti1p is mediated by the acidic C-terminus of Hsp104p and thus positions Sti1p at the potential exit site of unfolded polypeptides. Although Sti1p is not required in reconstituted Hsp104p-mediated protein refolding, it might play a role *in vivo* under undetermined conditions.

While the big picture of Hsp104p/ClpB function is rapidly emerging, the detailed mechanism of Hsp104p/ClpB-mediated disaggregation remains an enigma and represents a challenging problem for researchers in the field. One of the outstanding difficulties in studying these reactions is the non-uniform nature of the aggregated substrates generated for refolding *in vitro*. Aggregates of chemically or thermally denatured proteins are frequently amorphous. In large aggregates, some proteins are on the surface and chaperone accessible and others are buried and inaccessible in the initial stages of disaggregation. In part, this lack of substrate uniformity makes quantitative biochemistry on Hsp104p-mediated protein refolding qualitative at best. However, there is one aggregated substrate of Hsp104p, Sup35p, that displays remarkable structural uniformity and lies at the heart of one of the most interesting genetic and cell biological stories to emerge in the last decade.

6 Yeast prions and Hsp104p

In the 1960s, Dr. Brian Cox described a genetic element named [*PSI*$^+$] that caused omnipotent suppression (read-through of nonsense codons) in yeast (Cox 1965). Although suppressors with similar phenotypic consequences had been described many times, the unusual feature that distinguished this new element was that it was inherited as a dominant, non-Mendelian, cytoplasmic trait. Furthermore, the trait was metastable in that [*PSI*$^+$] could be eliminated from or "cured" in strains by certain treatments such as the exposure to a low concentration of guanidinium (Gdn). Despite being 'cured', the [*PSI*$^+$] element would reappear spontaneously with a low frequency in the resulting [*psi*$^-$] population (Tuite et al. 1981). Despite years of effort, few details emerged regarding the mechanism of [*PSI*$^+$] inheritance or clues that might explain its metastability.

Omnipotent suppression— read through of each of the three nonsense codons (UAA, UGA, UAG)— is essentially a manifestation of diminished fidelity of translation termination resulting in continuation of translation elongation. Indeed, evidence suggested a strong link between the [*PSI*$^+$] element and *SUP35*, the gene encoding a subunit of the translation termination factor in yeast. Most importantly, overexpression of Sup35 or only the N-terminal segment of the protein causes [*psi*$^-$] cells to become [*PSI*$^+$] with a high frequency (Ter-Avanesyan et al. 1993). Furthermore, altering the *SUP35* gene to delete all, or part, of the N-terminal segment of Sup35p prevents cells from converting to [*PSI*$^+$] (Ter-Avanesyan et al. 1994) and certain mutations in the Sup35 N-terminal domain perturb the stability of the [*PSI*$^+$] element (Doel et al. 1994). Lastly, mutations in Sup35 can also cause an omnipotent suppression phenotype, but in this case the phenotype is inherited as a simple Mendelian recessive trait (Wakem and Sherman 1990) (see Fig. 6).

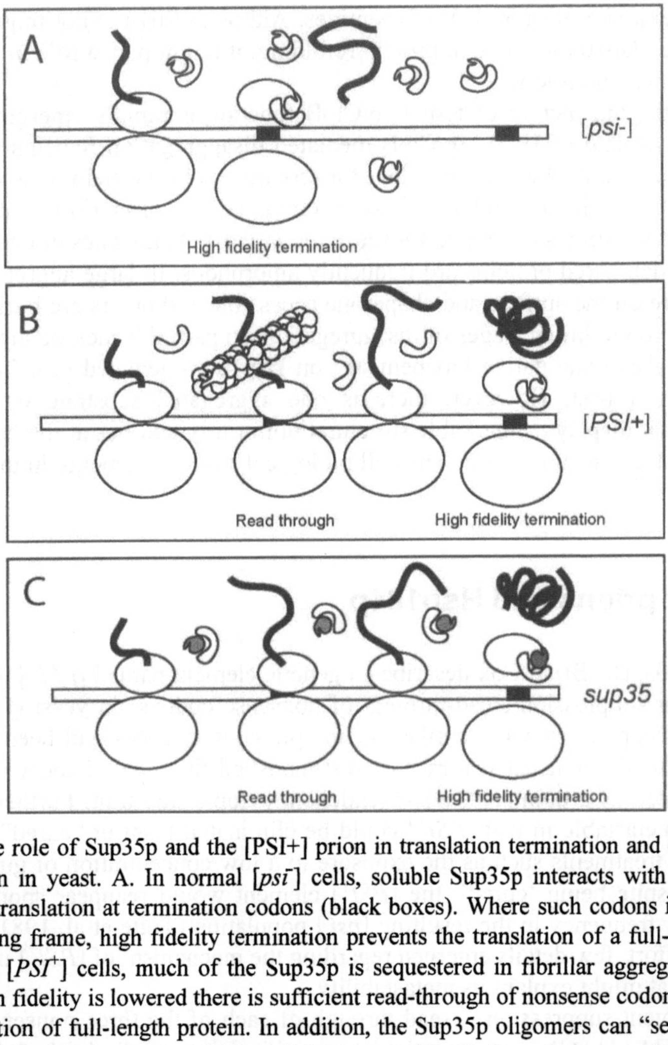

Fig. 6. The role of Sup35p and the [PSI+] prion in translation termination and omnipotent suppression in yeast. A. In normal [*psi*⁻] cells, soluble Sup35p interacts with Sup45p to terminate translation at termination codons (black boxes). Where such codons interrupt an open reading frame, high fidelity termination prevents the translation of a full-length protein. B. In [*PSI*⁺] cells, much of the Sup35p is sequestered in fibrillar aggregates. When termination fidelity is lowered there is sufficient read-through of nonsense codons to permit the translation of full-length protein. In addition, the Sup35p oligomers can "seed" the aggregation of newly synthesized Sup35p. Inheritance of seeds through the cytoplasm ensures stability of the trait during mitosis. C. In *sup35* mutant cells, where the termination function of Sup35p is reduced, omnipotent suppression of nonsense codons can also be observed. Since this defect is associated with a mutant nuclear gene, is inherited as a simple Mendelian trait.

In 1994 Dr. Reed Wickner proposed that the inheritance pattern of [*PSI*⁺] and another unusual genetic element, [*URE3*], could be explained by a biochemical mechanism that conceptually parallels the transmission of prions as the proposed by Stanley Prusiner (see Prusiner 1998). The prion hypothesis postulates that transmissible forms of spongiform encephalopathy can be transmitted from a diseased animal to a healthy one by a pathogenic conformer of the prion protein PrP. The pathogenic form of PrP acts as a template for the conversion of normally-folded PrP into the pathogenic form, thereby, causing disease and simultaneously propagating new infectious elements. Briefly, Wickner (1994) proposed three genetic criteria for prion-like processes in yeast. First, prion elements should be reversibly curable suggesting that if prion elements can be eliminated in a population of cells it should be reestablished through spontaneous misfolding at some appreciable frequency. Second, overexpression of a prionogenic protein in prion-free cells will kinetically favour *de novo* prion formation. Third, deletion of the gene encoding the prionogenic protein will result in complete abrogation of prion formation. In the case of [*PSI*$^+$] each criterion is broadly fulfilled with Sup35p being the underlying prionogenic protein.

Biochemical evidence for the prionogenic nature of Sup35p is abundant. Full-length Sup35p, or segments containing the N-terminal domain, which is crucial for induction and stability of [*PSI*$^+$], spontaneously form amyloid-like fibrils *in vitro* but only after a lag phase during which polymerization cannot be detected (Glover et al. 1997). The addition of preformed fibrils to solutions of unpolymerized protein eliminates the lag phase indicative of a seeded polymerization phenomenon. Extracts of [*PSI*$^+$] cells also eliminate the lag phase whereas extracts of [*psi*$^-$] cells cannot, suggesting that the [*PSI*$^+$] extracts contain "seeds". Finally, Sup35p in [*PSI*$^+$] cells is readily sedimentable by centrifugation whereas Sup35p in [*psi*$^-$] cells is mostly soluble (Patino et al. 1996). Taken together, a simple model in which preexisting aggregates of Sup35p recruit soluble Sup35p and are transferred during mitosis from mothers to daughters *via* the cytosol, explains the genetic behaviour of [*PSI*$^+$] (Fig. 6). In addition, the observation that the much of the cellular Sup35p is sequestered in an insoluble form in [*PSI*$^+$] cells and is thus unavailable to terminate translation with high fidelity, explains the defect in translation termination that is manifested as omnipotent suppression.

Around the same time that Wickner proposed that prion-like processes could explain the unusual pattern of [*PSI*$^+$] inheritance, a screen was underway in the lab of Sue Liebman for genes that, when overexpressed in yeast, would influence the mitotic stability of the [*PSI*$^+$]. Remarkably, multiple screens produced only one gene that efficiently induced the loss of [*PSI*$^+$] with high frequency— the gene encoding Hsp104p (Chernoff et al. 1995). Perhaps even more surprising was the fact that when the *HSP104* gene was knocked out in [*PSI*$^+$] cells, they also converted to [*psi*$^-$] and these cells could not be induced to revert to the [*PSI*$^+$] state even by the overexpression of Sup35p!

The role of Hsp104p in prion stability helps to explain one of the puzzling features of *S. cerevisiae* prions— the instability of prion elements in the presence of low millimolar concentration of guanidinium (Gdn). At first glance one might predict that the protein denaturing property of Gdn might directly destabilize Sup35p aggregates resulting in the curing of [*PSI*$^+$]. However, the Gdn concentrations required for prion curing (2 to 3 mM) are much too low to cause protein denaturation. An effect of low concentrations of Gdn on Hsp104p activity was first reported during the reconstitution of Hsp104p-dependent using Gdn denatured proteins as a substrate (Glover and Lindquist 1998). An initial investigation revealed that residual Gdn derived from the substrate denaturation step was inhibitory to Hsp104p-mediated refolding. It was consequently discovered that in

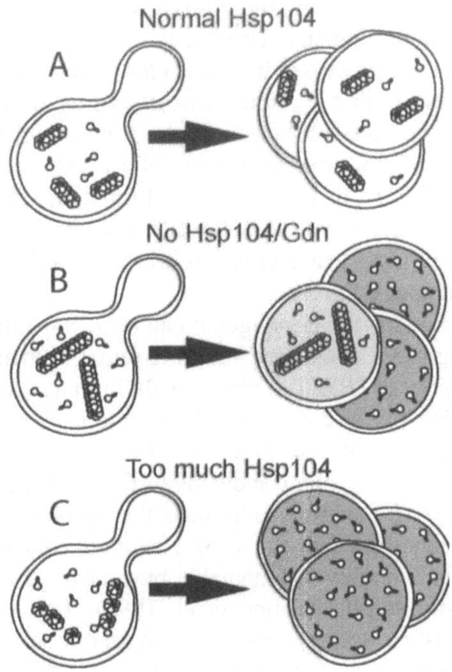

Fig. 7. The role of Hsp104p in [*PSI*$^+$] stability. A. In cells expressing a moderate amount of Hsp104p, severing of Sup35 fibrils generates sufficient new seeds to ensure stable inheritance of the [*PSI*$^+$] prion. B. When Hsp104p function is depleted by mutation of the *HSP104* gene or by growth of cells in guanidinium chloride (Gdn), fibrils are not severed. The reduced quantity and increased size of [*PSI*$^+$] prion seeds result in progressive loss of [*PSI*$^+$] cells in subsequent mitotic divisions. C. When Hsp104p is overexpressed, the severing properties of Hsp104p may reduce seeds to a size where they are no longer physically stable, essentially dissolving them.

addition to curing [*PSI*$^+$] cells, Gdn also inhibits the ability of cells to develop thermotolerance and refold thermally denatured proteins *in vivo* (Ferreira et al. 2001; Jung and Masison 2001). In addition, the ATPase activity of Hsp104p is strongly inhibited by the same concentrations of Gdn used to cure prions in yeast (Grimminger et al. 2004). Thus, Gdn most likely destabilizes [*PSI*$^+$] by inhibiting the normal function of Hsp104p.

The ability of Gdn to inhibit Hsp104p function is a convenient experimental tool. Blocking the function of Hsp104p with Gdn in cells and consequently quantitating the loss of [*PSI*$^+$] has been used to infer the number of prion seeds in cells at the time of Hsp104p inhibition (Eaglestone et al. 2000). For at least one yeast strain, this number was approximately 60. Furthermore, the size of Sup35p aggregates increases following exposure of cells to Gdn and begin to decrease to steady state size when Gdn is removed (Kryndushkin et al. 2003) suggesting a role for Hsp104p in the maintenance of the size (or length) and number of prion seeds during normal mitotic growth.

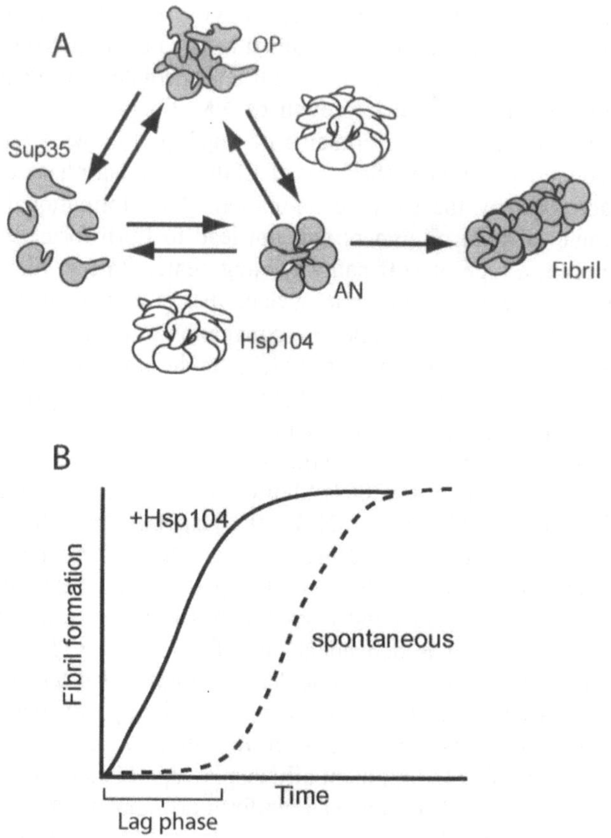

Fig. 8. A model for Hsp104p-mediated acceleration of fibril formation. A. In spontaneous reactions, fibril formation is preceded by the assembly of amyloidogenic nuclei (AN) that can initiate fibril formation. Other aggregates are off pathway (OP) and may become amyloidogenic only by slow rearrangement or by returning to the non-oligomeric state. Hsp104p may help to overcome rate-limiting steps in fibril formation by providing a scaffold or template on which prionogenic oligomers can be assembled (illustration based on Serio et al. 2000). B. Spontaneous fibril formation is preceded by a lag phase during which amyloidogenic nuclei are assembled. The addition of Hsp104 eliminates the lag phase (illustration based on Shorter and Lindquist 2004).

The role of Hsp104p in the mitotic stability of [*PSI*$^+$] can be accounted for based solely on the ability of Hsp104p to remodel protein aggregates (Fig. 7). The modest level of Hsp104p expression in unstressed cells is sufficient to sever longer fibrils and to generate the numbers of moderately sized "seeds" to ensure that they are transmitted to daughter cells during budding. Too much Hsp104p may create seeds that are too small to bephysically stable. Conversely, when Hsp104p is absent or inhibited by Gdn, the fibrils will not be severed to create new seeds and subsequently become too large and too few to be efficiently inherited.

As with the Hsp104p protein refolding activity, the function of Hsp104p in prion propagation and stability can be studied *in vitro* using purified components. Careful analysis of the kinetics and structural intermediates during spontaneous *in vitro* fibrillogenesis indicates that growth of NM fibers (a polypeptide derived from the N-terminal and middle domains of Sup35p) follows a "nucleated conformational conversion" model (Serio et al. 2000). The lag phase of polymerization is characterized by the slow accumulation of conformationally ambiguous oligomers. Some of these oligomers will not lead to fibril formation but instead probably represent a type of "off pathway" aggregate. Others will spontaneously convert to an intermediate oligomeric species that is competent to recruit additional Sup35pNM oligomers and undergo conversion to the high β-sheet structure conformation characteristic of the assembled, stable amyloid-like fibrils. Indeed the same oligomeric intermediates are likely important for the growth of fibrils in reactions containing preformed fibrils (Scheibel et al. 2004).

Only recently has the critical reconstitution of the Sup35p/Hsp104p interaction been accomplished (Inoue et al. 2004; Shorter and Lindquist 2004). Remarkably, at a relatively low Hsp104p/Sup35p-NM ratio, Hsp104p dramatically accelerates fibril formation completely eliminating the lag phase. These reactions initially contain no preexisting seeds suggesting that Hsp104p can accelerate the assembly of the oligomeric intermediates and may destabilize off-pathway oligomers. Importantly, this acceleration is dependent on ATP but not ATP hydrolysis since Hsp104p together with non-hydrolyzable ATP analogues also accelerates the reaction. This observation tends to support a model initially advanced in the early stages of Sup35p prion research (Patino et al. 1996). In this model, it was suggested that hexameric Hsp104p, potentially displaying six independent Sup35p interaction sites, is capable of serving as a platform or template upon which Sup35p oligomers can assemble (Fig. 8). Since Hsp104p would be generally passive in this role it is unlikely that ATP hydrolysis would be necessary.

At higher concentrations of Hsp104p, even spontaneous fibril formation is inhibited but more importantly, preformed fibrils are rapidly disassembled (Shorter and Lindquist 2004). In this instance, ATP hydrolysis is crucial suggesting that fiber depolymerization is more akin to Hsp104p-dependent protein refolding. Unlike protein refolding however, other chaperones do not influence the disassembly process. A different group (Inoue et al. 2004) however, demonstrated that although Hsp104p is indispensable for the depolymerization of preformed Sup35pNM fibrils, it does so in cooperation with a factor or factors present in yeast extracts prepared from *hsp104* mutant cells.

7 Implications for protein aggregation disease

The role of Hsp104p in the stability of yeast prions raises an interesting question regarding the influence of protein disaggregation on the progress of human diseases that are associated with the accumulation of aggregated protein. Many protein aggregation diseases— Alzheimer's Disease, Creutzfeld-Jacob Disease, and various systemic amyloidosis— are the result of aggregation that occurs outside the cell cytoplasm beyond the reach of cytosolic chaperones. However, other disease-associated aggregates accumulate in the cytoplasm or nucleoplasm of cells. Among these are the autosomal dominant polyglutamine (polyQ) repeat diseases including Huntington's Disease (HD), spinal and bulbar muscular atrophy (SBMA), dentorubral and pallidoluysian atrophy (DRPLA), and several forms of spinocerebellar ataxia (SCA1, SCA2, SCA3, SCA6, SCA7, and SCA17). Each of these diseases is associated with the expansion of CAG (glutamine codon) repeats in the coding regions genes that promote amyloid-like aggregation of the expressed protein. Molecular chaperones have been shown to colocalize with and in many cases, to alter the physical state of aggregates (for recent reviews see Opal and Zoghbi 2002; Sakahira et al. 2002; Wyttenbach 2004). Adding even more intrigue to the matter, animal cells lack Hsp104p orthologs and whether or not they possess an endogenous protein or protein complex that specializes in disaggregation is unknown.

A simple approach to understanding how Hsp104p might function in the aggregation of disease-associated proteins is to express them in yeast, perhaps generating a simplified model of the aggregation process (for review see Sherman and Muchowski 2003). In wild type yeast, the expression of segments of huntingtin with various lengths of polyglutamine fused to GFP results in modest aggregation of a 72 glutamine (72Q) fusion protein and extensive aggregation of a 103Q fusion protein (Krobitsch and Lindquist 2000). Hsp104p suppresses the formation of aggregates by the 72Q fusion protein when overexpressed but is only partially effective in suppressing aggregation by the 103Q fusion protein. In a striking parallel to the effect of Hsp104p expression on the stability of the [*PSI*$^+$] prion, the 103Q fusion protein does not form aggregates at all in *hsp104* deletion cells. Similar results were obtained by an independent group using 34Q and 80Q fusion proteins (Kimura et al. 2001).

Another more direct approach to understanding the potential of chaperone-mediated protein disaggregation in models of disease is to express Hsp104p in animal cells. To establish whether Hsp104p would function in animal cells as it does in yeast, we expressed Hsp104p in a human T-cell lymphoblast cell line (Mosser et al. 2004). Indeed, Hsp104p was able to assist in the refolding of thermally denatured firefly luciferase in human cells. *In vitro* experiments indicated that human Hsp70 and Hsp40 family members could substitute for their yeast counterparts in forming a hybrid bichaperone network for protein refolding. Importantly, Hsp104p apparently entered the nucleus of animal cells where aggregated polyglutamine repeat proteins frequently accumulate. Hsp104p also enhanced the ability of the cell line to survive heat-shock although its effect was not nearly so dramatic as its effects in yeast.

In an interesting *Caenorhabditis elegans* model of polyglutamine aggregation, a segment of huntingtin was expressed in the body wall muscle cells of transgenic nematodes where it formed aggregates and caused a moderate developmental delay (Satyal et al. 2000). The expression of Hsp104p in these animals diminished the aggregation of the polyglutamine-containing protein and ameliorated the delay. Likewise, in a tissue culture cell model, Hsp104p expression reduced aggregation of a huntingtin-GFP fusion and this was associated with a reduction in nuclear fragmentation used in these studies to signal progression into apoptotisis (Carmichael et al. 2000).

Whether the aggregates formed in polyglutamine repeat diseases are themselves toxic or if they are merely an end product of perturbed protein metabolism caused by soluble polyglutamine-expanded proteins is still somewhat controversial with supportive evidence on both sides. In addition, because the overexpression of some chaperones —notably Hsp70— inhibits proapoptotic signaling it is not clear whether reduced aggregation or simply the suppression of cell death signaling pathways is linked to improved cell survival in a cause and effect manner (for review see Kobayashi and Sobue 2001). We observed that Hsp104p expressed in human cells also inhibited proapoptotic signaling initiated by heat shock although in a manner that was distinct from that of overexpressed Hsp70 (Mosser et al. 2004). Understanding how aggregation-prone proteins cause cell death and how chaperones modulate aggregation and cell survival will be critical in establishing rational approaches to treating protein aggregation disease, perhaps by manipulating the molecular chaperone activity in cells pharmacologically.

8 Final Remarks

The interest in protein aggregation has in recent years moved into the limelight, particularly as aggregation is now associated with a large number of important human diseases. Whereas aggregation was once thought to be an irreversible process, studies on Hsp104p and its bacterial ortholog ClpB indicate that this is not always the case. Significant progress has been made in elucidating the basic mechanism of Hsp104/ClpB disaggregation but many important elements remain to be investigated more thoroughly. It will be interesting to know how Hsp104p recognizes and interacts with diverse yeast prions to promote mitotic stability, what characteristic of amorphous aggregates is recognized by Hsp104p, how Hsp104p "grabs on" to misfolded proteins to initiate unfolding and threading, and why Hsp104p functions in protein refolding only with yeast or human Hsp70s but not with bacterial chaperones. The answers to these and other important questions may inspire innovations in the treatment of protein aggregation disease in the distant future.

References

Abbas-Terki T, Donze O, Briand PA, Picard D (2001) Hsp104 interacts with hsp90 co-chaperones in respiring yeast. Mol Cell Biol 21:7569-7575

Attfield PV, Kletsas S, Hazell BW (1994) Concomitant appearance of intrinsic thermotolerance and storage of trehalose in *Saccharomyces cerevisiae* during early respiratory phase of batch-culture is CIF1-dependent. Microbiology 140:2625-2632

Beinker P, Schlee S, Groemping Y, Seidel R, Reinstein J (2002) The N terminus of ClpB from *Thermus thermophilus* is not essential for the chaperone activity. J Biol Chem 277:47160-47166

Bracken AP, Bond U (1999) Reassembly and protection of small nuclear ribonucleoprotein particles by heat shock proteins in yeast cells. RNA 5:1586-1596

Carmichael J, Chatellier J, Woolfson A, Milstein C, Fersht AR, Rubinsztein DC (2000) Bacterial and yeast chaperones reduce both aggregate formation and cell death in mammalian cell models of Huntington's Disease. Proc Natl Acad Sci USA 97:9701-9705

Cashikar AG, Schirmer EC, Hattendorf DA, Glover JR, Ramakrishnan MS, Ware DM, Lindquist SL (2002) Defining a pathway of communication from the C-terminal substrate binding domain to the N-terminal ATPase domain in Hsp104. Mol Cell 9:751-660

Chernoff YO, Lindquist SL, Ono B, Inge-Vechtomov SG, Liebman SW (1995) Role of the chaperone protein Hsp104 in propagation of the yeast prion- like factor [psi+]. Science 268:880-884

Cox BS (1965) [PSI], a cytoplasmic suppressor of super-suppressor in yeast. Heredity 20:505-521

Cyr DM (1995) Cooperation of the molecular chaperone Ydj1 with specific Hsp70 homologs to suppress protein aggregation. FEBS Lett 359:129-132

Doel SM, McCready SJ, Nierras CR, Cox BS (1994) The dominant PNM2- mutation, which eliminates the psi factor of *Saccharomyces cerevisiae* is the result of a missense mutation in the SUP35 gene. Genetics 137:659-670

Donaldson LW, Wojtyra U, Houry WA (2003) Solution structure of the dimeric zinc binding domain of the chaperone ClpX. J Biol Chem 278:48991-48996

Dougan DA, Reid BG, Horwich AL, Bukau B (2002) ClpS, a substrate modulator of the ClpAP machine. Mol Cell 9:673-683

Dougan DA, Weber-Ban E, Bukau B (2003) Targeted delivery of an *ssrA*-tagged substrate by the adaptor protein SspB to its cognate AAA^+ protein ClpX. Mol Cell 12:373-380

Eaglestone SS, Ruddock LW, Cox BS, Tuite MF (2000) Guanidine hydrochloride blocks a critical step in the propagation of the prion-like determinant [PSI^+] of *Saccharomyces cerevisiae*. Proc Natl Acad Sci USA 97:240-244

Elliott B, Haltiwanger RS, Futcher B (1996) Synergy between trehalose and Hsp104 for thermotolerance in *Saccharomyces cerevisiae*. Genetics 144:923-933

Escher A, O'Kane DJ, Lee J, Szalay AA (1989) Bacterial luciferase αβ fusion protein is fully active as a monomer and highly sensitive *in vivo* to elevated temperature. Proc Natl Acad Sci USA 86:6528-6532

Ferreira PC, Ness F, Edwards SR, Cox BS, Tuite MF (2001) The elimination of the yeast [PSI^+] prion by guanidine hydrochloride is the result of Hsp104 inactivation. Mol Microbiol 40:1357-1369

Glover JR, Kowal AS, Schirmer EC, Patino MM, Liu JJ, Lindquist S (1997) Self-seeded fibers formed by Sup35, the protein determinant of [PSI^+], a heritable prion-like factor of *S. cerevisiae*. Cell 89:811-819

Glover JR, Lindquist S (1998) Hsp104, Hsp70, and Hsp40: a novel chaperone system that rescues previously aggregated proteins. Cell 94:73-82

Glover JR, Tkach JM (2001) Crowbars and ratchets: hsp100 chaperones as tools in reversing protein aggregation. Biochem Cell Biol 79:557-568

Goloubinoff P, Mogk A, Zvi AP, Tomoyasu T, Bukau B (1999) Sequential mechanism of solubilization and refolding of stable protein aggregates by a bichaperone network. Proc Natl Acad Sci USA 96:13732-13737

Grimminger V, Richter K, Imhof A, Buchner J, Walter S (2004) The prion curing agent guanidinium chloride specifically inhibits ATP hydrolysis by Hsp104. J Biol Chem 279:7378-7383

Guo F, Esser L, Singh SK, Maurizi MR, Xia D (2002a) Crystal structure of the heterodimeric complex of the adaptor, ClpS, with the N-domain of the AAA+ chaperone, ClpA. J Biol Chem 277:46753-46762

Guo F, Maurizi MR, Esser L, Xia D (2002b) Crystal structure of ClpA, an Hsp100 chaperone and regulator of ClpAP protease. J Biol Chem 277:46743-46752

Hattendorf DA, Lindquist SL (2002a) Analysis of the AAA sensor-2 motif in the C-terminal ATPase domain of Hsp104 with a site-specific fluorescent probe of nucleotide binding. Proc Natl Acad Sci USA 99:2732-2737

Hattendorf DA, Lindquist SL (2002b) Cooperative kinetics of both Hsp104 ATPase domains and interdomain communication revealed by AAA sensor-1 mutants. EMBO J 21:12-21

Inoue Y, Taguchi H, Kishimoto A, Yoshida M (2004) Hsp104 binds to yeast sup35 prion fiber but needs other factor(s) to sever it. J Biol Chem 23:23

Ishikawa T, Beuron F, Kessel M, Wickner S, Maurizi MR, Steven AC (2001) Translocation pathway of protein substrates in ClpAP protease. Proc Natl Acad Sci USA 98:4328-4333

Jung G, Masison DC (2001) Guanidine hydrochloride inhibits Hsp104 activity *in vivo*: a possible explanation for its effect in curing yeast prions. Curr Microbiol 43:7-10

Kedzierska S, Akoev V, Barnett ME, Zolkiewski M (2003) Structure and function of the middle domain of ClpB from *Escherichia coli*. Biochemistry 42:14242-14248

Kim DY, Kim KK (2003) Crystal structure of ClpX molecular chaperone from *Helicobacter pylori*. J Biol Chem 278:50664-50670

Kimura Y, Koitabashi S, Kakizuka A, Fujita T (2001) Initial process of polyglutamine aggregate formation *in vivo*. Genes Cells 6:887-897

Kobayashi Y, Sobue G (2001) Protective effect of chaperones on polyglutamine diseases. Brain Res Bull 56:165-168

Krobitsch S, Lindquist S (2000) Aggregation of huntingtin in yeast varies with the length of the polyglutamine expansion and the expression of chaperone proteins. Proc Natl Acad Sci USA 97:1589-1594

Kryndushkin DS, Alexandrov IM, Ter-Avanesyan MD, Kushnirov VV (2003) Yeast [PSI^+] prion aggregates are formed by small Sup35 polymers fragmented by Hsp104. J Biol Chem 278:49636-49643

Lee GJ (1995) Assaying proteins for molecular chaperone activity. Meth Cell Biol 5:325-334

Lee P, Shabbir A, Cardozo C, Caplan AJ (2004) Sti1 and Cdc37 can stabilize Hsp90 in chaperone complexes with a protein kinase. Mol Biol Cell 15:1785-1792

Lee S, Sowa ME, Watanabe YH, Sigler PB, Chiu W, Yoshida M, Tsai FT (2003) The structure of ClpB: a molecular chaperone that rescues proteins from an aggregated state. Cell 115:229-240

Lindquist S, Kim G (1996) Heat-shock protein 104 expression is sufficient for thermotolerance in yeast. Proc Natl Acad Sci USA 93:5301-5306

Lum R, Tkach JM, Vierling E, Glover JR (2004) Evidence for an unfolding/threading mechanism for protein disaggregation by *Saccharomyces cerevisiae* Hsp104. J Biol Chem 279:29139-29146

Lupas AN, Martin J (2002) AAA proteins. Curr Opin Struct Biol 12:746-753

Mogk A, Schlieker C, Strub C, Rist W, Weibezahn J, Bukau B (2003) Roles of individual domains and conserved motifs of the AAA+ chaperone ClpB in oligomerization, ATP hydrolysis, and chaperone activity. J Biol Chem 278:17615-17624

Mosser DD, Ho S, Glover JR (2004) *Saccharomyces cerevisiae* Hsp104 enhances the chaperone capacity of human cells and inhibits heat stress-induced proapoptotic signaling. Biochemistry 43:8107-8115

Motohashi K, Watanabe Y, Yohda M, Yoshida M (1999) Heat-inactivated proteins are rescued by the DnaK.J-GrpE set and ClpB chaperones. Proc Natl Acad Sci USA 96:7184-7189

Neuwald AF, Aravind L, Spouge JL, Koonin EV (1999) AAA+: A class of chaperone-like ATPases associated with the assembly, operation, and disassembly of protein complexes. Genome Res 9:27-43

Ogura T, Wilkinson AJ (2001) AAA+ superfamily ATPases: common structure--diverse function. Genes Cells 6:575-597

Opal P, Zoghbi HY (2002) The role of chaperones in polyglutamine disease. Trends Mol Med 8:232-236

Park SK, Kim KI, Woo KM, Seol JH, Tanaka K, Ichihara A, Ha DB, Chung CH (1993) Site-directed mutagenesis of the dual translational initiation sites of the *clpB* gene of *Escherichia coli* and characterization of its gene products. J Biol Chem 268:20170-20174

Parsell DA, Sanchez Y, Stitzel JD, Lindquist S (1991) Hsp104 is a highly conserved protein with two essential nucleotide- binding sites. Nature 353:270-273

Parsell DA, Taulien J, Lindquist S (1993) The role of heat-shock proteins in thermotolerance. Philos Trans R Soc Lond B Biol Sci 339:279-285; discussion 285-276

Parsell DA, Kowal AS, Lindquist S (1994a) *Saccharomyces cerevisiae* Hsp104 protein. Purification and characterization of ATP-induced structural changes. J Biol Chem 269:4480-4487

Parsell DA, Kowal AS, Singer MA, Lindquist S (1994b) Protein disaggregation mediated by heat-shock protein Hsp104. Nature 372:475-478

Parsell DA, Lindquist S (1994) Heat shock proteins and stress tolerance. In: (eds) The bilogy of heat shock proteins and molecular chaperones. Cold Spring Harbor Laboratory Press, Cold Spring Harbor, pp 457-493

Patino MM, Liu JJ, Glover JR, Lindquist S (1996) Support for the prion hypothesis for inheritance of a phenotypic trait in yeast. Science 273:622-626

Prusiner SB (1998) Prions. Proc Natl Acad Sci USA 95:13363-13383

Reid BG, Fenton WA, Horwich AL, Weber-Ban EU (2001) ClpA mediates directional translocation of substrate proteins into the ClpP protease. Proc Natl Acad Sci USA 98:3768-3772

Sakahira H, Breuer P, Hayer-Hartl MK, Hartl FU (2002) Molecular chaperones as modulators of polyglutamine protein aggregation and toxicity. Proc Natl Acad Sci USA 99 Suppl 4:16412-16418

Sanchez Y, Lindquist SL (1990) HSP104 required for induced thermotolerance. Science 248:1112-1115

Satyal SH, Schmidt E, Kitagawa K, Sondheimer N, Lindquist S, Kramer JM, Morimoto RI (2000) Polyglutamine aggregates alter protein folding homeostasis in *Caenorhabditis elegans*. Proc Natl Acad Sci USA 97:5750-5755

Scheibel T, Bloom J, Lindquist SL (2004) The elongation of yeast prion fibers involves separable steps of association and conversion. Proc Natl Acad Sci USA 101:2287-2292

Schirmer EC, Glover JR, Singer MA, Lindquist S (1996) HSP100/Clp proteins: a common mechanism explains diverse functions. Trends Biochem Sci 21:289-296

Schirmer EC, Queitsch C, Kowal AS, Parsell DA, Lindquist S (1998) The ATPase activity of Hsp104, effects of environmental conditions and mutations. J Biol Chem 273:15546-15552

Schirmer EC, Ware DM, Queitsch C, Kowal AS, Lindquist SL (2001) Subunit interactions influence the biochemical and biological properties of Hsp104. Proc Natl Acad Sci USA 98:914-919

Schlee S, Beinker P, Akhrymuk A, Reinstein J (2004) A chaperone network for the resolubilization of protein aggregates: direct interaction of ClpB and DnaK. J Mol Biol 336:275-285

Schlieker C, Weibezahn J, Patzelt H, Tessarz P, Strub C, Zeth K, Erbse A, Schneider-Mergener J, Chin JW, Schultz PG, Bukau B, Mogk A (2004) Substrate recognition by the AAA+ chaperone ClpB. Nat Struct Mol Biol 11:607-615

Seol JH, Baek SH, Kang MS, Ha DB, Chung CH (1995) Distinctive roles of the two ATP-binding sites in ClpA, the ATPase component of protease Ti in *Escherichia coli*. J Biol Chem 270:8087-8092

Serio TR, Cashikar AG, Kowal AS, Sawicki GJ, Moslehi JJ, Serpell L, Arnsdorf MF, Lindquist SL (2000) Nucleated conformational conversion and the replication of conformational information by a prion determinant. Science 289:1317-1321

Sherman MY, Muchowski PJ (2003) Making yeast tremble: yeast models as tools to study neurodegenerative disorders. Neuromolecular Med 4:133-146

Shorter J, Lindquist S (2004) Hsp104 catalyzes formation and elimination of self-replicating Sup35 prion conformers. Science 304:1793-1797

Siddiqui SM, Sauer RT, Baker TA (2004) Role of the processing pore of the ClpX AAA+ ATPase in the recognition and engagement of specific protein substrates. Genes Dev 18:369-374

Singer MA, Lindquist S (1998) Multiple effects of trehalose on protein folding *in vitro* and *in vivo*. Mol Cell 1:639-648

Singh SK, Grimaud R, Hoskins JR, Wickner S, Maurizi MR (2000) Unfolding and internalization of proteins by the ATP-dependent proteases ClpXP and ClpAP. Proc Natl Acad Sci USA 97:8898-8903

Singh SK, Rozycki J, Ortega J, Ishikawa T, Lo J, Steven AC, Maurizi MR (2001) Functional domains of the ClpA and ClpX molecular chaperones identified by limited proteolysis and deletion analysis. J Biol Chem 276:29420-29429

Song HK, Hartmann C, Ramachandran R, Bochtler M, Behrendt R, Moroder L, Huber R (2000) Mutational studies on HslU and its docking mode with HslV. Proc Natl Acad Sci USA 97:14103-14108

Sousa MC, Trame CB, Tsuruta H, Wilbanks SM, Reddy VS, McKay DB (2000) Crystal and solution structures of an HslUV protease-chaperone complex. Cell 103:633-643

Swan TM, Watson K (1999) Stress tolerance in a yeast lipid mutant: membrane lipids influence tolerance to heat and ethanol independently of heat shock proteins and trehalose. Can J Microbiol 45:472-479

Ter-Avanesyan MD, Kushnirov VV, Dagkesamanskaya AR, Didichenko SA, Chernoff YO, Inge-Vechtomov SG, Smirnov VN (1993) Deletion analysis of the SUP35 gene of the yeast *Saccharomyces cerevisiae* reveals two non-overlapping functional regions in the encoded protein. Mol Microbiol 7:683-692

Ter-Avanesyan MD, Dagkesamanskaya AR, Kushnirov VV, Smirnov VN (1994) The SUP35 omnipotent suppressor gene is involved in the maintenance of the non-Mendelian determinant [*psi*$^+$] in the yeast *Saccharomyces cerevisiae*. Genetics 137:671-676

Tkach JM, Glover JR (2004) Amino acid substitutions in the C-terminal AAA+ module of Hsp104 prevent substrate recognition by disrupting oligomerization and Cause high temperature inactivation. J Biol Chem 279:35692-35701

Tuite MF, Mundy CR, Cox BS (1981) Agents that cause a high frequency of genetic change from [*psi*$^+$] to [*psi*$^-$] in *Saccharomyces cerevisiae*. Genetics 98:691-711

Vogel JL, Parsell DA, Lindquist S (1995) Heat-shock proteins Hsp104 and Hsp70 reactivate mRNA splicing after heat inactivation. Curr Biol 5:306-317

Wah DA, Levchenko I, Rieckhof GE, Bolon DN, Baker TA, Sauer RT (2003) Flexible linkers leash the substrate binding domain of SspB to a peptide module that stabilizes delivery complexes with the AAA+ ClpXP protease. Mol Cell 12:355-363

Wakem LP, Sherman F (1990) Isolation and characterization of omnipotent suppressors in the yeast *Saccharomyces cerevisiae*. Genetics 124:515-522

Wang J, Song JJ, Franklin MC, Kamtekar S, Im YJ, Rho SH, Seong IS, Lee CS, Chung CH, Eom SH (2001) Crystal structures of the HslVU peptidase-ATPase complex reveal an ATP- dependent proteolysis mechanism. Structure 9:177-184

Weibezahn J, Bukau B, Mogk A (2004a) Unscrambling an egg: protein disaggregation by AAA+ proteins. Microb Cell Fact 3:1

Weibezahn J, Tessarz P, Schlieker C, Zahn R, Maglica Z, Lee S, Zentgraf H, Weber-Ban EU, Dougan DA, Tsai FT, Mogk A, Bukau B (2004b) Thermotolerance requires refolding of aggregated proteins by substrate translocation through the central pore of ClpB. Cell 119:653-665

Wojtyra UA, Thibault G, Tuite A, Houry WA (2003) The N-terminal zinc binding domain of ClpX is a dimerization domain that modulates the chaperone function. J Biol Chem 278:48981-48990

Wyttenbach A (2004) Role of heat shock proteins during polyglutamine neurodegeneration: mechanisms and hypothesis. J Mol Neurosci 23:69-96

Xia D, Esser L, Singh SK, Guo F, Maurizi MR (2004) Crystallographic investigation of peptide binding sites in the N-domain of the ClpA chaperone. J Struct Biol 146:166-179

Zietkiewicz S, Krzewska J, Liberek K (2004) Successive and synergistic action of the Hsp70 and Hsp100 chaperones in protein disaggregation. J Biol Chem 279:44376-44383

Zolkiewski M (1999) ClpB cooperates with DnaK, DnaJ, and GrpE in suppressing protein aggregation. A novel multi-chaperone system from *Escherichia coli*. J Biol Chem 274:28083-28086

Glover, John R.
Department of Biochemistry, University of Toronto, 1 King's College Circle, Room 5302, Medical Sciences Building, Toronto, Ontario, Canada M5S 1A8
john.glover@utoronto.ca

Tkach, Johnny M.
Department of Biochemistry, University of Toronto, 1 King's College Circle, Room 5302, Medical Sciences Building, Toronto, Ontario, Canada M5S 1A8

Folding of newly synthesised proteins in the endoplasmic reticulum

Sanjika Dias-Gunasekara and Adam M. Benham

Abstract

The endoplasmic reticulum (ER) is a membranous compartment that can be found within any nucleated eukaryotic cell. Its job is to oversee the production of all the proteins that the cell secretes, or needs to express at the cell surface or within the secretory pathway itself. The type of proteins that pass through the ER is very varied, ranging from small, secreted peptide hormones, to large cell surface receptors. To the uninitiated, protein folding in the endoplasmic reticulum might seem straightforward. Unfortunately for biology students, but fortunately for researchers, it turns out that protein folding in the ER is a complex business, involving chaperones, quality control machinery and many accessory factors. These molecular helpers make sure that glycoproteins fold properly, and are directed to the right cellular location at the right time. Although many newly synthesised proteins follow a set of common "rules", some proteins require specific types of chaperones to assist them. In this review, recent advances in our knowledge of the early stages of ER protein folding will be discussed, focusing on the mammalian ER, but also drawing on examples of work in yeast.

1 The scope of protein folding in the ER

An important point to realise about ER protein folding is that it can go wrong, and does so quite often. At any given time, a proportion of polypeptides can become trapped in (relatively) energetically favourable, misfolded states (Schymkowitz et al. 2002). Alternatively, genetic mutations can give rise to changes in the amino acid sequence of a protein, causing the protein to become vulnerable to misfolding. Many diseases have an ER protein-misfolding component to them, including α1-antitrypsin deficiency (Lawless et al. 2004), Creuzfeld-Jakob disease (Aguzzi et al. 2004), the rheumatic disease ankylosing spondylitis (Ramos and Lopez de Castro 2002), and cystic fibrosis (Kopito 1999; Sharma et al. 2004). All of these diseases are at least partly caused by amino acid changes that cause the protein to misfold, or to be degraded by the ER quality control system. Misfolded proteins are dangerous to the cell, so if the ER fails to fold a protein correctly, a system called ER associated degradation (ERAD) exists to detect and remove the problematic protein from the ER. Thus an understanding of glycoprotein misfolding helps us towards the molecular mechanisms that underpin both the healthy and diseased state (Wigley et al. 2001).

Some cell types are dedicated to the production of particular proteins and have a huge stack of ER membranes to enable them to secrete many times their weight in protein every day. For example, a pancreatic exocrine cell will produce the digestive enzyme amylase whereas its companion islet cell will produce insulin, needed for the metabolic regulation of glucose. The activated plasma B cell secretes vast quantities of antibodies, whereas a fibroblast will secrete very little protein at all, in keeping with its role as a "structural" cell. To accommodate these differences, the ER needs a quality control system that is general, but that can be adapted to suit the needs of individual cell types and individual proteins.

In this chapter, we highlight the events that occur as a glycoprotein emerges from the ribosome and enters the ER environment. We will consider the array of protein chaperones and assistants that the glycoprotein meets, and describe how these important catalysts help to decide the fate of the young protein.

2 Entry into the ER

A polypeptide is targeted to the ER by virtue of its signal sequence, and is threaded from the ribosome into the ER through a channel called the translocon, or Sec61 complex (for reviews see Clemons et al. 2004; Johnson and van Waes 1999). As the nascent protein is being made, it will immediately start to encounter different local microenvironments, depending on whether it is destined to be a multi-membrane spanning protein, a soluble protein or a glycosylphosphatidyl inositol (GPI) linked protein, for example. We are now beginning to appreciate that these different proteins encounter distinct quality control issues even at this early stage of life. For example, Fluorescence Resonance Energy Transfer (FRET) studies have shown that both membrane and secretory proteins already differ in their folding within the ribosome itself, before they even enter the translocon. Hydrophobic trans-membrane sequences fold into tight helices that remain intact as they pass into the translocon, whereas secretory protein sequences take up more extended conformations (Woolhead et al. 2004).

The opening and closing of the translocon is tightly controlled to maintain selective permeability and this is partly regulated by binding to the ribosome. The lumenal side of the pore is sealed by BiP, and the seal must be in place before the cytosolic side opens, to maintain a permeability barrier (Hamman et al. 1998; Liao et al. 1997). BiP is a Ca^{2+} and ATP-dependent, Hsp70 family member of the ER lumen, possessing a KDEL ER retention signal. BiP is a protein that has contributed much to the understanding of ER biology: BiP was first characterised as an immunoglobulin binding chaperone (Haas and Wabl 1983), and as GRP78 it was one of the earliest "glucose regulated proteins" to be discovered (Munro and Pelham 1986). The versatile role of BiP as a chaperone will be discussed later on.

3 ER protein sorting

Different polypeptides must be targeted to different places in order to function properly. A protein may reside in the ER, assemble into complexes, remain as a monomer, or get directed to nuclear membranes. Alternatively, a protein may move apically or baso-laterally along the secretory pathway of polarised cells. Cellular polarity is essential for the proper function of many epithelial and endothelial cells (Nelson and Yeaman 2001). In epithelial cells, baso-lateral sorting is directed by signal motifs found in cytoplasmic domains. Usually, these motifs contain a conserved tyrosine residue (Y) and are of the form NPXY or YXXΦ (where X can be any amino acid and Φ is a hydrophobic residue) although some motifs are di-hydrophobic (e.g. the LDL receptor and Fc receptor) (Matter et al. 1994). Apical sorting signals are less well defined, but are usually present in the ER lumen rather than the cytoplasm. Targeting may also be achieved by a GPI anchor, as for decay accelerating factor (DAF) (Lisanti et al. 1989). Specific N-glycans can also be essential for apical sorting. For example, growth hormone, which is normally non-glycosylated and secreted both apically and baso-laterally from MDCK cells, is only secreted apically when glycosylated (Scheiffele et al. 1995). The lysosomal transmembrane protein endolyn also gets apically sorted in a glycan-dependent way, despite a cytoplasmic baso-lateral sorting motif that gets over-ridden by glycosylation (Potter et al. 2004). In general, the decision between apical and baso-lateral sorting occurs mainly in the trans-Golgi network (TGN). However, the relationship between translocation, "early" sorting events and ER protein folding is sure to develop in the coming years.

The sorting and quality control of inner nuclear membrane (INM) proteins often takes a back seat when compared to the quality control secreted proteins. This situation is now beginning to change (Mattaj 2004). Using proteins containing both cellular and viral INM sorting motifs, Art Johnson's group has used a photo cross-linking approach to show that an initial sorting decision is likely to occur early, within the translocon itself, based on preferential interaction of INM sequences with the TRAM protein (translocating chain-associated membrane protein, a component of the translocon) (Saksena et al. 2004). These results challenge the previous "passive" diffusion-retention model for sorting of inner nuclear membrane proteins, suggest that sorting of INMs is an active signal driven process, and point to TRAM as an important control point in nuclear membrane sorting. The emerging concept from these and other studies is that quality control and targeting decisions get made very early in the lifetime of a polypeptide, and are dependent on polypeptide folding.

4 Signal peptide cleavage

All glycoproteins that are inserted co-translationally into the ER through the translocon have an N-terminal signal sequence that is almost always cleaved from the mature polypeptide. The signal peptidase complex (SPC) of the ER was biochemically and genetically characterised in the 1990's, in both yeast, and in mammalian systems (Shelness et al. 1993). SPC is a hetero-oligomeric membrane complex comprised of 12, 18, 21, 22/23, and 25 kDa subunits. In *S. cerevisiae*, the Sec11 protein is necessary for signal peptide processing and cell viability and is homologous to the 18- and 21-kDa mammalian subunits. Details of how signal peptides influence glycoprotein folding and how the timing of signal peptide cleavage relates to glycoprotein maturation are beginning to emerge, particularly with respect to viral proteins. The subject is a particularly interesting area for virologists, because many viral proteins start their life as unprocessed polyproteins. Compact viral genomes use this trick to make efficient use of their limited complement of nucleotides, encoding proteins that must be processed from immature precursors. Processing by signal peptidase and related proteases has a critical impact upon whether an immature polyprotein goes on to fold productively. Semliki Forest virus (SFV), for example, encodes a p62-E1 precursor protein that is cleaved into the spike subunits p62 and E1 by signal peptidase. This step is required for p62 and E1 to form heterodimers from the same precursor protein, and is co-ordinated with the calnexin cycle (described below) (Andersson and Garoff 2003). Similarly, Hepatitis C virus (HCV) coat proteins E1 and E2 are also released from HCV polyprotein by signal peptidase cleavage, and require this step for maturation (Voisset and Dubuisson 2004). Thus efficient signal peptide proteolysis can be seen as a component of ER quality control for many viral glycoproteins.

In addition to the signal peptidase complex, a signal peptide peptidase (SPP) has been identified and characterised (Weihofen et al. 2002). SPP is an unusual, polytopic aspartic acid type intra-membrane protease with homology to presenilin (Xia and Wolfe 2003). This class of proteins is also attracting the interest of biologists interested in protein processing and misfolding because of the role of presenilin in generating amyloid β protein from amyloid precursor protein (APP) in Alzheimer's disease (Weihofen and Martoglio 2003).

Signal peptides have their uses after they have been clipped from the mature protein. One example can be seen in the realm of antigen presentation, in which the function of Major Histocompatibility Complex (MHC) molecules is usually to present peptide fragments, derived from pathogens, to patrolling T cells of the immune system. Classical MHC class I molecules bind to viral peptides in the ER as they are synthesised and start to assemble. It turns out that the non-classical class I molecule, HLA-E, does not bind to viral peptides. Instead, HLA-E binds to peptides derived from the highly conserved signal sequences of other classical MHC class I molecules. This self-recognition system is used to reassure the immune system, in particular natural killer cells, that the surveyed cell is healthy (Lemberg et al. 2001). The system is needed because a down regulation of classical class I molecules is a common ploy used by viruses to prevent the presentation of their own viral antigens by conventional MHC class I molecules. An immune response can thus be triggered based on the lack of displayed signal sequences. Ironically, viruses themselves also exploit SPP. In the case of Hepatitis C virus, SPP is essential for processing of its viral polyproteins (Okamoto et al. 2004). It seems that both viral and normal cellular glycoproteins require intra-membrane proteolytic enzymes as part of their quality control process, but make use of these enzymes in different ways.

5 The proline problem

Proline is a topological protein folding headache because as an imino acid it can exist in either a *cis* or *trans* form in a polypeptide chain. The orientation of proline can dramatically influence the secondary structure of the mature protein, and therefore determine whether a protein is properly folded and functional. *Trans*-proline is the most frequently encountered isomer in nature, but about 5% of prolines within β sheets can occur in the *cis* orientation. The proline *status quo* is surveyed by peptidyl-prolyl cis/trans isomerases (PPIases). PPIases are enzymes found in both the cytoplasm and the ER, whose job is to rearrange the *cis* to the *trans* form. Although these enzymes have been known for some time, their relative contribution to ER quality control is not clearly defined and there is scant data on how ER resident PPIases interact and communicate with the rest of the ER folding machinery. This is partly because PPIases are present not just in the ER, but in the cytosol and other compartments, and partly because PPIase inhibitors can have non-specific effects. Thus whether or not proline isomerisation limits ER glycoprotein folding is largely unknown. Nevertheless, there are some interesting reports linking the FK506 binding-protein (FKBP) family of PPIases to the control of calcium levels in the ER (or the sarcoplasmic reticulum in muscle). FKBPs are members of the extended PPIase family, and get their name from their ability to bind to the immunosuppressive drug FK506. Mutations in FKB12.6 have consequences for the activity of the calcium-pumping ryanodine receptors (RyR) localised at the luminal membranes of muscle (Avila et al. 2003). It has recently been shown that mice lacking FKB12.6 (and patients with the cardiac disorder catecholaminergic polymorphic ventricular tachycardia or CPVT) die from exercise induced sudden cardiac death because RyR is destabilised. In the unstable state, RyR cannot pump the calcium required for muscle contraction from the sarcoplasmic reticulum (Wehrens et al. 2004).

Other work has suggested that another family member, FKB23, can bind to the chaperone BiP in the ER and that this association is dependent on calcium levels (Zhang et al. 2004). Further investigation of this protein with respect to ER quality control is merited. However, it should be noted that numerous PPIases exist, some of which are ER/Golgi associated but not lumenally localised. This family of proteins has diversified to perform various roles in addition to and in place of a prolyl-isomerase function. The cytosolic cyclophilin protein, for example, is the famous target of the immunosuppressive drug cyclosporin A. Cyclosporin A blocks the function of cyclophilin in T cell signalling, preventing NF-AT transcription factor-dependent interleukin 2 secretion (see Matsuda et al. 2000) and references therein). Specific reagents and approaches that target ER resident PPIases are, therefore, required to conclusively demonstrate the physiological need for proline isomerisation in the folding of newly synthesised ER proteins.

6 Folding of ER glycoproteins

Most proteins that fold within the endoplasmic reticulum are glycoproteins, although it should be noted that not all ER proteins are glycosylated. Glycans play a critical role in determining how a nascent protein folds, and as such they have been under intense scrutiny from biochemists and cell biologists (Helenius and Aebi 2004). The mere presence of a glycan on a polypeptide emerging into the ER means that it is likely to attract the attention of the Ca^{2+} dependent, lectin-like chaperones calnexin (67 kD, membrane bound) (David et al. 1993; Ou et al. 1993) and calreticulin (55 kD, soluble) (Smith and Koch 1989). The thiol-disulfide oxidoreductase ERp57 also associates selectively with nascent glycoproteins in a complex with calnexin/calreticulin (Oliver et al. 1997), primarily interacting via its **b** domain (Russell et al. 2004).

Influenza virus hemagglutinin (HA) is a membrane fusion protein produced by the 'flu virus' and synthesised in the ER, and it has been used as a model for studying glycoprotein folding. Structures of the HA fusion protein have been solved (Bullough et al. 1994; Wilson et al. 1981) and the HA folding pathway is now relatively well understood (e.g. Braakman et al. 1991; Copeland et al. 1988). Mature influenza HA is a homo-trimer, with each monomer comprising an HA1 and HA2 chain linked by an intermolecular disulfide bond. N-linked glycans are strategically positioned to guide the interaction with calnexin (and the oxidoreductase ERp57), and help orchestrate the folding pathway prior to trimerisation (Daniels et al. 2003; Hebert et al. 1997). Although endogenous glycoproteins, such as the yeast acid phosphatase also require glycans for correct protein folding (Riederer and Hinnen 1991), not every glycoprotein relies on its sugars for this process. An explanation for these differences may lie in the position of the glycan on the polypeptide chain. If the sugar moiety is located towards the N-terminus and is attached as the nascent chain emerges from the ribosome, as in SFV p62 and influenza HA, the protein substrate is immediately seen by calnexin and ERp57, and is engaged in calnexin dependent quality control. If no "early" N-terminal sugars are present, as in SFV E1 and Vesicular Stomatitis virus (VSV) glycoprotein, the emerging polypeptide is not likely to be seen by calnexin. Instead, these proteins are more likely to interact with BiP and the oxidoreductase Protein Disulfide Isomerase (PDI) (Molinari and Helenius 2000). Although these experiments suggest that PDI/BiP and ERp57/calnexin could form distinct chaperone partnerships, it is not clear whether every viral and normal cellular glycoprotein will follow such a simple "rule". It is also uncertain whether PDI and ERp57 will perform exactly the same function (of disulfide bond oxidation) within the BiP and calnexin complexes (see section on disulfide bond formation). Future work will surely reveal more details of how chaperones select, or are selected by, both viral and cellular proteins.

Glycoproteins that interact with calnexin are subjected to a rigorous quality control inspection (Fig. 1). At first, a core oligosaccharide is transferred from the oligosaccharide transfer complex to the nascent protein at an NXS/T motif. This $Glc_3Man_9GlcNAc_2$ moiety has three terminal glucose residues that are trimmed by

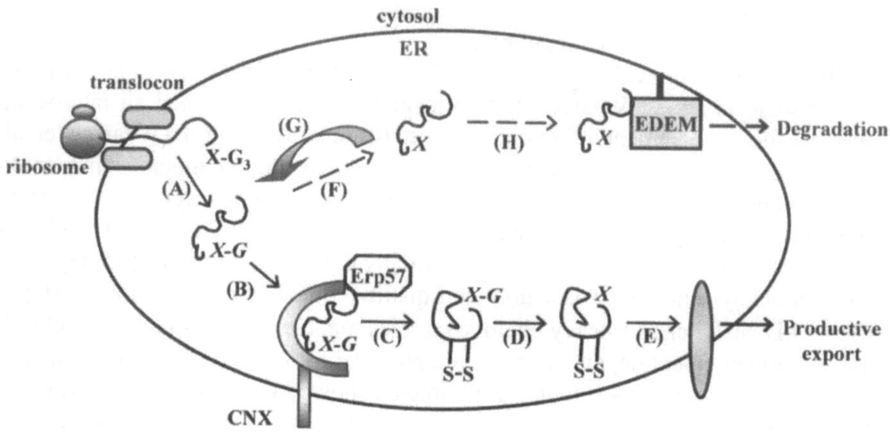

Fig. 1. The calnexin cycle and glycoprotein fate. Nascent polypeptides emerging from the polyribosome enter the ER lumen through a translocon complex and become N-glycosylated with the transfer of Glc_3-Man_9-$(GlcNAc)_2$ [X-G_3] core oligosaccharides from dolichol phosphate to an NXS/T motif by the oligosaccharide transferase complex. Glucosidase I and II remove the two terminal glucose residues (A), and the monoglucosylated (X-G) protein is able to bind CNX and ERp57. Upon release from this complex the protein will be either properly folded (B, solid arrows) or misfolded (F, dashed arrows). Deglucosylation of the remaining glucose residue by glucosidase II (C) targets properly folded disulfide bonded proteins for export out of the ER (D). For deglucosylated misfolded proteins (F), reglucosylation occurs via UGGT (UDP-glucose glycoprotein glucosyltransferase) (G). Alternatively, terminally misfolded proteins get hydrolysed by α-mannosidase I (H) and bind to ER degradation enhancing 1,2-mannosidase like protein (EDEM), which targets proteins for retrotranslocation and subsequent degradation.

glucosidases I and II to yield a monoglucosylated $GlcMan_9GlcNAc_2$ intermediate. (for review see Ellgaard and Helenius 2003). It is now well established that monoglucosylated proteins bind to both calnexin and calreticulin and are subsequently released following hydrolysis of the mono-glucose residue by glucosidase II (Hammond et al. 1994). These substrates may be re-glucosylated by UDP-glucose:glycoprotein glucosyltransferase (GT) allowing them to re-associate with calnexin (Caramelo et al. 2003). The idea behind this cycle of attachment and release to and from calnexin is to give the nascent polypeptide sufficient time to find a relatively folded conformation, or else be consigned to expulsion from the ER and subsequent degradation. The system relies on the detection of localised, disordered conformation and exposed hydrophobic patches by GT (Ritter and Helenius 2000; Trombetta and Helenius 2000). In this way, GT determines whether a protein is misfolded or not. A folded protein will not be reglucosylated, whereas a misfolded one will be sent back to calnexin for another try. Terminally misfolded glycoproteins are passed to EDEM (ER degradation enhancing 1,2-mannosidase like protein), another lectin like ER resident protein that receives misfolded substrates from calnexin and assists their degradation (Hosokawa et al. 2001; Molinari et al. 2003; Oda et al. 2003). ER degradation is the focus of another chapter and will not be discussed in detail here.

There is growing evidence that calnexin and calreticulin do more than just recognise sugar residues. Wearsch and colleagues recently published an interesting twist to the involvement of calreticulin in ER quality control (Wearsch et al. 2004). These investigators used *S. cerevisiae* glycosylation mutants that could only synthesise monoglucosylated glycoproteins. These strains were used to create MHC class I heavy chains with solely $GlcMan_9GlcNAc_2$ structures. MHC class I molecules are mammalian proteins not normally expressed by yeast, but they are a good model system whose assembly and quality control is quite well understood. The 45 kD MHC class I heavy chain assembles together with a small β_2m subunit and an antigenic peptide, and all three are required for the complex to pass the ER quality control apparatus. *In vitro* experiments showed that calreticulin bound efficiently to both assembled, peptide-loaded MHC class I complexes as well as class I folding intermediates, including free heavy chains and empty HC-β_2m heterodimers. This suggested that calreticulin cannot differentiate between native and non-native MHC class I conformations and questions its role in the quality control of peptide-loaded MHC class I complexes. However, some caution is required in interpreting folding differences between yeast and mammalian cells, since there are some differences between the two organisms. Notably, *S. cerevisiae* expresses only one calnexin/calreticulin type protein, most closely related to calnexin (Parlati et al. 1995). Nevertheless, the results of Wearsch *et al.* are thought provoking and further use of this type of expression system using other model cellular and viral ER substrates may be very informative.

Structural studies have started to reveal how calnexin/calreticulin and ERp57 come together to form a complex that can recognise a wide range of glycans and hold them in a scaffold suitable for protein folding. A combination of NMR and a new membrane-based yeast two-hybrid system (MYTHS) have been used to show that direct interactions occur between the tip of the calnexin P-domain (a stable, non-helical fold towards the C-terminus of the protein) and the basic carboxy-terminus of ERp57 (Pollock et al. 2004). Other work, based on biochemistry and domain swapping experiments, suggests that the association between calnexin and ERp57 also requires other protein determinants (Silvennoinen et al. 2004). The NMR structure of the calreticulin P-domain (from rat) has also been studied in detail. The P domain has a hairpin fold that encompasses the whole polypeptide chain, with the two chain termini in close spatial proximity (Ellgaard et al. 2001). The crystal structure of the ER lumenal portion of calnexin has now been solved to 2.9 Å and has along extended 140 Å β-sandwich arm, including two proline-based motifs, topped with a globular lectin domain. This structure is ideal for positioning substrates for interaction with ERp57 (Schrag et al. 2001).

Calnexin and calreticulin are sometimes thought of as interchangeable proteins that perform essentially the same function. The work on MHC class I molecules suggests that this is not the case, and it seems that calnexin may be able to perform additional tasks by virtue of its transmembrane domain. Although the weight of evidence suggests that calnexin primarily interacts with mono-glucosylated sugar residues, *in vitro* studies do suggest that calnexin can recognise non-glycosylated proteins as well as glycoproteins (Ihara et al. 1999; Saito et al. 1999). In particular, calnexin can interact with the tetraspanin glycoprotein CD82 in a transmembrane-dependent manner (Cannon and Cresswell 2001). CD82 is a lymphoid-specific partner of the immune accessory molecules CD4 and CD8. Truncated CD82 molecules lacking one or more transmembrane segments are retained in the ER. This occurs even when CD82 mutants have a properly folded ER lumenal domain. Transport of these proteins can be restored with co-expression of the CD82 transmembrane segments only, suggesting that these hydrophobic sequences are important for quality control of CD82. Although CD82 is glycosylated, it transiently associates with calnexin (but not with calreticulin) in a transmembrane dependent manner, implying that calnexin can sense misfolded conformations even within a bilayer.

Studies using the multi-membrane spanning, non-glycosylated ER substrate proteolipid protein (PLP), also suggest that calnexin can interact with transmembrane segments in the ER (Swanton et al. 2003). Calnexin binds to PLP transmembrane domain constructs, as well as a misfolded mutant of PLP. PLP is an important protein, being a major component of myelin. Mutations in PLP lead to X-linked Pelizaeus-Merzbacher disease and spastic paraplegia, in which motor development and co-ordination are severely impaired. Although the molecular details of how calnexin interacts with PLP (and CD82) need to be further resolved, it would be no surprise if calnexin is found to be important for the quality control of even more multi-membrane spanning non-glycosylated proteins.

The calnexin cycle is not the be-all and end-all of quality control in the ER. Other decisions must be made and checkpoints passed for a protein to reach its final destination. Important areas of research that will be discussed only briefly here include the role of lipids and membranes in quality control (Bogdanov and Dowhan 1999), and how proteins are targeted to different ER exit points. The function of lipids as chaperones has been quite well studied in prokaryotes, but has not been extensively addressed in the eukaryotic ER. In contrast, the assembly of cargo at the correct ER exit point(s) has been more intensively worked on in eukaryotes (Watanabe and Riezman 2004). For misfolded, membrane-spanning proteins, it appears that there are at least two checkpoints that have to be passed. The cell first examines the cytosolic portion, and if it is misfolded the protein is retained in the ER, then degraded. The second checkpoint, which can involve the calnexin cycle and/or other players, monitors the status of the luminal domains (Vashist and Ng 2004). Life and death decisions for ER proteins can be both multi-factorial and multi-compartmental.

Another area that should not be overlooked is the role that ions and small molecules play in the ER. Calcium levels are obviously important to the function of calcium dependent proteins like calnexin and calreticulin. In the sarcoplasmic reticulum, ERp57 plays a critical role in calcium dependent redox regulation of SERCA2b, an ER ATPase that regulates cytosolic calcium levels and maintains the ER calcium store (Li and Camacho 2004). Likewise, zinc is an important cation in the ER. Certain ER chaperones depend on Zn^{2+} for their function, including the yeast DnaJ homologue Scj1p (Silberstein et al. 1998). PDI has also been shown to require Zn^{2+} for dimerisation, though the physiological significance of this is unclear (Solovyov and Gilbert 2004). It has also been shown that depletion of Zn^{2+} induces an unfolded protein response, and the Zn^{2+} transporter Msc2, localised to the ER, is required for appropriate function of the compartment (Ellis et al. 2004). It will be interesting to examine the specific effects of Zn^{2+} depletion and Msc2 transporter loss on the quality control of specific protein substrates.

7 BiP

We have already discussed how BiP gates the translocon during the entry of nascent polypeptides into the ER. BiP is a member of the Hsp70 class of ATP dependent chaperones, is the homologue of yeast Kar2p, and has a number of other functions in the ER. BiP has been well studied as the antibody heavy chain binding protein, helping retain incompletely folded immunoglobulins in the ER until they assemble with the relevant light chains (Vanhove et al. 2001). The peptide motif that BiP recognises is a heptameric one, with an alternating hydrophobic residue signature (Gierasch 1994). Mammalian BiP can also associate with cofactors, including BAP (BiP-associated Protein). BAP acts as a nucleotide exchange factor to regulate BiPs ATPase activity (Chung et al. 2002). A further role for BiP is in ERAD. BiP, along with the translocon, has been genetically linked to the retrograde transport step of ER associated degradation (Plemper et al. 1997), although the precise involvement of the translocon in this process is still unclear.

BiP also plays an important part in the unfolded protein response (UPR) in the ER through its affinity for exposed hydrophobic sequences in unfolded proteins. A pool of BiP is normally associated with IRE1 and PERK, the membrane associated kinases that signal the UPR. As the concentration of unfolded proteins increases, BiP binds to them, and is released from its interactions with IRE1 and PERK. PERK and IRE1 are then free to oligomerise, and by doing so become active, signalling the presence of unfolded proteins in the ER to the nucleus (Bertolotti et al. 2000). For a thorough description of the unfolded protein response, see the accompanying chapter by Maho Niwa.

8 Disulfide bond formation

Disulfide bond formation is an integral part of the protein folding process in the endoplasmic reticulum (Tu and Weissman 2004). A disulfide bond is a covalent modification that brings two cysteine residue sulfhydryls together to form an S-S bridge. This bridge can be either intra-molecular (between two cysteines in the same protein) or inter-molecular (between two cysteines in two different polypeptide chains). Disulfide bridges can be structural, or required for an active site, and they can form co-translationally during protein folding in the ER (Bergman and Kuehl 1979; Peters and Davidson 1982). Most disulfides are found in secreted or membrane-bound proteins, but there are some important examples of cytosolic

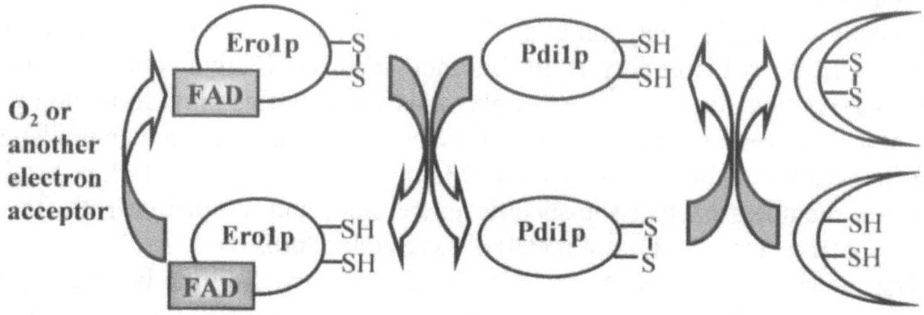

Fig. 2. Oxidising equivalents flow from Ero1p to Pdi1p and target proteins. In *S. cerevisiae* (and other eukaryotes), Pdi1p is oxidised through dithiol-disulfide exchange with Ero1p. Proteins folding in the ER and requiring disulfide bonds can then be oxidised through thiol-disulfide exchange with Pdi1p.

heat shock proteins and transcription factors that are subject to regulated disulfide bond switching (Linke and Jakob 2003).

In yeast, the essential gene required for the catalysis of disulfide bonds in the ER is *ERO1* (Pollard et al. 1998; Frand and Kaiser 1998). The gene product, Ero1p, is an ER resident flavoprotein that uses O_2 as its terminal electron acceptor and passes oxidising equivalents on to Pdi1p (Fig. 2), the yeast protein disulfide isomerase (Frand and Kaiser 1999; Tu et al. 2000). There are two Ero1p counterparts in man, Ero1α and Ero1β (Cabibbo et al. 2000; Pagani et al. 2000). Ero1α is induced by hypoxia (Gess et al. 2003) and Ero1β is induced by the unfolded protein response (Pagani al. 2000). Ero1α and Ero1β show some differences in their tissue specific distribution, with Ero1β being notably highly expressed in secretory tissues such as the pancreas, stomach, and pituitary gland. Two clusters of conserved cysteine residues are important for Ero structure and function (Benham et al. 2000; Frand and Kaiser 2000). These are a CxxxxC sequence towards the N-terminus that interacts with PDI and a more C-terminal CxxCxxC sequence that accepts electrons from the N-terminal cysteines and passes them on to FAD. The 2.2 Å and 2.8 Å crystal structures of a truncated form of Ero1p revealed that the N and C terminal cysteine clusters can communicate via a flexible hinge, with FAD being partially buried in the centre of the protein in an unconventional binding site (Gross et al. 2004). The adenine ring system of FAD is exposed to solvent, with the isoalloxazine ring able to contact the CxxCxxC motif.

Flexible hinges are a recurring theme in ER redox proteins. Erv2p, which can operate in parallel to Ero1p in yeast to support disulfide bond formation, uses a similar principle (Sevier et al. 2001). The 1.5Å crystal structure of Erv2p strongly suggests that two redox active cysteine rich sites communicate via a flexible hinge to exchange electrons (Gross et al. 2002). In the case of Erv2p, a central CGEC sequence and a more carboxy-terminal CGC motif are separated by 52 amino acids. Although Erv2p can replace Ero1p function and transfer oxidising equivalents to Pdi1p, there is no direct Erv2p homologue in the mammalian ER. The so-called mammalian quiescin-sulfhydryl oxidase (QSOX) proteins, including ALR and Q6, are related to Erv2p and posses thioredoxin domains, but these proteins are probably mostly secreted (Benayoun et al. 2001). At present, there is no evidence for a role of QSOX in ER quality control, although FMOs (Flavin-containing Monooxygenases) have been suggested to play a part in disulfide bond formation in the ER, at least in yeast (Suh et al. 1999). Further work is required in mammalian systems to establish a function for the FMO family in ER quality control, particularly as some members have tissue-specific expression, and alternative roles in substrate-specific detoxification of metabolites (Zeigler 2002).

The focus of Ero1ps attention is Pdi1p, the primary disulfide bond catalyst in the yeast ER. Pdi1p is required for oxidation and isomerisation (the reorganisation or "unscrambling") of disulfide bonds. The precise contribution that Pdi1p makes to oxidation and isomerisation in yeast has been open to debate. Using active site mutants of Pdi1p, it was claimed that isomerisation was the essential function of Pdi1p (Laboissiere et al. 1995). However, Pdi1p deficient yeast can be reconstituted with an inducible active site **a** domain construct essentially lacking isomerase activity (Solovyov et al. 2004). This module restores both viability and oxidative protein folding to the Pdi1p deficient yeast. These results suggest that oxidation, rather than isomerisation, is the essential activity of Pdi1p, although both oxidation and isomerisation reactions are likely to be undertaken by Pdi1p *in vivo.* Work in mammalian cells suggests that PDI is at least partly oxidised, in keeping with its role as a primary protein oxidant (Mezghrani et al. 2001).

Pdi1p and its human counterpart PDI can be thought of as renaissance proteins, accomplished at many tasks. PDI works as an oxidoreductase (Bulleid and Freedman 1988), a peptide-binding protein (Klappa et al. 1998), a component of ERAD (Gillece et al. 1999) and as a multi-protein complex in the assembly and retention of procollagen (Bottomley et al. 2001). PDI is also required for the unfolding and export of cholera toxin from the ER to the cytosol (Tsai et al. 2001) and for the reduction prior to export of ricin toxin (Spooner et al. 2004). However, a full discussion of the role of PDI in protein export and ER degradation is beyond the scope of this review.

PDI has a modular structure that is based on the prototypical cytoplasmic redox protein thioredoxin (Darby et al. 1996). PDI modules are denoted by bold lower case letters, and PDI has the domain structure **abb'a'**. The **a** and **a'** domains are catalytically active and contain a WCGHC motif. This motif has a strong tendency to donate its intra-motif disulfide bond to a substrate, and this is reflected in the high redox potential of PDI (E^o = -0.18 V for rat PDI) (Lundstrom and Holmgren 1993). The PDI **b** domains lack the WCGHC motif and are catalytically inactive. The C-terminal domain is acidic and retention in the ER is mediated by a KDEL sequence. The structural basis of the thioredoxin fold in PDI has been demonstrated by NMR studies of individual domains (Kemmink et al. 1995, 1996, 1997), but a crystal structure of this highly dynamic protein has not been forthcoming. All PDI homologues are recognisable by the modular arrangement of their thioredoxin domains, although these have been shuffled and expanded during evolution (Alanen et al. 2003a).

Much recent work has focused on protein-mediated pathways for disulfide bond formation in the ER, but oxidising and reducing equivalents can come from other sources. Glutathione should not be forgotten as an important source of sulfhydryl groups in the luminal environment. Reduced glutathione (GSH) is a tripeptide composed of the amino acids γ-Glu-CysH-Gly and is involved in a number of essential biosynthetic and metabolic pathways (Meister and Anderson 1983; Ziegler 1985). The ER lumen contains relatively more oxidised glutathione (GSSG) than reduced glutathione (GSH), in contrast to the cytosol. It has been estimated that the ratio changes from between 30-100 GSH: 1 GSSG in the cytosol to 3 GSH: 1 GSSG in the ER (Hwang et al. 1992). A large proportion of glutathione is also conjugated to ER resident proteins, and held as a "reservoir" in the ER (Bass et al. 2004). The full extent of these bound thiols has only just been appreciated.

Because glutathione can switch between oxidised and reduced states, it was assumed to be the primary oxidising agent of the ER prior to the discovery of Ero1p. Recent work in mammalian cells shows that glutathione still has an important role to play in the ER. Lowering intracellular GSH levels using specific inhibitors of GSH synthesis causes model proteins such as tPA (tissue-type plasminogen activator) and immunoglobulin J chain to undergo oxidative misfolding. Correct disulfide bond formation requires both Ero1 and GSH. This work demonstrates that GSH controls the reductive pathway of the endoplasmic reticulum, rather than the oxidative pathway (Chakravarthi and Bulleid 2004) and that the reducing environment of the cytosol is actually important for the oxidative function of the endoplasmic reticulum (Molteni et al. 2004). These ideas could have profound implications for our understanding of ER glycoprotein folding, suggesting that separate pathways exist for protein oxidation and isomerisation that may be subject to different forms of control, regulation and therefore exploitation.

A notable example of the importance of isomerisation, as well as oxidation, in glycoprotein folding protein comes from the low-density lipoprotein receptor (LDL-R). Mutations in LDL-R, of which there are many, are associated with familial hypercholesterolemia, a disease that results in increased cholesterol deposition (Fass et al. 1997). LDL-R is a large, cholesterol binding, N and O-glycosylated lipoprotein, with 3 Epidermal Growth Factor (EGF) domains, 13 LDLRA/B domain repeats and 30 predicted disulfide bonds. This protein folds by initially forming collapsed, non-native disulfide bonded intermediates between distant cysteines. It appears that the LDL-R folding pathway is post-translational and not vectorial, involving long-range non-native disulfides that subsequently require isomerisation (Jansens et al. 2002). It will be very interesting to find out the relative contribution that the isomerase and oxidase functions of PDI and its homologues make to the productive folding of proteins like LDL-R.

PDI is not the only protein disulfide isomerase in the ER. In yeast, Pdi1p homologues include Mpd1p, Mpd2p, Eug1p and Eps1p. To some extent, the functions of these proteins are overlapping and redundant, although Mpd1p is the only homologue capable of performing all the essential functions of Pdi1p (Norgaard et al. 2001). Eps1p is unusual in that it is membrane associated, and has been implicated in the quality control and trafficking of the yeast H^+-ATPase Pma1p. Eps1p is required for Pma1p cell surface expression (Wang and Chang 1999). In mammals, the most well known examples of PDI homologues are ERp57 and ERp72.

ERp57 can catalyse oxidation, reduction, and isomerisation reactions *in vitro* (Frickel et al. 2004) and is an important component of the calnexin cycle, as we have already discussed. Both PDI and ERp57 can catalyse viral S-S bond formation in living cells (Molinari and Helenius 1999). ERp57 is also a component of the MHC class I peptide loading complex (Cresswell et al. 1999), although whether it is involved in oxidation, isomerisation or reduction reactions within the complex is not clear (Dick et al. 2002; Antoniou et al. 2002). Less is known about the function of ERp72, other than that it can be selectively found in certain chaperone/substrate complexes, for example in the folding of thyroglobulin (Sargsyan et al. 2002a).

9 Introducing more ER chaperones

One of the most striking observations to emerge from ongoing proteomic and database analysis of the ER is that there are a very large number of putative ER chaperones and folding factors, particularly PDI homologues, about which we know very little. Some are cell and/or tissue specific, but many of these proteins are likely to be involved in the quality control of newly synthesised glycoproteins. A brief description of some of these PDI proteins is given in Table 1, but is not exhaustive. In the coming years more folding assistants are likely to emerge, and the challenge is to work out the detail of how these proteins function, and how they are co-ordinated, both in general and in a tissue or substrate specific way.

10 Folding of specialised proteins in the ER

The PDI homologue ERp28 (Table 1) is interesting because it lacks a conventional CxxC motif, and therefore has no oxidoreductase activity. Instead, this protein may have evolved a chaperone-only function (Ferrari et al. 1998). The *D. melanogaster* orthologue of ERp28 is called *wind.* By contrast, *wind* does have a CxxC motif and is the only PDI protein so far to have been crystallised (Ma et al. 2003). *Wind* ensures that a glycosaminoglycan-modifying enzyme called *pipe* is appropriately localised within the secretory pathway. In this capacity, *wind* is critical for dorso-ventral polarity (Sen et al. 2000). The function of *wind* suggests that this PDI protein has evolved a substrate specific developmental function.

A number of other proteins also require particular help from specialised chaperones. Apo-lipoprotein B (Apo B) is an interesting example of a protein that must be lipidated prior to export from the ER and seems to be associated with a complex pool of chaperones both in the ER and the trans-golgi apparatus, including PDI, BiP, gp96, ERp72, calreticulin, and the PPIase cyclophilin B (Zhang and Herscovitz 2003). Apo B plays a key role in the formation of high and very low

Table 1. Recently discovered PDI homologues likely to be involved in oxidative ER protein folding.

Novel PDI homologues.	**Characteristics**	**Postulated function**	**References**
ERp28 (ERp29 in rat)	Lacks CXXC motif. Thioredoxin fold acts as a specific homodimerization module	Non-redox chaperone	(Ferrari et al. 1998)
EndoPDI/ ERp46	Three CXXC motifs	Protective role in endothelium	(Sullivan et al. 2003; Knoblach et al. 2003)
ERp18	Contains a CGAC motif and probable insertion between β3 and α3 of the thioredoxin fold	Potential protein disulfide isomerase	(Alanen et al. 2003b)
ERp44	Contains a CRFS motif	Key player in thiol-mediated retention	(Anelli et al. 2002, 2003)
ERdJ5/JPDI	J domain with 4 potentially active thioredoxin domains	Likely to combine redox function with Hsp70-dependent folding	(Cunnea et al. 2003; Hosoda et al. 2003)
PDILT	Variant SXXC and SXXS motifs.	Testis specific chaperone	(Van Lith et al. 2005)
TMX3	CXXC motif. Type1 membrane protein with KKSS retrieval sequence.	Likely dithiol oxidase	(Haugstetter et al. 2005)

density lipoprotein particles, which are assembled in the ER, and it will be interesting to determine precisely what role these ER chaperones play in this protein complex as it travels along the secretory pathway (Stillemark et al. 2000).

The immunoglobulin heavy chain provides an example of how chaperone complexes can be spatially organised to achieve the productive folding of a substrate. Many chaperones and folding factors can be found in a multiprotein complex with unassembled heavy chains that is independent of nascent protein synthesis (Meunier et al. 2002). The components of the complex include BiP, the Hsp90 family member gp96 (GRP94), PDI, the PDI homologue P5 (PDA6), the Hsp40 co-chaperone ERdj3, cyclophilin B, ERp72, and GRP170 (ORP150 or HYOU1 in humans), an ATP-dependent Hsp70 family member up regulated by hypoxia. Interestingly, ERdj3 is absent from the complex when protein translation is inhibited, and calnexin and calreticulin are absent from the complex entirely. The data suggest that ER networks of chaperones can form and then bind to unfolded protein substrates, rather than free pools of chaperones assembling independently onto substrates. The spatial separation of at least two types of chaperone system may relate to the specificity of the calnexin cycle, and we can expect further definition of the molecular basis for separate pools of "chaperones" in the ER.

The cystic fibrosis transmembrane receptor (CFTR), LDL-R and thyroglobulin have very different functions, being involved in chloride transport, cholesterol metabolism and thyroid development, respectively. Yet these three proteins all start their lives in the ER, and can be used to illustrate the point that different substrates with different features can require different types of chaperone to fold properly. CFTR, with its large cytoplasmic domain, interacts not only with ER chaperones, but also with cytosolic chaperones including the ATP-dependent Hsc70 and Hsp90 (Loo et al. 1998; Strickland et al. 1997). This adds another layer of complexity to the quality control of this trans-membrane ion channel. The "foldedness" of CFTR cytosolic domains is monitored by CHIP (carboxyl terminus of Hsc70-interacting protein), a U-box ubiquitin ligase that can team up with Hsc70 and target misfolded CFTR for proteasomal degradation (Meacham et al. 2001). CHIPs task is helped by BAG-1, a cytosolic factor that binds to the ATPase domain of Hsc70 whilst CHIP occupies itself with the carboxy terminus (Alberti et al. 2002). By stimulating the nucleotide exchange activity of Hsc70, BAG-1 accelerates the delivery of misfolded CFTR to the proteasome. This system is subject to regulation. The ubiquitin ligase activity of CHIP is inhibited by HspBP1, a cytosolic accessory factor whose expression therefore helps to stabilise CFTR as it folds in the ER (Alberti et al. 2004). CFTR is, therefore, subjected to quality control steps on both sides of the ER membrane. Although the cytosolic chaperone machinery is not necessarily unique to CFTR, only a restricted set of secretory pathway proteins with large cytosolic domains will be able to interact with these cytosolic folding/degradation factors.

Many LDL-R family proteins interact with RAP (low density lipoprotein receptor-related protein-associated protein or α2-macroglobulin receptor-associated protein) in the ER (Bu et al. 1995). RAP seems to protect LDLR family members from spurious ligand interactions and is a specialised 40 kD ER resident chaperone with an unusual HNEL ER retention sequence.

Thyroglobulin is an interesting example of a substrate that is produced by a dedicated cell type. The precursor of thyroid hormone, thyroglobulin is a major product of the thyrocyte, and failure of thyroglobulin secretion can lead to hypothyroidism. Certain types of genetic thyroid deficiencies can thus be classified as ER storage diseases (Medeiros-Neto et al. 1996; Kim et al. 1996). Thyroglobulin is associated with a number of chaperones as it folds in the ER, including BiP, ERp72, gp96, calnexin and PDI (Kim and Arvan 1995; Kuznetsov et al. 1994). ERp29 (see Table 1) was subsequently identified as a novel component of thyroglobulin complexes in the rat thyrocyte ER (Sargsyan et al. 2002). Given subsequent developments in the PDI field, it will be interesting to discover the temporal/spatial nature of these associations, and to determine just how substrate specific ERp29 is for thyroglobulin.

11 Techniques, model systems and what's next

The analysis of protein folding and oxidation in the endoplasmic reticulum has traditionally relied on careful biochemistry combined with cell biological techniques, using purified proteins *in vitro*, cultured cells, structural investigation and yeast genetics. The identification and scrutiny of mutations causing certain human genetic diseases has led to an appreciation that misfolding in the ER can have serious consequences for human health. In turn, this has advanced our understanding of how the healthy ER works. The completion of the human, mouse, and rat genomes, coupled with sophisticated proteomic approaches, is turning up more ER chaperones than previously appreciated. The coming years should see the development of techniques to study protein folding in more appropriate physiological environments. 3-D culture techniques (Luo and Shoichet 2004), improvements in fluorescent/redox imaging (Dooley et al. 2004; Ostergaard et al. 2001), secretion/stress inducible animal models (Iwawaki et al. 2004) and mammalian RNAi technology (Mittal 2004) will all help the field to develop. Today we understand how the folding of many individual ER proteins occurs and how many individual chaperones work. For tomorrow, we need to address how ER protein folding is coordinated and controlled in time and space within complex physiological systems.

Acknowledgements

Work in our laboratory is generously funded by the Wellcome Trust, the arthritis research campaign, BBSRC, ONE North-East and the EU. S D-G is an Overseas Research Scholar. We thank Dr. Arto Määttä for critical reading of the manuscript and JAG Williams and R Kataky for support.

References

Aguzzi A, Heikenwalder M, Miele G (2004) Progress and problems in the biology, diagnostics, and therapeutics of prion diseases. J Clin Invest 114:153-160

Alanen HI, Salo KE, Pekkala M, Siekkinen HM, Pirneskoski A, Ruddock LW (2003a) Defining the domain boundaries of the human protein disulfide isomerases. Antioxid Redox Signal 5:367-374

Alanen HI, Williamson RA, Howard MJ, Lappi A-K, Jantti HP, Rautio SM, Kellokumpu S, Ruddock LW (2003b) Functional Characterization of ERp18, a new endoplasmic reticulum-located thioredoxin superfamily member. J Biol Chem 278:28912-28920

Alberti S, Bohse K, Arndt V, Schmitz A, Hohfeld J (2004) The cochaperone HspBP1 inhibits the CHIP ubiquitin ligase and stimulates the maturation of the cystic fibrosis transmembrane conductance regulator. Mol Biol Cell 15:4003-4010

Alberti S, Demand J, Esser C, Emmerich N, Schild H, Hohfeld J (2002) Ubiquitylation of BAG-1 suggests a novel regulatory mechanism during the sorting of chaperone substrates to the proteasome. J Biol Chem 277:45920-45927

Andersson H, Garoff H (2003) Lectin-mediated retention of p62 facilitates p62-E1 heterodimerization in endoplasmic reticulum of Semliki Forest virus-infected cells. J Virol 77:6676-6682

Anelli T, Alessio M, Bachi A, Bergamelli L, Bertoli G, Camerinin S, Mezghrani A, Ruffato E, Simmen T, Sitia R (2003) Thiol-mediated protein retention in the endoplasmic reticulum: the role of ERp44. EMBO J 22:5015-5022

Anelli T, Alessio M, Mezghrani A, Simmen T, Talamo F, Bachi A, Sitia R (2002) ERp44, a novel endoplasmic reticulum folding assistant of the thioredoxin family. EMBO J 21:835-844

Antoniou AN, Ford S, Alphey M, Osborne A, Elliott T, Powis SJ (2002) The oxidoreductase ERp57 efficiently reduces partially folded in preference to fully folded MHC class I molecules. EMBO J 21:2655-2663

Avila G, Lee EH, Perez CF, Allen PD, Dirksen RJ (2003) FKBP12 binding to RyR1 modulates excitation-contraction coupling in mouse skeletal myotubes. J Biol Chem 278:22600-22608

Bass R, Ruddock LW, Klappa P, Freedman RB (2004) A major fraction of endoplasmic reticulum-located glutathione is present as mixed disulfides with protein. J Biol Chem 279:5257-5262

Benayoun B, Esnard-Feve A, Castella S, Courty Y, Esnard F (2001) Rat seminal vesicle FAD-dependent sulfhydryl oxidase. Biochemical characterization and molecular cloning of a member of the new sulfhydryl oxidase/quiescin Q6 gene family. J Biol Chem 276:13830-13837

Benham AM, Cabibbo A, Fassio A, Bulleid N, Sitia R, Braakman I (2000) The CXXCXXC motif determines the folding, structure and stability of human Ero1-Lalpha. EMBO J 19:4493-4502

Bergman LW, Kuehl WM (1979) Formation of an intrachain disulfide bond on nascent immunoglobulin light chains. J Biol Chem 254:8869-8876

Bertolotti A, Zhang Y, Hendershot LM, Harding HP, Ron D (2000) Dynamic interaction of BiP and ER stress transducers in the unfolded-protein response. Nat Cell Biol 2:326-332

Bogdanov M, Dowhan W (1999) Lipid-assisted Protein Folding. J Biol Chem 274:36827-36830

Bottomley MJ, Batten MR, Lumb RA, Bulleid NJ (2001) Quality control in the endoplasmic reticulum. PDI mediates the ER retention of unassembled procollagen C-propeptides. Curr Biol 11:1114-1118

Braakman I, Hoover-Litty H, Wagner KR, Helenius A (1991) Folding of influenza hemagglutinin in the endoplasmic reticulum. J Cell Biol 114:401-411

Bu G, Geuze H, Strous G, Schwartz A (1995) 39 kDa receptor-associated protein is an ER resident protein and molecular chaperone for LDL receptor-related protein. EMBO J 14:2269-2280

Bulleid NJ, Freedman RB (1988) Defective co-translational formation of disulphide bonds in protein disulphide-isomerase-deficient microsomes. Nature 335:649-651

Bullough PA, Hughson FM, Skehel JJ, Wiley DC (1994) Structure of influenza haemagglutinin at the pH of membrane fusion. Nature 371:37-43

Cabibbo A, Pagani M, Fabbri M, Rocchi M, Farmery MR, Bulleid NJ, Sitia R (2000) ERO1-L, a human protein that favors disulfide bond formation in the endoplasmic reticulum. J Biol Chem 275:4827-4833

Cannon KS, Cresswell P (2001) Quality control of transmembrane domain assembly in the tetraspanin CD82. EMBO J 20:2443-2453

Caramelo JJ, Castro OA, Alonso LG, De Prat-Gay G, Parodi AJ (2003) Inaugural article: UDP-Glc:glycoprotein glucosyltransferase recognizes structured and solvent accessible hydrophobic patches in molten globule-like folding intermediates. Proc Natl Acad Sci USA 100:86-91

Chakravarthi S, Bulleid NJ (2004) Glutathione is required to regulate the formation of native disulfide bonds within proteins entering the secretory pathway. J Biol Chem 279:39872-39879

Chung KT, Shen Y, Hendershot LM (2002) BAP, a mammalian BiP-associated protein, is a nucleotide exchange factor that regulates the ATPase activity of BiP. J Biol Chem 277:47557-47563

Clemons J, William M, Menetret J-F, Akey CW, Rapoport TA (2004) Structural insight into the protein translocation channel. Curr Opin Struct Biol 14:390-396

Copeland CS, Zimmer KP, Wagner KR, Healey GA, Mellman I, Helenius A (1988) Folding, trimerization, and transport are sequential events in the biogenesis of influenza virus hemagglutinin. Cell 53:197-209

Cresswell P, Bangia N, Dick T, Diedrich G (1999) The nature of the MHC class I peptide loading complex. Immunol Rev 172:21-28

Cunnea PM, Miranda-Vizuete A, Bertoli G, Simmen T, Damdimopoulos AE, Hermann S, Leinonen S, Huikko MP, Gustafsson J-A, Sitia R, Spyrou G (2003) ERdj5, an endoplasmic reticulum (ER)-resident protein containing DnaJ and thioredoxin domains, is expressed in secretory cells or following ER stress. J Biol Chem 278:1059-1066

Daniels R, Kurowski B, Johnson A, Hebert DN (2003) N-linked glycans direct the cotranslational folding pathway of influenza hemagglutinin. Mol Cell 11:79-90

Darby NJ, Kemmink J, Creighton TE (1996) Identifying and characterizing a structural domain of protein disulfide isomerase. Biochemistry 35:10517-10528

David V, Hochstenbach F, Rajagopalan S, Brenner M (1993) Interaction with newly synthesized and retained proteins in the endoplasmic reticulum suggests a chaperone function for human integral membrane protein IP90 (calnexin). J Biol Chem 268:9585-9592

Dick TP, Bangia N, Peaper DR, Cresswell P (2002) Disulfide bond isomerization and the assembly of MHC class I-peptide complexes. Immunity 16:87-98

Dooley CT, Dore TM, Hanson GT, Jackson WC, Remington SJ, Tsien RY (2004) Imaging dynamic redox changes in mammalian cells with green fluorescent protein indicators. J Biol Chem 279:22284-22293

Ellgaard L, Helenius A (2003) Quality control in the endoplasmic reticulum. Nat Rev Mol Cell Biol 4:181-191

Ellgaard L, Riek R, Herrmann T, Guntert P, Braun D, Helenius A, Wuthrich K (2001) NMR structure of the calreticulin P-domain. Proc Natl Acad Sci USA 98:3133-3138

Ellis CD, Wang F, Macdiarmid CW, Clark S, Lyons T, Eide DJ (2004) Zinc and the Msc2 zinc transporter protein are required for endoplasmic reticulum function. J Cell Biol 166:325-335

Fass D, Blacklow S, Kim PS, Berger JM (1997) Molecular basis of familial hypercholesterolaemia from structure of LDL receptor module. Nature 388:691-693

Ferrari DM, Nguyen Van P, Kratzin HD, Soling H-D (1998) ERp28, a human endoplasmic-reticulum-lumenal protein, is a member of the protein disulfide isomerase family but lacks a CXXC thioredoxin-box motif. Eur J Biochem 255:570-579

Frand AR, Kaiser CA (1998) The ERO1 gene of yeast is required for oxidation of protein dithiols in the endoplasmic reticulum. Mol Cell 1:161-170

Frand AR, Kaiser CA (1999) Ero1p oxidizes protein disulfide isomerase in a pathway for disulfide bond formation in the endoplasmic reticulum. Mol Cell 4:469-477

Frand AR, Kaiser CA (2000) Two pairs of conserved cysteines are required for the oxidative activity of Ero1p in protein disulfide bond formation in the endoplasmic reticulum. Mol Biol Cell 11:2833-2843

Frickel E-M, Frei P, Bouvier M, Stafford WF, Helenius A, Glockshuber R, Ellgaard L (2004) ERp57 is a multifunctional thiol-disulfide oxidoreductase. J Biol Chem 279:18277-18287

Gess B, Hofbauer K-H, Wenger RH, Lohaus C, Meyer HE, Kurtz A (2003) The cellular oxygen tension regulates expression of the endoplasmic oxidoreductase ERO1-Lalpha. Eur J Biochem 270:2228-2235

Gierasch LM (1994) Molecular chaperones. Panning for chaperone-binding peptides. Curr Biol 4:173-174

Gillece P, Luz JM, Lennarz WJ, De La Cruz FJ, Romisch K (1999) Export of a cysteine-free misfolded secretory protein from the endoplasmic reticulum for degradation requires interaction with protein disulfide isomerase. J Cell Biol 147:1443-1456

Gross E, Kastner DB, Kaiser CA, Fass D (2004) Structure of Ero1p, source of disulfide bonds for oxidative protein folding in the cell. Cell 117:601-610

Gross E, Sevier CS, Vala A, Kaiser CA, Fass D (2002) A new FAD-binding fold and intersubunit disulfide shuttle in the thiol oxidase Erv2p. Nat Struct Biol 9:61-67

Haas IG, Wabl M (1983) Immunoglobulin heavy chain binding protein. Nature 306:387-389

Hamman BD, Hendershot LM, Johnson AE (1998) BiP maintains the permeability barrier of the ER membrane by sealing the lumenal end of the translocon pore before and early in translocation. Cell 92:747-758

Hammond C, Braakman I, Helenius A (1994) Role of N-linked oligosaccharide recognition, glucose trimming, and calnexin in glycoprotein folding and quality control. Proc Natl Acad Sci USA 91:913-917

Haugstetter J, Blicher T, Ellgaard L (2005) Identification and characterisation of a novel thioredoxin-related transmembrane protein of the endoplasmic reticulum. J Biol Chem 280:8371-8380

Hebert DN, Zhang J-X, Chen W, Foellmer B, Helenius A (1997) The number and location of glycans on influenza hemagglutinin determine folding and association with calnexin and calreticulin. J Cell Biol 139:613-623

Helenius A, Aebi M (2004) Roles of N-linked glycans in the endoplasmic reticulum. Annu Rev Biochem 73:1019-1049

Hosoda A, Kimata Y, Tsuru A, Kohno K (2003) JPDI, a novel endoplasmic reticulum-resident protein containing both a BiP-interacting J-domain and thioredoxin-like motifs. J Biol Chem 278:2669-2676

Hosokawa N, Wada I, Hasegawa K, Yorihuzi T, Tremblay LO, Herscovics A, Nagata K (2001) A novel ER alpha-mannosidase-like protein accelerates ER-associated degradation. EMBO Rep 2:415-422

Hwang C, Sinskey AJ, Lodish HF (1992) Oxidized redox state of glutathione in the endoplasmic reticulum. Science 257:1496-1502

Ihara Y, Cohen-Doyle MF, Saito Y, Williams DB (1999) Calnexin discriminates between protein conformational states and functions as a molecular chaperone *in vitro*. Mol Cell 4:331-341

Iwawaki T, Akai R, Kohno K, Miura M (2004) A transgenic mouse model for monitoring endoplasmic reticulum stress. Nat Med 10:98-102

Jansens A, Braakman I (2002) Coordinated non-vectorial folding in a newly synthesized multidomain protein. Science 298:2401-2403

Johnson AE, Van Waes MA (1999) The translocon: a dynamic gateway at the ER membrane. Annu Rev Cell Dev Biol 15:799-842

Kemmink J, Darby NJ, Dijkstra K, Nilges M, Creighton TE (1996) Structure determination of the N-terminal thioredoxin-like domain of protein disulfide isomerase using multidimensional heteronuclear 13C/15N NMR spectroscopy. Biochemistry 35:7684-7691

Kemmink J, Darby NJ, Dijkstra K, Nilges M, Creighton TE (1997) The folding catalyst protein disulfide isomerase is constructed of active and inactive thioredoxin modules. Curr Biol 7:239-245

Kemmink J, Darby NJ, Dijkstra K, Scheek RM, Creighton TE (1995) Nuclear magnetic resonance characterization of the N-terminal thioredoxin-like domain of protein disulfide isomerase. Protein Sci 4:2587-2593

Kim P, Arvan P (1995) Calnexin and BiP act as sequential molecular chaperones during thyroglobulin folding in the endoplasmic reticulum. J Cell Biol 128:29-38

Kim P, Kwon O, Arvan P (1996) An endoplasmic reticulum storage disease causing congenital goiter with hypothyroidism. J Cell Biol 133:517-527

Klappa P, Ruddock LW, Darby NJ, Freedman RB (1998) The b' domain provides the principal peptide-binding site of protein disulfide isomerase but all domains contribute to binding of misfolded proteins. EMBO J 17:927-935

Knoblach B, Keller BO, Groenendyk J, Aldred S, Zheng J, Lemire BD, Li L, Michalak M (2003) ERp19 and ERp46, new members of the thioredoxin family of endoplasmic reticulum proteins. Mol Cell Proteomics 2:1104-119

Kopito RR (1999) Biosynthesis and degradation of CFTR. Physiol Rev 79:167-173

Kuznetsov G, Chen L, Nigam S (1994) Several endoplasmic reticulum stress proteins, including ERp72, interact with thyroglobulin during its maturation. J Biol Chem 269:22990-22995

Laboissiere MC, Sturley SL, Raines RT (1995) The essential function of protein-disulfide isomerase is to unscramble non-native disulfide bonds. J Biol Chem 270:28006-28009

Lawless MW, Greene CM, Mulgrew A, Taggart CC, O'Neill SJ, McElvaney NG (2004) Activation of endoplasmic reticulum-specific stress responses associated with the conformational disease Z alpha 1-antitrypsin deficiency. J Immunol 172:5722-5726

Lemberg MK, Bland FA, Weihofen A, Braud VM, Martoglio B (2001) Intramembrane proteolysis of signal peptides: an essential step in the generation of HLA-E epitopes. J Immunol 167:6441-6446

Li Y, Camacho P (2004) Ca2+-dependent redox modulation of SERCA 2b by ERp57. J Cell Biol 164:35-46

Liao S, Lin J, Do H, Johnson AE (1997) Both lumenal and cytosolic gating of the aqueous ER translocon pore are regulated from inside the ribosome during membrane protein integration. Cell 90:31-41

Linke K, Jakob U (2003) Not every disulfide lasts forever: disulfide bond formation as a redox switch. Antioxid Redox Signal 5:425-434

Lisanti MP, Caras IW, Davitz MA, Rodriguez-Boulan E (1989) A glycophospholipid membrane anchor acts as an apical targeting signal in polarized epithelial cells. J Cell Biol 109:2145-2156

Loo MA, Jensen TJ, Cui L, Hou Y, Chang XB, Riordan JR (1998) Perturbation of Hsp90 interaction with nascent CFTR prevents its maturation and accelerates its degradation by the proteasome. EMBO J 17:6879-6887

Lundstrom J, Holmgren A (1993) Determination of the reduction-oxidation potential of the thioredoxin-like domains of protein disulfide-isomerase from the equilibrium with glutathione and thioredoxin. Biochemistry 32:6649-6655

Luo L, Shoichet MS (2004) A photolabile hydrogel for guided three-dimensional cell growth and migration. Nat Mater 3:249-253

Ma Q, Guo C, Barnewitz K, Sheldrick GM, Soling HD, Uson I, Ferrari DM (2003) Crystal structure and functional analysis of *Drosophila Wind*, a protein-disulfide isomerase-related protein. J Biol Chem 278:44600-44607

Matsuda S, Shibasaki F, Takehana K, Mori H, Nishida E, Koyasu S (2000) Two distinct action mechanisms of immunophilin-ligand complexes for the blockade of T-cell activation. EMBO Rep 1:428-434

Mattaj IW (2004) Sorting out the nuclear envelope from the endoplasmic reticulum. Nat Rev Mol Cell Biol 5:65-69

Matter K, Yamamoto EM, Mellman I (1994) Structural requirements and sequence motifs for polarized sorting and endocytosis of LDL and Fc receptors in MDCK cells. J Cell Biol 126:991-1004

Meacham GC, Patterson C, Zhang W, Younger JM, Cyr DM (2001) The Hsc70 co-chaperone CHIP targets immature CFTR for proteasomal degradation. Nat Cell Biol 3:100-105

Medeiros-Neto G, Kim PS, Yoo SE, Vono J, Targovnik HM, Camargo R, Hossain SA, Arvan P (1996) Congenital hypothyroid goiter with deficient thyroglobulin. Identification of an endoplasmic reticulum storage disease with induction of molecular chaperones. J Clin Invest 98:2838-2844

Meister A, Anderson ME (1983) Glutathione. Annu Rev Biochem 52:711-760

Meunier L, Usherwood Y-K, Chung KT, Hendershot LM (2002) A subset of chaperones and folding enzymes form multiprotein complexes in endoplasmic reticulum to bind nascent proteins. Mol Biol Cell 13:4456-4469

Mezghrani A, Fassio A, Benham A, Simmen T, Braakman I, Sitia R (2001) Manipulation of oxidative protein folding and PDI redox state in mammalian cells. EMBO J 20:6288-6296

Mittal V (2004) Improving the efficiency of RNA interference in mammals. Nat Rev Genet 5:355-365

Molinari M, Calanca V, Galli C, Lucca P, Paganetti P (2003) Role of EDEM in the release of misfolded glycoproteins from the calnexin cycle. Science 299:1397-1400

Molinari M, Helenius A (1999) Glycoproteins form mixed disulphides with oxidoreductases during folding in living cells. Nature 402:90-93

Molinari M, Helenius A (2000) Chaperone selection during glycoprotein translocation into the endoplasmic reticulum. Science 288:331-333

Molteni Sn, Fassio A, Ciriolo Mr, Filomeni G, Pasqualetto E, Fagioli C, Sitia R (2004) Glutathione limits Ero1-dependent oxidation in the endoplasmic reticulum. J Biol Chem 279:32667-32673

Munro S, Pelham H (1986) An Hsp70-like protein in the ER: identity with the 78 kd glucose-regulated protein and immunoglobulin heavy chain binding protein. Cell 46:291-300

Nelson WJ, Yeaman C (2001) Protein trafficking in the exocytic pathway of polarized epithelial cells. Trends Cell Biol 1:483-486

Norgaard P, Westphal V, Tachibana C, Alsoe L, Holst B, Winther JR (2001) Functional differences in yeast protein disulfide isomerases. J Cell Biol 152:553-562

Oda Y, Hosokawa N, Wada I, Nagata K (2003) EDEM as an acceptor of terminally misfolded glycoproteins released from calnexin. Science 299:1394-1397

Okamoto K, Moriishi K, Miyamura T, Matsuura Y (2004) Intramembrane proteolysis and endoplasmic reticulum retention of Hepatitis C virus core protein. J Virol 78:6370-6380

Oliver JD, Van Der Wal FJ, Bulleid NJ, High S (1997) Interaction of the thiol-dependent reductase ERp57 with nascent glycoproteins. Science 275:86-88

Ostergaard H, Henriksen A, Hansen FG, Winther JR (2001) Shedding light on disulfide bond formation: engineering a redox switch in green fluorescent protein. EMBO J 20:5853-5862

Ou WJ, Cameron PH, Thomas DY, Bergeron JJ (1993) Association of folding intermediates of glycoproteins with calnexin during protein maturation. Nature 364:771-776

Pagani M, Fabbri M, Benedetti C, Fassio A, Pilati S, Bulleid NJ, Cabibbo A, Sitia R (2000) Endoplasmic reticulum oxidoreductin 1-lbeta (ERO1-Lbeta), a human gene induced in the course of the unfolded protein response. J Biol Chem 275:23685-23692

Parlati F, Dominguez M, Bergeron JJM, Thomas DY (1995) *Saccharomyces cerevisiae* CNE1 encodes an endoplasmic reticulum (ER) membrane protein with sequence similarity to calnexin and calreticulin and functions as a constituent of the ER quality control apparatus. J Biol Chem 270:244-253

Peters T Jr, Davidson LK (1982) The biosynthesis of rat serum albumin. *In vivo* studies on the formation of the disulfide bonds. J Biol Chem 257:8847-8853

Plemper RK, Bohmler S, Bordallo J, Sommer T, Wolf DH (1997) Mutant analysis links the translocon and BiP to retrograde protein transport for ER degradation. Nature 388:891-895

Pollard MG, Travers KJ, Weissman JS (1998) Ero1p: a novel and ubiquitous protein with an essential role in oxidative protein folding in the endoplasmic reticulum. Mol Cell 1:171-182

Pollock S, Kozlov G, Pelletier MF, Trempe JF, Jansen G, Sitnikov D, Bergeron JJ, Gehring K, Ekiel I, Thomas DY (2004) Specific interaction of ERp57 and calnexin determined by NMR spectroscopy and an ER two-hybrid system. EMBO J 23:1020-1029

Potter BA, Ihrke G, Bruns JR, Weixel KM, Weisz OA (2004) Specific N-Glycans direct apical delivery of transmembrane, but not soluble or glycosylphosphatidylinositol-anchored forms of endolyn in Madin-Darby canine kidney cells. Mol Biol Cell 15:1407-1416

Ramos M, Lopez De Castro JA (2002) HLA-B27 and the pathogenesis of spondyloarthritis. Tissue Antigens 60:191-205

Riederer MA, Hinnen A (1991) Removal of N-glycosylation sites of the yeast acid phosphatase severely affects protein folding. J Bacteriol 173:3539-3546

Ritter C, Helenius A (2000) Recognition of local glycoprotein misfolding by the ER folding sensor UDP-glucose:glycoprotein glucosyltransferase. Nat Struct Biol 7:278-280

Russell SJ, Ruddock LW, Salo KEH, Oliver JD, Roebuck QP, Llewellyn DH, Roderick Hl, Koivunen P, Myllyharju J, High S (2004) The primary substrate binding site in the b' domain of ERp57 is adapted for endoplasmic reticulum lectin association. J Biol Chem 279:18861-18869

Saito Y, Ihara Y, Leach MR, Cohen-Doyle MF, Williams DB (1999) Calreticulin functions *in vitro* as a molecular chaperone for both glycosylated and non-glycosylated proteins. EMBO J 18:6718-6729

Saksena S, Shao Y, Braunagel SC, Summers MD, Johnson AE (2004) Cotranslational integration and initial sorting at the endoplasmic reticulum translocon of proteins destined for the inner nuclear membrane. Proc Natl Acad Sci USA 101:12537-12542

Sargsyan E, Baryshev M, Szekely L, Sharipo A, Mkrtchian S (2002) Identification of ERp29, an endoplasmic reticulum lumenal protein, as a new member of the thyroglobulin folding complex. J Biol Chem 277:17009-17015

Scheiffele P, Peranen J, Simons K (1995) N-glycans as apical sorting signals in epithelial cells. Nature 378:96-98

Schrag JD, Bergeron JJM, Li Y, Borisova S, Hahn M, Thomas DY, Cygler M (2001) The structure of calnexin, an ER chaperone involved in quality control of protein folding. Mol Cell 8:633-644

Schymkowitz JWH, Rousseau F, Serrano L (2002) Surfing on protein folding energy landscapes. Proc Natl Acad Sci USA 99:15846-15848

Sen J, Goltz J, Konsolaki M, Schupbach T, Stein D (2000) Windbeutel is required for function and correct subcellular localization of the *Drosophila* patterning protein *Pipe*. Development 127:5541-5550

Sevier CS, Cuozzo JW, Vala A, Aslund F, Kaiser CA (2001) A flavoprotein oxidase defines a new endoplasmic reticulum pathway for biosynthetic disulphide bond formation. Nat Cell Biol 3:874-882

Sharma M, Pampinella F, Nemes C, Benharouga M, So J, Du K, Bache Kg, Papsin B, Zerangue N, Stenmark H, Lukacs Gl (2004) Misfolding diverts CFTR from recycling to degradation: quality control at early endosomes. J Cell Biol 164:923-933

Shelness G, Lin L, Nicchitta C (1993) Membrane topology and biogenesis of eukaryotic signal peptidase. J Biol Chem 268:5201-5208

Silberstein S, Schlenstedt G, Silver Pa, Gilmore R (1998) A role for the DnaJ homologue Scj1p in protein folding in the yeast endoplasmic reticulum. J Cell Biol 143:921-933

Silvennoinen L, Myllyharju J, Ruoppolo M, Orru S, Caterino M, Kivirikko KI, Koivunen P (2004) Identification and characterization of structural domains of human ERp57: association with calreticulin requires several domains. J Biol Chem 279:13607-13615

Smith M, Koch G (1989) Multiple zones in the sequence of calreticulin (CRP55, calregulin, HACBP), a major calcium binding ER/SR protein. EMBO J 8:3581-3586

Solovyov A, Gilbert HF (2004) Zinc-dependent dimerization of the folding catalyst, protein disulfide isomerase. Protein Sci 13:1902-1907

Solovyov A, Xiao R, Gilbert Hf (2004) Sulfhydryl oxidation, not disulfide isomerization, is the principal function of protein disulfide isomerase in yeast *Saccharomyces cerevisiae*. J Biol Chem 279:34095-34100

Spooner RA, Watson PD, Marsden CJ, Smith DC, Moore KA, Cook JP, Lord JM, Roberts LM (2004) Protein disulphide isomerase reduces ricin to its A and B chains in the endoplasmic reticulum. Biochem J 383:285-293

Stillemark P, Boren J, Andersson M, Larsson T, Rustaeus S, Karlsson K-A, Olofsson S-O (2000) The assembly and secretion of Apolipoprotein B-48-containing very low density lipoproteins in McA-RH7777 cells. J Biol Chem 275:10506-10513

Strickland E, Qu B-H, Millen L, Thomas PJ (1997) The molecular chaperone Hsc70 assists the *in vitro* folding of the N-terminal nucleotide-binding domain of the cystic fibrosis transmembrane conductance regulator. J Biol Chem 272:25421-25424

Suh J-K, Poulsen LL, Ziegler DM, Robertus JD (1999) Yeast flavin-containing monooxygenase generates oxidizing equivalents that control protein folding in the endoplasmic reticulum. Proc Natl Acad Sci USA 96:2687-2691

Sullivan DC, Huminiecki L, Moore JW, Boyle JJ, Poulsom R, Creamer D, Barker J, Bicknell R (2003) EndoPDI, a novel protein-disulfide isomerase-like protein that is preferentially expressed in endothelial cells acts as a stress survival factor. J Biol Chem 278:47079-47088

Swanton E, High S, Woodman P (2003) Role of calnexin in the glycan-independent quality control of proteolipid protein. EMBO J 22:2948-2958

Trombetta ES, Helenius A (2000) Conformational requirements for glycoprotein reglucosylation in the endoplasmic reticulum. J Cell Biol 148:1123-1129

Tsai B, Rodighiero C, Lencer WI, Rapoport TA (2001) Protein disulfide isomerase acts as a redox-dependent chaperone to unfold cholera toxin. Cell 104:937-948

Tu BP, Ho-Schleyer SC, Travers KJ, Weissman JS (2000) Biochemical basis of oxidative protein folding in the endoplasmic reticulum. Science 290:1571-1574

Tu BP, Weissman JS (2004) Oxidative protein folding in eukaryotes: mechanisms and consequneces. J Cell Biol 164:341-346

Vanhove M, Usherwood YK, Hendershot LM (2001) Unassembled Ig heavy chains do not cycle from BiP *in vivo* but require light chains to trigger their release. Immunity 15:105-114

van Lith M, Hartigan N, Hatch J, Benham AM (2005) PDILT, a divergent testis-specific protein disulfide isomerase with a non-classical SXXC motif that engages in disulfide-dependent interactions in the endoplasmic reticulum. J Biol Chem 280:1376-1383

Vashist S, Ng DTW (2004) Misfolded proteins are sorted by a sequential checkpoint mechanism of ER quality control. J Cell Biol 165:41-52

Voisset C, Dubuisson J (2004) Functional hepatitis C virus envelope glycoproteins. Biol Cell 96:413-420

Wang Q, Chang A (1999) Eps1, a novel PDI-related protein involved in ER quality control in yeast. EMBO J 18:5972-5982

Watanabe R, Riezman H (2004) Differential ER exit in yeast and mammalian cells. Curr Opin Cell Biol 16:350-355

Wearsch PA, Jakob CA, Vallin A, Dwek RA, Rudd PM, Cresswell P (2004) Major histocompatibility complex class I molecules expressed with monoglucosylated N-linked glycans bind calreticulin independently of their assembly status. J Biol Chem 279:25112-25121

Wehrens XHT, Lehnart SE, Reiken SR, Deng S-X, Vest JA, Cervantes D, Coromilas J, Landry DW, Marks AR (2004) Protection from cardiac arrhythmia through ryanodine receptor-stabilizing protein calstabin2. Science 304:292-296

Weihofen A, Binns K, Lemberg MK, Ashman K, Martoglio B (2002) Identification of signal peptide peptidase, a presenilin-type aspartic protease. Science 296:2215-2218

Weihofen A, Martoglio B (2003) Intramembrane-cleaving proteases: controlled liberation of proteins and bioactive peptides. Trends Cell Biol 13:71-78

Wigley WC, Stidham RD, Smith NM, Hunt JF, Thomas PJ (2001) Protein solubility and folding monitored *in vivo* by structural complementation of a genetic marker protein. Nat Biotechnol 19:131-136

Wilson IA, Skehel JJ, Wiley DC (1981) Structure of the haemagglutinin membrane glycoprotein of influenza virus at 3 A resolution. Nature 289:366-373

Woolhead CA, Mccormick PJ, Johnson AE (2004) Nascent membrane and secretory proteins differ in FRET-detected folding far inside the ribosome and in their exposure to ribosomal proteins. Cell 116:725-736

Xia W, Wolfe MS (2003) Intramembrane proteolysis by presenilin and presenilin-like proteases. J Cell Sci 116:2839-2844

Zeigler DM (2002) An overview of the mechanism, substrate specificities, and structure of FMOs. Drug Metab Rev 34:503-511

Zhang J, Herscovitz H (2003) Nascent lipidated apolipoprotein B is transported to the golgi as an incompletely folded intermediate as probed by its association with network of endoplasmic reticulum molecular chaperones, GRP94, ERp72, BiP, calreticulin, and cyclophilin B. J Biol Chem 278:7459-7468

Zhang X, Wang Y, Li H, Zhang W, Wu D, Mi H (2004) The mouse FKBP23 binds to BiP in ER and the binding of C-terminal domain is interrelated with Ca2+ concentration. FEBS Lett 559:57-60

Ziegler DM (1985) Role of reversible oxidation-reduction of enzyme thiols-disulfides in metabolic regulation. Annu Rev Biochem 54:305-329

Abbreviations

Apo B: apo-lipoprotein B
APP: amyloid precursor protein
ATP: adenosine triphosphate
BAP: BiP-associated protein
BiP: [Immunuglobulin] Binding Protein
CFTR: cystic fibrosis transmembrane receptor
CHIP: carboxyl terminus of Hsc70-interacting protein
CPVT: catecholaminergic polymorphic ventricular tachycardia
DAF: decay accelerating factor
EDEM: ER degradation enhancing 1,2-mannosidase like protein
EGF: epidermal growth factor
ER: endoplasmic reticulum
ERAD: ER associated degradation
ERO: endoplasmic reticulum oxidoreductin
FAD: flavin adenine dinucleotide
FKBP: FK506 binding-protein
FMO: flavin-containing mono-oxygenases
FRET: fluorescence resonance energy transfer
GPI: glycosylphosphatidyl inositol
GRP: glucose regulated protein
GSH: glutathione, reduced
GSSG: glutathione, oxidised
GT: UDP-glucose:glycoprotein glucosyltransferase
HA: hemagglutinin
HCV: hepatitis C virus
INM: inner nuclear membrane
kDa: kilodalton
LDL-R: low-density lipoprotein receptor
MDCK cells: Madin-Darby canine kidney cells

MHC: major histocompatibility complex
MYTHS: membrane-based yeast two-hybrid system
NF-AT: nuclear factor-activated T cells
PDI: protein disulfide isomerase
PLP: proteolipid protein
PPIase: peptidyl-prolyl cis/trans isomerase
QSOX: quiescin-sulfhydryl oxidase
RAP: low density lipoprotein receptor-related protein-associated protein
RyR: ryanodine receptor
SFV: Semliki Forest virus
SPC: signal peptidase complex
SPP: signal peptide peptidase
TGN: trans-Golgi network
tPA: tissue-type plasminogen activator
TRAM: translocating chain-associated membrane protein
UPR: unfolded protein response
VSV: vesicular stomatitis virus

Benham, Adam M.
Department of Biological Sciences, University of Durham, South Road, Durham, DH1 3LE, England.
Adam.benham@durham.ac.uk

Dias-Gunasekara, Sanjika
Department of Biological Sciences, University of Durham, South Road, Durham, DH1 3LE, England.

Quality control of proteins in the mitochondrion

Mark Nolden, Brigitte Kisters-Woike, Thomas Langer, and Martin Graef

Abstract

The quality control of proteins within mitochondria is ensured by conserved and ubiquitous ATP-dependent molecular chaperones and proteases, present in various subcompartments of the organelle. Hsp70 chaperones drive protein import and facilitate folding of newly imported preproteins, but are also required for proteolysis of misfolded polypeptides by ATP-dependent proteases. Energy dependent proteases in mitochondria include Lon and Clp proteases in the matrix space and two AAA proteases in the inner membrane, all of them compartmental proteases of the AAA^+ family with chaperone-like properties. Studies in yeast identify essential regulatory roles of these proteases for mitochondrial genome integrity, gene expression, the assembly of the respiratory chain, and mitochondrial morphology. An impaired proteolytic system in mitochondria has been identified as a cause for neurodegeneration in human. The present review summarizes the current understanding of the protein quality system in mitochondria and discusses the molecular action of protein machineries involved.

1 Stability of mitochondria

Mitochondria are dynamic organelles whose number, shape, and protein composition varies in different metabolic and differentiation states (Shaw and Nunnari 2002). The proteomic analysis of murine mitochondria isolated from different tissues revealed a drastically altered protein composition affecting about half of all mitochondrial proteins (Mootha et al. 2003). These alterations appear to be largely triggered by changes of gene expression programmes, which allow the adaptation of mitochondrial activities and regulate the biogenesis of the organelle. Similarly, synthesis of respiratory chain complexes and enzymes of the TCA-cycle is upregulated in yeast cells shifted from fermenting to respiring carbon sources (Hughes et al. 2000; Epstein et al. 2001; Ohlmeier et al. 2004). On the other hand, proteolytic processes, either affecting individual mitochondrial proteins or the whole organelle, can contribute to an altered abundance of mitochondrial proteins.

Mitochondria can be rapidly removed from the cell by autophagy which is considered as a mechanism to replenish the cellular amino acid pool (Abeliovich and Klionsky 2001; Huang and Klionsky 2002; Rodriguez-Enriquez et al. 2004). An increased degradation of mitochondria in vacuoles has been observed in nitrogen-starved yeast cells (Takeshige et al. 1992). Similarly, inactivation of the *i*-AAA protease in yeast mitochondria appears to induce their vacuolar degradation suggesting that mitochondrial dysfunction can trigger autophagy (Campbell and Thorsness 1998). The autophagic removal of mitochondria has recently also been linked to apoptosis and might exert a regulatory role in this process by the selected removal of impaired mitochondria (Gozuacik and Kimchi 2004).

Whereas an increasing number of stimuli are identified which lead to the autophagic degradation of mitochondria, next to nothing is known about the half-lives of individual mitochondrial proteins. Early experiments in mammalian cells pointed to different half-lives of mitochondrial proteins localised in different subcompartments of the organelle already suggesting the presence of various proteases within mitochondria (Russel et al. 1980). However, more general studies on the stability of mitochondrial proteins in different cells and tissues or under varying environmental conditions have not been performed. Recent labelling experiments in logarithmically growing yeast cells revealed that only ~5% of the mitochondrial proteins are degraded per hour indicating a high stability of the mitochondrial proteome under these conditions (Augustin et al. 2005). Similarly, 5-10% of newly imported mitochondrial preproteins were found to be subject to proteolysis suggesting the efficiency of mitochondrial biogenesis within the organelle to be >90%.

The turnover of mitochondrial proteins is mediated by an independent proteolytic system within the organelle. Components of this system are mostly ubiquitous and conserved; their inactivation often leads to pleiotropic phenotypes in various organisms indicating essential functions. Polypeptides can be degraded completely to amino acids presumably by the concerted action of ATP-dependent proteases and oligopeptidases present in different mitochondrial subcompartments (Desautels and Goldberg 1982). Whereas oligopeptidases are only poorly characterized, studies in yeast point to a dual function of ATP-dependent proteases within mitochondria: on one hand, they comprise a quality control system and degrade non-assembled or misfolded polypeptides; on the other hand, they appear to specifically regulate essential steps in mitochondrial biogenesis. These ATP-dependent proteases show high similarity to prokaryotic enzymes and share a conserved ATPase-module characteristic of the AAA^+ family of Walker-type ATPases (Neuwald et al. 1999; Ogura and Wilkinson 2001; Lupas and Martin 2002; Frickey and Lupas 2004; Iyer et al. 2004). This module exerts chaperone-like activities and specifically binds to non-native substrate polypeptides. It is, therefore, crucial for the selective removal of non-native polypeptides by the ATP-dependent proteases in mitochondria. Nevertheless, molecular chaperones were found to be required for the degradation of various misfolded substrate proteins and therefore represent further components of the mitochondrial quality control system. They are thought to maintain aggregation-prone polypeptides in a soluble state allowing their proteolysis, but may also play a role for the specific recognition of substrate proteins.

2 Molecular chaperone proteins and mitochondrial proteolysis

First evidence for a role of molecular chaperone proteins in mitochondrial proteolysis was reported for the major Hsp70 protein (mtHsp70) located in the mitochondrial matrix space (Craig et al. 1989; Kang et al. 1990). This conserved chaperone protein is the central part of the mitochondrial protein import motor and drives the unfolding of preproteins bound to the mitochondrial surface and their subsequent vectorial translocation across both membranes (Neupert and Brunner 2002; Rehling et al. 2004). Moreover, folding of newly imported preproteins after completion of import is facilitated by mtHsp70 proteins (Craig et al. 1989; Kang et al. 1990). Whereas the folding of some preproteins depends on mtHsp70 proteins only, others need to be transferred to the Hsp60 chaperonin system (Manning-Krieg et al. 1991; Martin 1997; Voos and Röttgers 2002).

Like other Hsp70 proteins (Bukau and Horwich 1998; Hartl and Hayer-Hartl 2002), mtHsp70 exerts its function by repeated binding-release cycles to unfolded polypeptides. Several regulatory proteins confer specificity to mtHsp70 for certain substrates and modulate their ATP-dependent interaction with the chaperone (Fig. 1). During protein import, mtHsp70 is brought in close proximity to the import sites by binding to Tim44, a peripheral component of the protein translocase in the inner membrane (TIM complex) (Schneider et al. 1994). The association of incoming preproteins with mtHsp70 is facilitated by Tim14 (Pam18), a membrane-bound DnaJ-like molecular chaperone which stimulates the ATPase activity of mtHsp70 (Mokranjac et al. 2003). Another component of the mitochondrial import motor is Tim16 (Pam16) which is distantly related to J-like proteins (Frazier et al. 2004; Kozany et al. 2004). It antagonizes Tim14 and thereby modulates the interaction of Tim14 with mtHsp70 (Li et al. 2004). Finally, substrate release from mtHsp70 is triggered by the nucleotide-release factor Mge1, which is homologous to bacterial GrpE (Ikeda et al. 1994; Laloraya et al. 1994; Nakai et al. 1994). Mge1 does also cooperate with mtHsp70 during the folding of newly imported proteins in the matrix space (Westermann et al. 1995). In contrast, the interaction of mtHsp70 with folding polypeptides is regulated by another J-like protein, Mdj1, which is not involved in mitochondrial protein import (Rowley et al. 1994; Laloraya et al. 1995). Thus, different co-chaperones modulate the binding of mtHsp70 to newly imported preproteins and folding polypeptides.

The role of mtHsp70 during proteolysis in mitochondria is far less characterized. The degradation of misfolded or missorted polypeptides in the matrix space was found to be impaired in yeast mitochondria carrying a temperature-sensitive allele of *SSC1* encoding mtHsp70 (Wagner et al. 1994). Similarly, proteolysis depended on Mge1 and Mdj1 demonstrating that the same co-chaperones modulate the interaction of mtHsp70 with folding polypeptides and polypeptides prone to degradation. According to the currently favoured model, the fate of a polypeptide associated with mtHsp70 is determined by kinetic partitioning (Wagner et al. 1994). Repeated binding-release cycles of mtHsp70 facilitate the folding of polypeptides, which do not rebind to mtHsp70 in their native conformation. Irreversibly misfolded polypeptides, however, remain bound to mtHsp70 for a prolonged time period allowing their proteolytic destruction. In agreement with this scenario, the mtHsp70 machinery can cooperate during proteolysis with different ATP-dependent proteases, the membrane-bound *m*-AAA protease as well as the Lon-like PIM1 protease in the matrix (see below) (Savel′ev et al. 1998). Although most of the current evidence is in support of the kinetic partitioning model, some puzzling observations have been made. Deletion of *MDJ1* in yeast did not only impair the binding of a misfolded model substrate to mtHsp70, as expected from the regulatory role of other J-like proteins in the Hsp70-cycle, but inhibited also the release of bound polypeptides from the chaperone (Wagner et al. 1994). Moreover, as protein aggregates cannot be degraded by ATP-dependent proteases in mitochondria, the sole function of the mtHsp70 machinery during proteolysis would be to prevent the aggregation of misfolded proteins. However, suppression of protein aggregation in *ssc1* mutant mitochondria by overexpression of the ClpB/Hsp100 protein Hsp78 did not restore the degradation of misfolded polypeptides (Schmitt et al. 1995). It is, therefore, possible that molecular chaperones confer specificity to the proteolysis of misfolded proteins, reminiscent of the co-factor-regulated action of Hsp70 proteins during protein degradation in the mammalian cytosol (Hohfeld et al. 2001).

Two recent studies suggest that Hsp78 affects proteolytic processes not only by suppressing substrate aggregation. Hsp78 was found to be required for the degradation of soluble misfolded polypeptides by the Lon-like PIM1 protease in the matrix space if these substrates were overexpressed within mitochondria (Rottgers et al. 2002). Similarly, proteolysis of a mutant Ilv5 protein in the matrix was triggered by Hsp78 in glucose-repressed cells even when the Ilv5 variant was present in a disaggregated form (Bateman et al. 2002). How does Hsp78 affect these proteolytic processes if not by suppressing protein aggregation? Hsp78 has overlapping functions with mtHsp70 in mitochondria and cooperates with this chaperone machinery in the disassembly of insoluble aggregates of heat-denatured proteins (Krzewska et al. 2001). Hsp78 is required to maintain mitochondrial thermotolerance and allows the restoration of the mitochondrial protein synthesis apparatus after heat stress (Fig. 1) (Schmitt et al. 1996; Germaniuk et al. 2002). Thus, its function appears to be similar to the bacterial homologue ClpB and cytosolic Hsp104 (Horwich et al. 1999). It is therefore conceivable that the chaperone activity of Hsp78 alters the conformation of soluble substrate, which thereby becomes accessible for Lon protease. Although further experiments have to be awaited to elucidate the molecular function of Hsp78 as well as mtHsp70 in proteolysis, our current understanding suggests already a complex interplay between the proteolytic and folding machineries within mitochondria.

3 ATP-dependent proteases of mitochondria

Energy dependent proteases reshape the cellular proteome by the selective degradation of proteins in different cellular conditions or in response to an altered

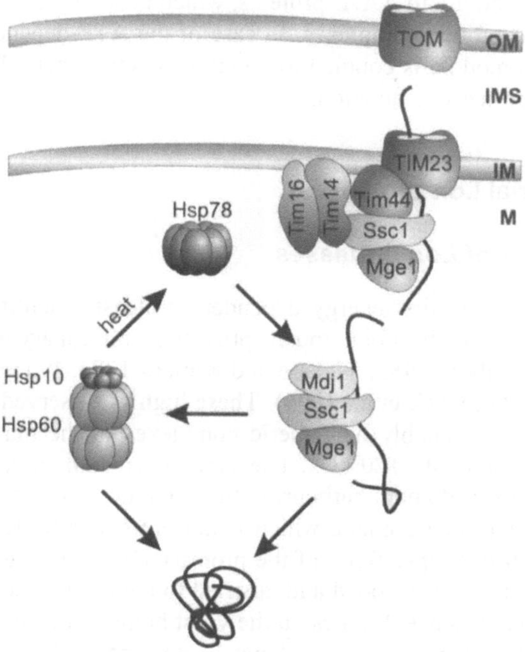

Fig. 1. Mitochondrial chaperone network for protein import and folding. mtHsp70 (Ssc1) drives the unfolding of preproteins and their subsequent vectorial translocation across both membranes. The folding of newly imported preproteins is either mediated by mtHsp70 or requires transfer to the chaperonin system Hsp60/Hsp10. Substrate interaction of mtHsp70 in these processes is regulated by different cofactors, including the J-proteins Mdj1, Tim14 (Pam18) and Tim16 (Pam16) and the nucleotide release factor Mge1. Hsp78 reactivates mitochondrial protein aggregates, which are denatured by heat stress, and transfers these proteins to mtHsp70 for refolding. TOM, translocase of outer membrane; TIM, translocase of inner membrane containing Tim23 and Tim17.

external environment (Sauer et al. 2004). They comprise conserved and widespread protein families, members of which are present in the cytosol and mitochondria of eukaryotic cells. Whereas proteolysis in the cytosol is largely mediated by 26S proteasomes (Goldberg 2003; Pickart and Cohen 2004), mitochondria harbour several ATP-dependent proteases localised in different subcompartments of the organelle. Similar to their bacterial homologues, these peptidases form large protein complexes. Despite the lack of any structural information for the mitochondrial enzymes, they are most likely compartmental peptidases providing an internal cavity for proteolysis by a ring-shaped assembly of their subunits. Entry into this cavity is thought to require the processive unfolding of substrate polypeptides consuming part of the energy spent for proteolysis. All known energy dependent proteases are AAA^{+} ATPases and are characterized by a conserved P-loop ATPase domain (Neuwald et al. 1999; Ogura and Wilkinson 2001; Lupas and Martin 2002; Frickey and Lupas 2004; Iyer et al. 2004). Numerous studies assigned chaperone activity to AAA$^+$ proteins, which generally function in the disassembly of cellular protein complexes. In case of the ATP-dependent proteases of mitochondria this module is coupled to metallo- or serine peptidase domains and thereby serves a proteolytic function.

3.1 Mitochondrial Lon proteases

3.1.1 Biogenesis of Lon proteases

Lon proteases were the first energy dependent proteases identified within mitochondria and appear to be ubiquitously present in all eukaryotic cells (Fig. 2) (Desautels and Goldberg 1982; Watabe and Kimura 1985; Wang et al. 1993; Suzuki et al. 1994; Van Dyck et al. 1994). These highly conserved peptidases form homo-oligomeric, presumably heptameric complexes in the matrix space which are built up of subunits of ~120 kDa. The subunits contain an AAA$^+$ domain followed by a proteolytic domain harbouring the catalytic serine-lysine dyad. A less conserved amino terminal domain, which is not present in bacterial homologues, may affect the substrate specificity of the protease (Van Dyck and Langer 1999). Lon proteases are nuclear-encoded and targeted to the mitochondrial matrix by an amino terminal presequence. Studies on the yeast homologue, termed PIM1 protease, revealed a two-step processing of newly imported subunits (Wagner et al. 1997). The N-terminal matrix targeting sequence is cleaved off upon import into mitochondria presumably by the mitochondrial processing peptidase. Efficient sorting to mitochondria is supported by a pro-region following the presequence, which is removed within mitochondria autocatalytically. Maturation depends on the ATP-dependent assembly and concomitant activation of the Lon protease in the matrix. Purified PIM1 protease subunits assemble in a ~800 kDa complex *in vitro* (Kutejová et al. 1993; Van Dijl et al. 1998). Electron microscopic analysis revealed a sevenfold symmetry of this homo-oligomeric complex and a striking asymmetry in the presence of non-hydrolysable ATP-analogon which might indicate a lateral opening of the ring structure under these conditions (Stahlberg et al. 1999). Notably, Pim1 subunits were detected in a complex of ~1600 kDa in mitochondrial extracts suggesting that the protease is part of a large assembly in the mitochondrial matrix space (Wagner et al. 1997).

3.1.2 Functions of Lon proteases within mitochondria

Lon proteases represent a central component of the quality control system in mitochondria and mediate the degradation of misfolded polypeptides in the matrix space (Fig. 2). Misfolding is induced under heat stress conditions providing a rationale for the increased expression of yeast Pim1 subunits under these conditions (Van Dyck et al. 1994). Similarly, oxidatively damaged aconitase is degraded by Lon protease in human mitochondria (Bota and Davies 2002). In agreement with its quality control function, electron dense particles which may be composed of aggregated polypeptides have been observed to accumulate in yeast mitochondria

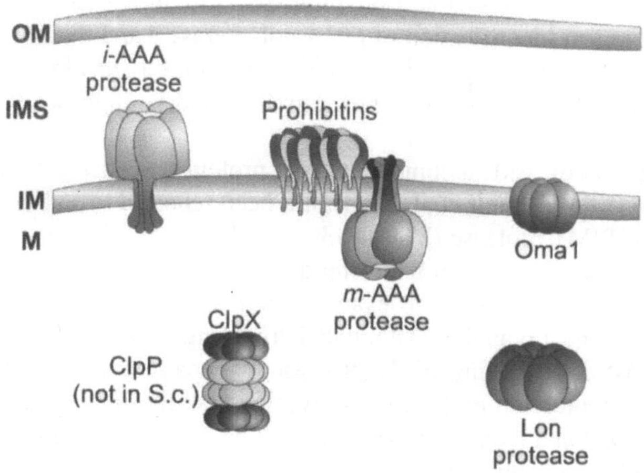

Fig. 2. Mitochondrial proteases involved in protein quality control. Lon protease appears to be ubiquitously present in all eukaryotic cells, whereas the protease ClpP is not present in *S. cerevisiae.* Two AAA protease complexes have been identified in the inner mitochondrial membrane. The *i*-AAA protease is active on the intermembrane side of the membrane, the *m*-AAA protease on the matrix side. The *m*-AAA protease and prohibitins build up a super-complex of ~ 2 MDa. Oma1 represents a conserved ATP-independent metallopeptidase in the inner membrane.

lacking PIM1 protease (Suzuki et al. 1994). These aggregates might impair diverse processes within mitochondria thereby explaining the pleiotropic phenotypes associated with mutations in the yeast *PIM1* gene. However, as the expression of mutant PIM1 variants with low proteolytic activity are sufficient to restore mitochondrial defects in the absence of PIM1 protease, it appears likely that PIM1 protease affects multiple processes during mitochondrial biogenesis by affecting the stability of regulatory proteins.

PIM1 protease is required for the maintenance of functional mitochondrial DNA (mtDNA) in yeast (Suzuki et al. 1994; Van Dyck et al. 1994). As essential subunits of respiratory chain complexes are encoded by mtDNA (Burger et al. 2003), yeast cells lacking Pim1 cannot utilize non-fermentable substrates as the sole carbon source. The role of PIM1 protease in mitochondrial DNA metabolism is still unclear, however, it is linked to its proteolytic activity. It might be related to the peculiar property of Lon proteases to bind nucleic acids. *E. coli* Lon protease was found to bind single- and double-stranded DNA resulting in the stimulation of its ATPase activity (Zehnbauer et al. 1981; Chung and Goldberg 1982; Charette et al. 1984). Subsequent *in vitro* studies revealed that bacterial and murine Lon proteases bind specifically to GT-rich DNA sequences (Fu et al. 1997; Fu and Markovitz 1998; Lu et al. 2003). The binding affinity, however, is apparently not restricted to DNA as human Lon protease was recently found to bind also GU-rich RNA oligonucleotides (Liu et al. 2004). DNA binding is stimulated by protein substrates but inhibited by ATP, suggesting that conformational changes of Lon proteases modulate the affinity for nucleic acids. The physiological relevance of

DNA binding by mitochondrial Lon proteases still needs to be substantiated. However, the recent identification of DNA polymerase γ and the mtDNA helicase Twinkle, both components of mitochondrial nucleoids (Garrido et al. 2003), as binding partners of Lon protease in human mitochondria might lead the way (Liu et al. 2004).

Overexpression of Sss1, a component of a protein translocase in the endoplasmic reticulum (ER) (Esnault et al. 1994), has been reported to stabilize mtDNA in the absence of PIM1 protease (Van Dyck et al. 1998). While the molecular basis of this suppressive effect is not understood, the maintenance of mtDNA in Pim1-deficient mitochondria allowed to examine further regulatory functions of PIM1 protease within the organelle. Although harbouring intact mtDNA, respiratory growth of yeast cells lacking PIM1 protease is impaired. The analysis of mitochondrial gene expression in these cells revealed that the splicing (and thereby the stability) of *COXI* and *COB* transcripts as well as the translation of mature *COXI* mRNA depends on the proteolytic activity of Pim1 (Van Dyck et al. 1998).

The phenotypic analysis of yeast PIM1 mutants points to multiple functions of Lon proteases in mitochondrial gene expression and the biogenesis of the respiratory chain, although the identification of short-lived substrates remains to be awaited. Overexpression of a proteolytically inactive PIM1 protease was found to suppress defects in the assembly of respiratory chain complexes in mitochondria lacking the membrane-bound *m*-AAA protease (see below) (Rep et al. 1996). This observation was taken as evidence for a chaperone function of PIM1 protease independent of its proteolytic activity. However, proteolytically inactive PIM1 subunits were demonstrated to assemble with wild type subunits in these cells forming a hetero-oligomeric PIM1 protease with residual proteolytic activity (Wagner et al. 1997). Thus, although a chaperone-like activity of the AAA^+ domain of mitochondrial Lon proteases during proteolysis is likely, mitochondrial defects in the absence of PIM1 protease appear to reflect essential proteolytic functions during mitochondrial biogenesis. Whatever the substrates are, a number of complementation studies suggested functionally conserved roles of Lon proteases within mitochondria (Teichmann et al. 1996; Barakat et al. 1998; Lu et al. 2003).

3.2 Mitochondrial Clp proteases

Clp proteases comprise another conserved family of energy dependent proteases with representatives identified within mitochondria. They form hetero-oligomeric, barrel-like complexes composed of one or two hexameric rings of AAA^+ ATPase subunits with chaperone activity and a heptameric double ring of proteolytic ClpP subunits (Fig. 2) (Horwich et al. 1999; Kang et al. 2002; Sauer et al. 2004). In contrast to other ATP-dependent proteases, Clp proteases are not ubiquitously present in mitochondria of eukaryotic cells. Whereas mitochondria appear to contain generally ClpX-like ATPase subunits (Corydon et al. 1998; Van Dyck et al. 1998; Halperin et al. 2001), ClpP-like proteins are present in mitochondria of mammals and plants but not in yeast. This has hampered the characterization of the role of Clp proteases within mitochondria in recent years. *In vitro* studies demonstrated that human ClpX and ClpP form proteolytically active ring complexes (Corydon et al. 1998, 2000). Like in *E. coli*, the human ClpXP protease is able to degrade different proteins and ClpX is responsible for the substrate selection (Kang et al. 2002). Similarly, a ClpXP protease was identified in plant mitochondria and shown to degrade α-casein in mitochondrial extracts (Halperin et al. 2001). Whereas only one ClpP protein is present in mammals, plants express a large variety of ClpP homologues localised in plastids and mitochondria raising the possibility that hetero-oligomeric ClpP complexes may exist (Adam et al. 2001). In tomato, however, mitochondria harbour only homo-tetradecameric ClpP2 complexes reminiscent of bacterial ClpP core complexes (Peltier et al. 2004).

3.3 AAA proteases in the inner membrane

3.3.1 The family of AAA proteases

Central components of the quality control system for mitochondrial inner membrane proteins are AAA proteases, membrane-embedded ATP-dependent metallopeptidases homologous to bacterial FtsH (Langer 2000). Their highly conserved subunits harbour an AAA domain typically spanning 200-250 amino acids, which defines a subfamily of the larger AAA^+ superfamily of Walker-type ATPases. In addition to the Walker A (or P-loop) and Walker B motifs essential for nucleotide binding and hydrolysis of all Walker-type ATPases, AAA proteins are characterized by the so-called second region of homology (SRH) within the AAA module (Neuwald et al. 1999; Ogura and Wilkinson 2001; Lupas and Martin 2002; Frickey and Lupas 2004; Iyer et al. 2004). The crystal structures of various AAA proteins revealed that a conserved arginine residue within the SRH-region is essential for ATP hydrolysis by protruding into neighbouring AAA subunits (Karata et al. 1999, 2001). In agreement with such an intermolecular catalytic role replacement of this arginine residue in bacterial and mitochondrial AAA protease subunits by site-directed mutagenesis inactivated the enzymes (Karata et al. 1999, 2001; Korbel et al. 2004).

As other energy dependent proteases, AAA proteases constitute large protein complexes, which are composed of identical or closely related subunits with molecular masses of 70-80 kDa (Fig. 3). One or two transmembrane segments in the N-terminal region anchor the proteins to the inner membrane. This region is followed by the AAA domain and the proteolytic domain which harbours a consensus metal binding site characteristic of metallopeptidases of the thermolysin family (Rawlings and Barrett 1995). A coiled-coil region is found at the C-terminal end. Modelling of the AAA domain of Yme1 based on the crystal structure of p97 revealed the typical fold of AAA domains (Fig. 4A). AAA protease subunits assemble into large complexes in the inner membrane with a native molecular mass of ~1 MDa (Arlt et al. 1996; Leonhard et al. 1996). Although next to nothing is known about the stoichiometry of the subunits in an AAA protease complex, crystal structures of other AAA proteins suggest a hexameric ring assembly as a

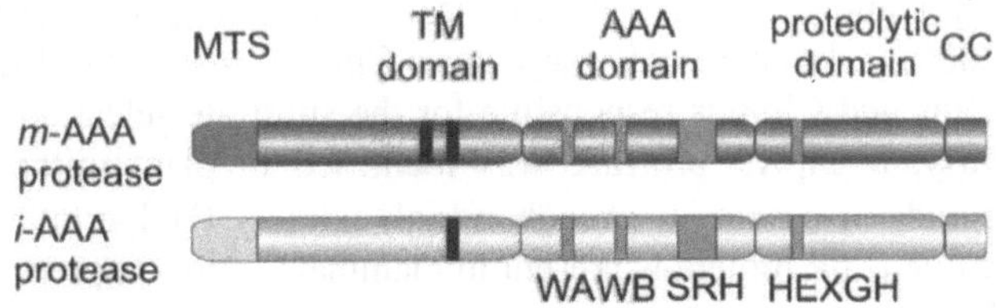

Fig. 3. Domain structure of AAA protease subunits. Linear presentation of the domain structure of the *m*- and *i*-AAA protease. AAA, ATPase domain; CC, coiled-coil region; HEXGH, consensus metal binding site representing the proteolytic centre; MTS, mitochondrial targeting signal; SRH, second region of homology; WA, Walker A box; WB, Walker B box; TM, transmembrane.

general feature of AAA proteins (Lenzen et al. 1998; Yu et al. 1998; Zhang et al. 2000). Consistently, the predicted structure of the AAA domain of Yme1 can be fitted to the hexameric ring structure of p97 (Fig. 4B). It is noteworthy, however, that the observed native molecular mass of AAA proteases in detergent extracts is difficult to reconcile with a hexameric assembly of known AAA protease subunits.

Two AAA proteases have been identified in yeast, *N. crassa* and human mitochondria which are termed *m*- and *i*- AAA protease to indicate their different topology in the inner membrane (Fig. 2): the *m*-AAA protease is active on the matrix, the *i*-AAA protease on the intermembrane side of the membrane. The *m*-AAA protease is composed of Yta10 (Afg3) and Yta12 (Rca1) subunits in yeast (Arlt et al. 1996) and of Afg3L2 and paraplegin in human (Atorino et al. 2003), whereas MAP-1 represents the only *m*-AAA protease subunit identified in *N. crassa* (Klanner et al. 2001). On the other hand, proteins homologous to yeast Yme1 are the only known subunits of *i*-AAA proteases (Leonhard et al. 1996; Shah et al. 2000). The inspection of known genome sequences indicates that two AAA proteases with catalytic sites at opposite membrane surfaces are indeed ubiquitously present in the mitochondrial inner membrane of eukaryotic cells.

3.3.2 Roles of AAA proteases in mitochondria

The deletion or inactivation of mitochondrial AAA proteases causes severe pleiotropic phenotypes in various systems, which are best characterized in yeast. Yeast cells lacking both AAA proteases are not viable (Lemaire et al. 2000; Leonhard et al. 2000). Biochemical experiments demonstrate overlapping substrate specificity of both enzymes (Leonhard et al. 2000), but additive effects of both mutations may also contribute. In general, identical phenotypes have been observed after deletion of AAA proteases or mutating the proteolytic sites of all subunits of AAA protease complexes, suggesting that the observed functional defects reflect proteolytic functions of AAA proteases within mitochondria (Weber et al. 1996; Arlt et al. 1998).

The *i*-AAA protease subunit Yme1 was first identified in a genetic screen for yeast mutants resulting in an increased rate of DNA escape from mitochondria to

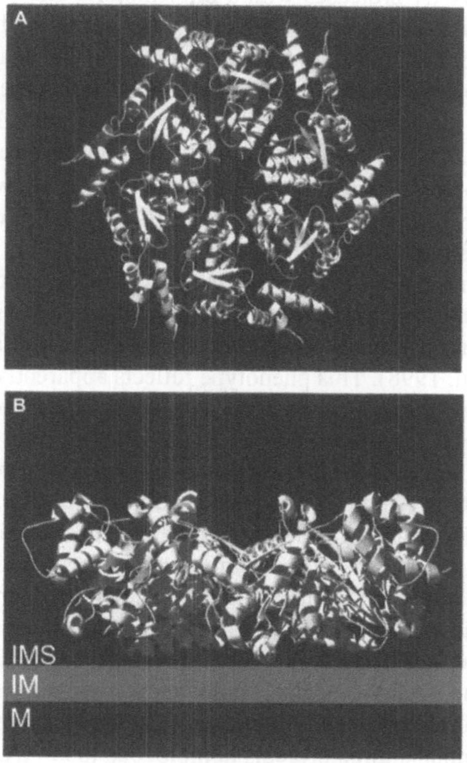

Fig. 4. Structural hexameric ring model of Yme1 AAA domains. (A) Top view of Yme1 AAA domains modelled into a hexameric ring complex. (B) Side view of the hexameric ring complex relative to the inner mitochondrial membrane. The transmembrane and proteolytic domains of Yme1 subunits are not shown. The structural model for the Yme1 AAA domain is based on the crystal structure of the AAA domain of p97 (Zhang et al. 2000), and was generated using SWISS-MODEL, Swiss-Pdb Viewer (Guex and Peitsch 1997) and AMBER. IMS, intermembrane space; IM, inner membrane; M, matrix.

the nucleus (Thorsness and Fox 1993). The *i*-AAA protease is required for respiratory growth of yeast cells at high temperatures and for fermenting growth at low temperature. Moreover, *yme1* mutant cells are petite negative, i.e. exhibit an strong growth phenotype when the mitochondrial DNA is deleted (Thorsness et al. 1993). Similarly, inactivation of the *i*-AAA protease subunit IAP-1 of *N. crassa* causes impaired respiratory growth at high temperature pointing to conserved functions within mitochondria (Klanner et al. 2001). This conclusion is further substantiated by complementation studies in yeast: expression of IAP-1 restored fermentative growth at low temperature and converted *Δyme1* into petite positive cells, whereas the expression of the human homologue hYme1 allowed growth on non-fermentable carbon sources at high temperature (Shah et al. 2000; Klanner et al. 2001). While these studies demonstrate functional conservation, differences in both the roles of *i*-AAA proteases within mitochondria and their substrate specificity appear to exist. Deletion of *YME1* in yeast leads to the accumulation of punctate mitochondria suggesting a role of Yme1 in the maintenance of mitochondrial morphology (Campbell et al. 1994). This phenotype can be suppressed by IAP-1 although the morphology of mitochondria is not affected in *N. crassa* cells lacking IAP-1. On the other hand, IAP-1 does not restore respiratory growth at elevated temperature in yeast *Δyme1* cells nor does it allow the proteolysis of non-assembled cytochrome oxidase subunit 2 (Cox2), a known substrate of the yeast *i*-AAA protease (Klanner et al. 2001).

Yeast cells lacking the *m*-AAA protease or expressing proteolytically inactive variants of both Yta10 and Yta12 are respiratory deficient and not able to grow on non-fermentable carbon sources (Guélin et al. 1994; Tauer et al. 1994; Tzagoloff et al. 1994; Arlt et al. 1998). This phenotype reflects apparently multiple functions of the protease during the biogenesis of the organelle. The expression of two respiratory chain subunits, cytochrome oxidase subunit 1 (Cox1) and cytochrome *b* (Cob), which both are encoded by intron-containing genes in the mitochondrial genome of yeast (Burger et al. 2003), is under the control of the *m*-AAA protease (Arlt et al. 1998). The *m*-AAA protease is required for the splicing of *COX1* and *COB* transcripts (Arlt et al. 1998) and thus acts similarly to the matrix-localised PIM1 protease (see above). The inactivation of the *m*-AAA protease also impairs the posttranslational assembly of respiratory chain complexes and of the F_1F_O-ATP synthase (Paul and Tzagoloff 1995; Arlt et al. 1998; Galluhn and Langer 2004). Similar to the *i*-AAA protease, complementation studies in yeast suggest a functional conservation of *m*-AAA proteases (Kolodziejczak et al. 2002; Atorino et al. 2003). In mammals, three proteins homologous to *m*-AAA protease subunits have been identified and were named AFG3L1, AFG3L2 and paraplegin (Casari et al. 1998; Shah et al. 1998; Banfi et al. 1999; Kremmidiotis et al. 2001). AFG3L1 is encoded by a pseudogene in humans but expressed in mice (Kremmidiotis et al. 2001). AFG3L2 and paraplegin have been demonstrated to assemble into a large hetero-oligomeric complex in the mitochondrial inner membrane with a native molecular mass similar to the yeast *m*-AAA protease (Atorino et al. 2003). Expression of both AFG3L2 and paraplegin in *Δyta10Δyta12* yeast cells restores their respiratory competence demonstrating functional conservation between human and yeast *m*-AAA proteases. Loss of paraplegin causes neurodegeneration in an autosomal recessive form of hereditary spastic paraplegia (HSP) in humans (Casari et al. 1998) indicating important functions of the human *m*-AAA protease in mitochondrial biogenesis. HSP patient fibroblasts exhibit an increased sensitivity towards oxidative stress and complex I deficiencies reminiscent of defects observed in *m*-AAA protease deficient yeast cells (Atorino et al. 2003). The analysis of a recently developed mouse model revealed the appearance of abnormal, giant mitochondria with rearranged cristae early in development suggesting that it is a primary cause triggering the disease (Ferreirinha et al. 2004).

Despite the extensive characterization of phenotypes associated with AAA protease mutants, the molecular basis of the observed defects is only poorly understood. The accumulation of non-native polypeptides in AAA protease deficient cells may impair mitochondrial activities and cause pleiotropic phenotypes. However, protein aggregates have not been detected in mitochondria lacking AAA proteases. It is therefore conceivable that the turnover of short-lived regulatory proteins by AAA proteases is required for the biogenesis of mitochondria and the maintenance of their activities. The current understanding of the role of AAA proteases is clearly limited by the fact that short-lived substrates of AAA proteases, whose impaired proteolysis would explain the observed defects in AAA protease deficient cells, have not been identified.

3.3.3 Prohibitins

Studies in yeast identify the *m*-AAA protease as part of a large supercomplex in the inner membrane containing prohibitins (Steglich et al. 1999). Similarly, the bacterial AAA protease FtsH assembles into a large complex with proteins distantly related to prohibitins suggesting that this complex may represent the holoenzyme (Kihara et al. 1996; Saikawa et al. 2004). Prohibitins form a conserved protein family, which appears to be ubiquitously present in the inner membrane of mitochondria. Prohibitins have been linked to diverse cellular processes such as cellular signalling and transcriptional control, senescence, phospholipid metabolism and organelle inheritance (Nijtmans et al. 2002). Two homologous proteins, Phb1 and Phb2, assemble into a large protein complex with a native molecular mass of ~1.2 MDa (Steglich et al. 1999; Nijtmans et al. 2000; Artal-Sanz et al. 2003). Although the yeast *m*-AAA protease binds quantitatively to the prohibitin complex, prohibitins are not essential for its proteolytic activity. Rather, an accelerated proteolysis of non-assembled membrane proteins was observed in yeast cells lacking the prohibitin complex (Steglich et al. 1999). It is presently unclear how prohibitins exert their apparently regulatory role during proteolysis by the *m*-AAA protease. The interaction of non-assembled respiratory chain subunits with the prohibitin complex has lead to the proposal of a chaperone activity of prohbitins during the biogenesis of the respiratory chain (Nijtmans et al. 2000). This activity, however, is apparently not essential, as the assembly of respiratory chain complexes is not affected in yeast cells lacking prohibitins. Electron microscopic analysis of purified prohibitin complexes revealed a ring-like assembly of prohibitins with outer dimensions of ~270 Å x 200 Å (Tatsuta et al. 2005). It appears therefore conceivable that prohibitins exert a scaffolding function in the inner membrane, which ensures the organisation and integrity of the inner membrane. This would provide an explanation for the functional interaction of prohibitins with various mitochondrial proteins and would be in agreement with their proposed role for the maintenance of mitochondrial morphology.

4 Quality control of inner membrane proteins

The proteolytic breakdown of membrane proteins represents a major energetic challenge for any cellular degradation system as hydrophobic, membrane-embedded segments have to be removed from the bilayer for proteolysis. In case of the endoplasmic reticulum, proteins are dislocated from the membrane by protein conducting channels allowing the proteolysis by 26S proteasomes in the cytosol (Hampton 2002). The inner membrane of mitochondria is characterised by a high protein content emphasising the need for an efficient quality control system in this subcompartment. However, it is not accessible for cytosolic proteasomes, which could explain why an independent proteolytic system derived from prokaryotic ancestors was maintained during evolution. AAA proteases with catalytic sites exposed to opposite membrane surfaces are central components of this system. As the turnover of substrate proteins can be easily monitored after their import into isolated mitochondria, the quality control function of AAA proteases in yeast has been exploited in recent years to understand general principles of membrane protein degradation by AAA proteases.

4.1 Substrate recognition by AAA proteases

A large variety of non-assembled inner membrane proteins are degraded by AAA proteases in yeast indicating broad substrate specificity (Nakai et al. 1994, 1995; Pajic et al. 1994; Pearce and Sherman 1995; Arlt et al. 1996; Guélin et al. 1996). Studies on the yeast *i*-AAA protease provided first evidence for a chaperone-like activity of the AAA domain of AAA proteases (Leonhard et al. 1999). Model substrates with unfolded domains protruding from the membrane are bound by the AAA domain whereas folding prevents recognition. The AAA domain thus serves as a sensor of the folding state of solvent-exposed domains of membrane proteins and thereby ensures the specificity of proteolysis. These observations also raise the intriguing question how misfolded membrane proteins devoid of large solvent-exposed domains are recognized by the proteolytic system. N-terminal helices of the AAA domain of Yme1, which were found to be sufficient for binding unfolded polypeptides, are close to the membrane as indicated by the crystal structure of the AAA domain of bacterial FtsH (Leonhard et al. 1999; Niwa et al. 2002). Furthermore, binding and proteolysis by AAA proteases requires only short N- or C-terminal tails of substrate proteins composed of at least 20 amino acids protruding from the membrane (Leonhard et al. 2000). Thus, recognition can apparently occur in close proximity to the membrane surface. Moreover, loop structures might also be recognized by AAA proteases. A mutant variant of the polytopic inner membrane protein Oxa1 is degraded by the *m*-AAA protease due to a point mutation present in an intermembrane space loop of Oxa1 while only a folded C-terminal domain and internal loops are exposed to the matrix (Käser et al. 2003).

These experiments suggest that unfolding or disassembly and thereby the exposure of hydrophobic surfaces is sufficient to explain their recognition by the AAA domains of AAA proteases. However, additional signals might exist triggering proteolysis by AAA proteases. Such signals might be of relevance for short-lived regulatory proteins, which are present in the native state within mitochondria but need to be degraded under certain conditions. Notably, the *m*-AAA protease was recently found to act as a processing peptidase for yeast cytochrome *c* peroxidase

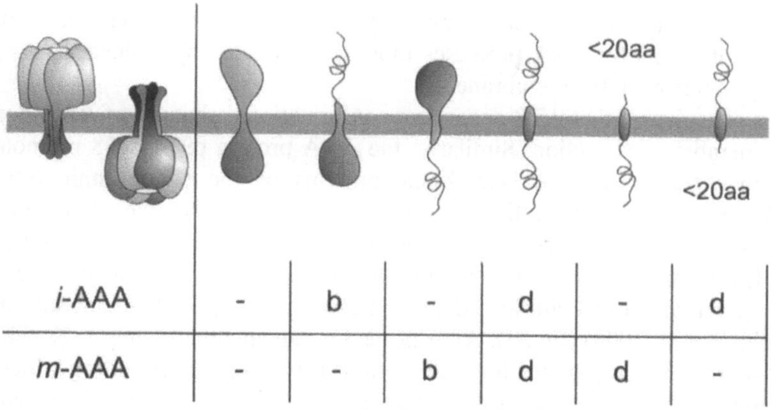

Fig. 5. The folding state of solvent exposed domains of membrane proteins determines the recognition and degradation by *m*-AAA and *i*-AAA proteases; b, binding to the respective AAA protease(s); d, degradation mediated by the respective AAA protease(s).

(Ccp1) (Esser et al. 2002). The precursor form of Ccp1 carries a bipartite presequence which is sequentially cleaved off by the *m*-AAA protease and Pcp1 (Rbd1), a rhomboid-like peptidase in the inner membrane, resulting in sorting of newly imported Ccp1 to the intermembrane space. It is currently not understood how the precursor form of Ccp1 is recognized by the *m*-AAA protease and how its complete degradation is avoided. Thus, it appears that we are just beginning to understand which characteristics of a protein drive its degradation by AAA proteases.

4.2 Substrate dislocation during proteolysis by AAA proteases

AAA proteases with catalytic sites exposed to opposite membrane surfaces are apparently ubiquitously present in the mitochondrial inner membrane of eukaryotic cells. It is therefore conceivable that the conserved membrane topology of AAA proteases is essential to ensure the degradation of membrane proteins with different topologies.

According to the current view, the involvement of one or the other AAA protease is mainly determined by the membrane topology of substrate proteins, as AAA proteases recognize protein domains exposed to the same side of the inner membrane. In a systematic analysis of the degradation of a single membrane spanning protein, Yme2, containing hydrophilic domains on both sides of the membrane, properties of a substrate protein have been elucidated which trigger its degradation (Fig. 5) (Leonhard et al. 2000). Proteolysis was found to require a minimal length of tails of >20 amino acid residues protruding from the membrane surface and unfolding of solvent-exposed domains on both sides of the membrane. If solvent-exposed domains are unfolded, both AAA proteases can mediate the degradation from either membrane surface. This implies that protein domains are translocated across the membrane bilayer for proteolysis at the opposite membrane surface. Thus, proteolysis by AAA proteases most likely occurs by dislocation of membrane proteins from the membrane.

It is conceivable that ATP-dependent conformational changes of AAA-domains drive membrane extraction. Similarly, the AAA protein p97/Cdc48 promotes the retrograde translocation of membrane proteins of the endoplasmic reticulum which are prone to degradation by 26S proteasomes in the cytosol (Zhang et al. 2000; Ye et al. 2001; Jarosch et al. 2003). It is noteworthy that the presence of a folded domain on one membrane side prevents proteolysis even when an AAA protease can bind to an unfolded domain present at the opposite membrane surface (Leonhard et al. 2000) (Fig. 5). AAA proteases can apparently only exert a limited force on membrane proteins not sufficient to drive domain unfolding which is in agreement with studies suggesting a limited unfoldase activity of the bacterial AAA protease FtsH (Herman et al. 2003).

The dislocation of membrane proteins raises the intriguing question whether protein translocases, which mediate import and membrane insertion of newly synthesised mitochondrial proteins, are required for proteolysis. Saturation of TIM23 translocases, a multicomponent protein complex in the inner membrane for the translocation of presequence-containing proteins, did not impair the degradation of model substrate proteins (Korbel et al. 2004). On the other hand, transmembrane segments of *m*-AAA protease subunits are essential for the degradation of integral membrane proteins (Korbel et al. 2004). Mutant yeast *m*-AAA proteases lacking the transmembrane domain of either Yta10 or Yta12 are proteolytically active. They maintain respiratory growth, degrade the peripheral membrane protein Atp7 and cleave the presequence of Ccp1. However, proteolysis of various integral membrane proteins was impaired suggesting a pivotal role of membrane-embedded parts of AAA proteases during substrate dislocation.

It remains to be determined whether hydrophilic pores are formed by the transmembrane segments of AAA protease subunits or whether membrane extraction is facilitated by another mechanism. A pore-like assembly of transmembrane segments would be in line with the predicted ring assembly of AAA protease subunits which, by analogy to other compartmental proteases, provides an internal cavity for proteolysis (Sauer et al. 2004). However, the formation of ring complexes by AAA proteases poses another caveat. How can integral membrane protein substrates access this cavity considering that AAA protease subunits are anchored in the membrane? Three scenarios can be envisioned each of them remaining highly speculative and posing additional questions (Fig. 6): First, substrate proteins might be extracted from the membrane and directly enter the formed cavity via the proteolytic domains (apical insertion model). As substrate polypeptides would not traverse the AAA ring, it remains unclear how the uptake of substrates would be energized. Moreover, accessory proteins should be required to facilitate membrane dislocation. Second, substrate polypeptides might be squeezed in-between the membrane and the membrane-exposed surface of the

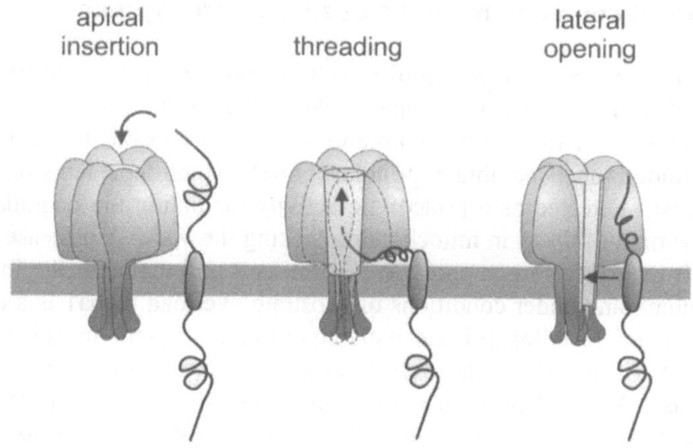

Fig. 6. How can membrane-bound substrates enter a proteolytic cavity in AAA proteases? Apical insertion model: the substrate prone to degradation is extracted from the membrane and inserted apically into the central cavity. In contrast to the general model for AAA$^+$ protein action the substrate would not pass the pore of the AAA domain ring. Threading model: the substrate protein is inserted into the central pore formed by AAA domains between the membrane and the membrane-exposed surface of the AAA domains. Lateral opening model: the AAA protease ring opens laterally to allow access of the substrate protein to the central cavity of the AAA protease. After closure of the ring the substrate is transported vectorially into the proteolytic chamber.

AAA domains to enter the pore formed by AAA domains (threading model). According to this model, substrates would be in close proximity to N-terminal helices of the AAA domains which were found to be involved in substrate binding (Leonhard et al. 1999). Translocation into the proteolytic cavity might be facilitated by alternate conformational changes of N-terminal regions altering the distance between the AAA ring and the membrane surface. Third, AAA proteases may form ring-like structures, which laterally open during substrate binding and enclose the substrate for subsequent degradation (lateral opening model). This mechanism would be reminiscent of DNA-binding by ring-shaped helicases (Ahnert et al. 2000). Initial interaction of substrate polypeptides may trigger the conformational changes of the proteolytic complex. However, a lateral opening of AAA proteases would imply an asymmetry in the ring complex, which is difficult to envision for homo-oligomeric complexes. The threading and the lateral opening model are not mutually exclusive and intermediate forms can be envisioned. For example, unfolded solvent exposed domains could insert into the central pore of the AAA ring according to the threading model. The substrate protein would be pulled into the proteolytic chamber until transmembrane domains have to be extracted from the membrane to allow further pulling/degradation. Substrate dislocation could then be facilitated by lateral opening of membrane-embedded parts of AAA protease subunits only. Clearly, further experiments are required to elucidate the mechanism of substrate dislocation by AAA proteases.

4.3 Additional components of the quality control system

While the complete degradation of membrane proteins appears to be generally ATP-dependent and mediated by energy dependent proteases, additional peptidases are present in the inner membrane which are capable to recognize and cleave misfolded inner membrane proteins. Oma1, which represents one of these peptidases, was detected as a proteolytic activity mediating the degradation of a misfolded variant of Oxa1 in mitochondria lacking the *m*-AAA protease (Käser et al. 2003). It may therefore support AAA proteases and maintain the integrity of the inner membrane under conditions of substrate overload. Oma1 is a conserved membrane-bound metallopeptidase with homologues present in eukaryotes, including plants, in eubacteria as well as in archaebacteria. In contrast to the *m*-AAA protease, ATP-independent proteolysis by Oma1 does not lead to the complete degradation of the polytopic membrane protein Oxa1 but results in the accumulation of proteolytic fragments (Käser et al. 2003). It is conceivable that complete turnover of membrane proteins depends on their membrane extraction and is therefore ATP-dependent. Notably, misfolded Oxa1 was still cleaved in mitochondria lacking both *m*-AAA protease and Oma1 indicating the existence of further peptidases in the inner membrane (Käser et al. 2003). One candidate could be the rhomboid-like peptidase Pcp1 (Rbd1) (Esser et al. 2002; Herlan et al. 2003; Sesaki et al. 2003a; 2003b), an intramembrane cleaving peptidase which acts as a processing enzyme for Ccp1 and Mgm1, but others may await their identification.

5 Regulation of quality control systems of mitochondria

Although an elaborate quality control system facilitates protein folding and ensures the removal of misfolded proteins, a number of stress conditions can interfere with protein folding and result in an overload of misfolded polypeptides within mitochondria, including thermal or oxidative stress or an imbalance in mitochondrial and nuclear gene expression. Therefore, an adaptation of the mitochondrial quality system to altered cellular conditions is needed to avoid perturbation of mitochondrial functions. Many mitochondrial chaperone proteins and proteases are encoded by heat-shock inducible genes and expressed at increased levels at high temperatures to cope with thermally destabilised proteins. Moreover, compartment-specific stress responses exist which enable cells to adjust the organellar protein folding and degradation capacity to stress conditions affecting only individual organelles.

The accumulation of unfolded proteins in the mitochondrial matrix was found to lead to the transcriptional upregulation of nuclear genes encoding mitochondrial chaperones and proteases, demonstrating the existence of a mitochondrial specific stress response in mammalian cells (Martinus et al. 1996; Zhao et al. 2002; Yoneda et al. 2004). Similarly, protein processing was impaired in mitochondria of *C. elegans* by a systematic RNAi approach and shown to cause the increased expression of nuclear-encoded components of the mitochondrial quality control system (Kuzmin et al. 2004; Yoneda et al. 2004). It remains to be determined whether the mitochondrial specific stress response is indeed triggered by the initial recognition of misfolded polypeptides within mitochondria, similar to the unfolded proteins response in the ER (Zhang and Kaufman 2004). Alternatively, it might be indirectly caused by the deleterious effect of non-native polypeptides on mitochondrial activities and the concomitant reduced transmembrane potential (Kuzmin et al. 2004). Pleiotropic effects on nuclear gene expression were indeed observed in respiratory deficient yeast cells (Traven et al. 2001). This includes the induction of non-mitochondrial anaplerotic pathways in frame of the so-called retrograde response (Butow and Avadhani 2004).

The induction of mitochondrial chaperones and proteases upon accumulation of misfolded proteins within the organelle appears to be compartment-specific and does not lead to an altered expression of stress proteins located in the endoplasmic reticulum (Zhao et al. 2002; Yoneda et al. 2004). A regulatory element was identified in the promoters of Cpn60, Cpn10, ClpP and mtDnaJ which includes a binding site for the transcription factor CHOP (Zhao et al. 2002). CHOP transcription is also induced by the unfolded protein response in the ER (Li et al. 2000; Yoshida et al. 2000) indicating that a complex interplay may exist between different organelle specific stress responses. Interestingly, ER stress was found to result in reduced steady-state levels of cytochrome oxidase subunits and the upregulation of mitochondrial Lon protease in cultured astrocytes (Hori et al. 2002). On the other hand, ER stress did not affect the expression of mitochondrial quality control components in *C. elegans* (Yoneda et al. 2004). Thus, although the existence of a mitochondrial specific stress response has been established, much remains to be learned about this complex regulatory circuit.

References

Abeliovich H, Klionsky DJ (2001) Autophagy in yeast: mechanistic insights and physiological function. Microbiol Mol Biol Rev 65:463-479

Adam Z, Adamska I, Nakabayashi K, Ostersetzer O, Haussuhl K, Manuell A, Zheng B, Vallon O, Rodermel SR, Shinozaki K, Clarke AK (2001) Chloroplast and mitochondrial proteases an *Arabidopsis*. A proposed nomenclature. Plant Physiol 125:1912-1918

Ahnert P, Picha KM, Patel SS (2000) A ring-opening mechanism for DNA binding in the central channel of the T7 helicase-primase protein. EMBO J 19:3418-3427

Arlt H, Steglich G, Perryman R, Guiard B, Neupert W, Langer T (1998) The formation of respiratory chain complexes in mitochondria is under the proteolytic control of the *m*-AAA protease. EMBO J 17:4837-4847

Arlt H, Tauer R, Feldmann H, Neupert W, Langer T (1996) The YTA10-12-complex, an AAA protease with chaperone-like activity in the inner membrane of mitochondria. Cell 85:875-885

Artal-Sanz M, Tsang WY, Willems EM, Grivell LA, Lemire BD, van der Spek H, Nijtmans LG, Sanz MA (2003) The mitochondrial prohibitin complex is essential for embryonic viability and germline function in *Caenorhabditis elegans*. J Biol Chem 278:32091-32099

Atorino L, Silvestri L, Koppen M, Cassina L, Ballabio A, Marconi R, Langer T, Casari G (2003) Loss of *m*-AAA protease in mitochondria causes complex I deficiency and increased sensitivity to oxidative stress in hereditary spastic paraplegia. J Cell Biol 163:777-787

Augustin S, Nolden M, Müller S, Hardt O, Arnold I, Langer T (2005) Characterization of peptides released from mitochondria: evidence for constant proteolysis and peptide efflux. J Biol Chem 280:2691-2699

Banfi S, Bassi MT, Andolfi G, Marchitiello A, Zanotta S, Ballabio A, Casari G, Franco B (1999) Identification and characterization of AFG3L2, a novel paraplegin-related gene. Genomics 59:51-58

Barakat S, Pearce DA, Sherman F, Rapp WD (1998) Maize contains a Lon protease gene that can partially complement a yeast *pim1*-deletion mutant. Plant Mol Biol 37:141-154

Bateman JM, Iacovino M, Perlman PS, Butow RA (2002) Mitochondrial DNA instability mutants of the bifunctional protein Ilv5p have altered organization in mitochondria and are targeted for degradation by Hsp78 and the Pim1p protease. J Biol Chem 277:47946-47953

Bota DA, Davies KJ (2002) Lon protease preferentially degrades oxidized mitochondrial aconitase by an ATP-stimulated mechanism. Nat Cell Biol 4:674-680

Bukau B, Horwich AL (1998) The Hsp70 and Hsp60 chaperone machines. Cell 92:351-366

Burger G, Gray MW, Lang BF (2003) Mitochondrial genomes: anything goes. Trends Genet 19:709-716

Butow RA, Avadhani NG (2004) Mitochondrial signaling: the retrograde response. Mol Cell 14:1-15

Campbell CL, Tanaka N, White KH, Thorsness PE (1994) Mitochondrial morphological and functional defects in yeast caused by *yme1* are suppressed by mutation of a 26S protease subunit homologue. Mol Biol Cell 5:899-905

Campbell CL, Thorsness PE (1998) Escape of mitochondrial DNA to the nucleus in *yme1* yeast is mediated by vacuolar-dependent turnover of abnormal mitochondrial compartments. J Cell Sci 111:2455-2464

Casari G, De-Fusco M, Ciarmatori S, Zeviani M, Mora M, Fernandez P, DeMichele G, Filla A, Cocozza S, Marconi R, Durr A, Fontaine B, Ballabio A (1998) Spastic paraplegia and OXPHOS impairment caused by mutations in paraplegin, a nuclear-encoded mitochondrial metalloprotease. Cell 93:973-983

Charette MF, Henderson GW, Doane LL, Markovitz A (1984) DNA-stimulated ATPase activity on the lon (CapR) protein. J Bacteriol 158:195-201

Chung CH, Goldberg AL (1982) DNA stimulates ATP-dependent proteolysis and protein-dependent ATPase activity of protease La from *Escherichia coli*. Proc Natl Acad Sci USA 79:795-799

Corydon TJ, Bross P, Holst HU, Neve S, Kristiansen K, Gregersen N, Bolund L (1998) A human homologue of *Escherichia coli* ClpP caseinolytic protease: recombinant expression, intracellular processing and subcellular localization. Biochem J 331:309-316

Corydon TJ, Wilsbech M, Jespersgaard C, Andresen BS, Borglum AD, Pederson S, Bolund L, Gregersen N, Bross P (2000) Human and mouse mitochondrial orthologs of bacterial clpX. Mamm Genome 11:899-905

Craig EA, Kramer J, Shilling J, Werner-Washburne M, Holmes S, Kosic-Smithers J, Nicolet CM (1989) *SSC1*, an essential member of the yeast HSP70 multigene family, encodes a mitochondrial protein. Mol Cell Biol 9:3000-3008

Desautels M, Goldberg AL (1982) Liver mitochondria contain an ATP-dependent, vanadate-sensitive pathway for the degradation of proteins. Proc Natl Acad Sci USA 79:1869-1873

Epstein CB, Waddle JA, Hale W, Davé V, Thornton J, Macatee TL, Garner HR, Butow RA (2001) Genome-wide responses to mitochondrial dysfunction. Mol Biol Cell 12:297-308

Esnault Y, Feldheim D, Blondel MO, Schekman R, Képès F (1994) *SSS1* encodes a stabilizing component of the Sec61 subcomplex of the yeast protein translocation apparatus. J Biol Chem 269:27478-27485

Esser K, Tursun B, Ingenhoven M, Michaelis G, Pratje E (2002) A novel two-step mechanism for removal of a mitochondrial signal sequence involves the *m*-AAA complex and the putative rhomboid protease Pcp1. J Mol Biol 323:835-843

Ferreirinha F, Quattrini A, Priozzi M, Valsecchi V, Dina G, Broccoli V, Auricchio A, Piemonte F, Tozzi G, Gaeta L, Casari G, Ballabio A, Rugarli EI (2004) Axonal degeneration in paraplegin-deficient mice is associated with abnormal mitochondria and impairment of axonal transport. J Clin Invest 113:231-242

Frazier AE, Dudek J, Guiard B, Voos W, Li Y, Lind M, Meisinger C, Geissler A, Sickmann A, Meyer HE, Bilanchone V, Cumsky MG, Truscott KN, Pfanner N, Rehling P (2004) Pam16 has an essential role in the mitochondrial protein import motor. Nat Struct Mol Biol 11:226-233

Frickey T, Lupas AN (2004) Phylogenetic analysis of AAA proteins. J Struct Biol 146:2-10

Fu GK, Markovitz DM (1998) The human Lon protease binds to mitochondrial promoters in a single-stranded, site-specific, strand-specific manner. Biochemistry 37:1905-1909

Fu GK, Smith MJ, Markovitz DM (1997) Bacterial protease Lon is a site-specific DNA-binding protein. J Biol Chem 272:534-538

Galluhn D, Langer T (2004) Reversible assembly of the ATP-binding cassette transporter Mdl1 with the F_1F_O-ATP synthase in mitochondria. J Biol Chem 279:38338-38345

Garrido N, Griparic L, Jokitalo E, Wartiovaara J, van der Bliek AM, Spelbrink JN (2003) Composition and dynamics of human mitochondrial nucleoids. Mol Biol Cell 14:1583-1596

Germaniuk A, Liberek K, Marszalek J (2002) A bichaperone (Hsp70-Hsp78) system restores mitochondrial DNA synthesis following thermal inactivation of Mip1p polymerase. J Biol Chem 277:27801-27808

Goldberg AL (2003) Protein degradation and protection against misfolded or damaged proteins. Nature 426:895-899

Gozuacik D, Kimchi A (2004) Autophagy as a cell death and tumor suppressor mechanism. Oncogene 23:2891-2906

Guélin E, Rep M, Grivell LA (1994) Sequence of the *AFG3* gene encoding a new member of the FtsH/Yme1/Tma subfamily of the AAA-protein family. Yeast 10:1389-1394

Guélin E, Rep M, Grivell LA (1996) Afg3p, a mitochondrial ATP-dependent metalloprotease, is involved in the degradation of mitochondrially-encoded Cox1, Cox3, Cob, Su6, Su8, and Su9 subunits of the inner membrane complexes III, IV, and V. FEBS Lett 381:42-46

Guex N, Peitsch MC (1997) SWISS-MODEL and the Swiss-PdbViewer: an environment for comparative protein modeling. Electrophoresis 18:2714-2723

Halperin T, Zheng B, Itzhaki H, Clarke AK, Adam Z (2001) Plant mitochondria contain proteolytic and regulatory subunits of the ATP-dependent Clp protease. Plant Mol Biol 45:461-468

Hampton RY (2002) ER-associated degradation in protein quality control and cellular regulation. Curr Opin Cell Biol 14:476-482

Hartl FU, Hayer-Hartl M (2002) Molecular chaperones in the cytosol: from nascent chain to folded protein. Science 295:1852-1858

Herlan M, Vogel F, Bornhövd C, Neupert W, Reichert AS (2003) Processing of Mgm1 by the rhomboid-type protease Pcp1 is required for maintenance of mitochondrial morphology and of mitochondrial DNA. J Biol Chem 278:27781-27788

Herman C, Prakash S, Lu CZ, Matouschek A, Gross CA (2003) Lack of a robust unfoldase activity confers a unique level of substrate specificity to the universal AAA protease FtsH. Mol Cell 11:659-669

Hohfeld J, Cyr DM, Patterson C (2001) From the cradle to the grave: molecular chaperones that may choose between folding and degradation. EMBO Rep 2:885-890

Hori O, Ichinoda F, Tamatani T, Yamaguchi A, Sato N, Ozawa K, Kitao Y, Miyazaki M, Harding HP, Ron D, Tohyama M, Stern DM, Ogawa S (2002) Transmission of cell stress from endoplasmic reticulum to mitochondria: enhanced expression of Lon protease. J Cell Biol 157:1151-1160

Horwich AL, Weber-Ban EU, Finley D (1999) Chaperone rings in protein folding and degradation. Proc Natl Acad Sci USA 96:11033-11040

Huang WP, Klionsky DJ (2002) Autophagy in yeast: a review of the molecular machinery. Cell Struct Funct 27:409-420

Hughes TR, Marton MJ, Jones AR, Roberts CJ, Stoughton R, Armour CD, Bennett HA, Coffey E, Dai H, He YD, Kidd MJ, King AM, Meyer MR, Slade D, Lum PY, Stepaniants SB, Shoemaker DD, Gachotte D, Chakraburtty K, Simon J, Bard M, Friend SH (2000) Functional discovery via a compendium of expression profiles. Cell 102:109-126

Ikeda E, Yoshida S, Mitsuzawa H, Uno I, Toh-e A (1994) YGE1 is a yeast homolog of *Escherichia coli* grpE and is required for maintenance of mitochondrial functions. FEBS Lett 339:265-268

Iyer LM, Leipe DD, Koonin EV, Aravind L (2004) Evolutionary history and higher order classification of AAA+ ATPases. J Struct Biol 146:11-31

Jarosch E, Lenk U, Sommer T (2003) Endoplasmic reticulum-associated protein degradation. Int Rev Cytol 223:39-81

Kang PJ, Ostermann J, Shilling J, Neupert W, Craig EA, Pfanner N (1990) Requirement for hsp70 in the mitochondrial matrix for translocation and folding of precursor proteins. Nature 348:137-143

Kang SG, Ortega J, Singh SK, Wang N, Huang NN, Steven AC, Maurizi MR (2002) Functional proteolytic complexes of the human mitochondrial ATP-dependent protease, hClpXP. J Biol Chem 277:21095-21102

Karata K, Inagawa T, Wilkinson AJ, Tatsuta T, Ogura T (1999) Dissecting the role of a conserved motif (the second region of homology) in the AAA family of ATPases. Site-directed mutagenesis of the ATP-dependent protease FtsH. J Biol Chem 274:26225-26232

Karata K, Verma CS, Wilkinson AJ, Ogura T (2001) Probing the mechanism of ATP hydrolysis and substrate translocation in the AAA protease FtsH by modelling and mutagenesis. Mol Microbiol 39:890-903

Käser M, Kambacheld M, Kisters-Woike B, Langer T (2003) Oma1, a novel membrane-bound metallopeptidase in mitochondria with activities overlapping with the *m*-AAA protease. J Biol Chem 278:46414-46423

Kihara A, Akiyama Y, Ito K (1996) A protease complex in the *Escherichia coli* plasma membrane: HflKC (HflA) forms a complex with FtsH (HflB), regulating its proteolytic activity against SecY. EMBO J 15:6122-6131

Klanner C, Prokisch H, Langer T (2001) MAP-1 and IAP-1, two novel AAA proteases with catalytic sites on opposite membrane surfaces in the mitochondrial inner membrane of *Neurospora crassa*. Mol Biol Cell 12:2858-2869

Kolodziejczak M, Kolaczkowska A, Szczesny B, Urantowka A, Knorpp C, Kieleczawa J, Janska H (2002) A higher plant mitochondrial homologue of the yeast *m*-AAA protease. Molecular cloning, localization, and putative function. J Biol Chem 277:43792-43798

Korbel D, Wurth S, Kaser M, Langer T (2004) Membrane protein turnover by the *m*-AAA protease in mitochondria depends on the transmembrane domains of its subunits. EMBO Rep 5:698-703

Kozany C, Mokranjac D, Sichting M, Neupert W, Hell K (2004) The J domain-related co-chaperone Tim16 is a constituent of the mitochondrial TIM23 preprotein translocase. Nat Struct Mol Biol 11:234-241

Kremmidiotis G, Gardner AE, Settasatian C, Savoia A, Sutherland GR, Callen DF (2001) Molecular and functional analyses of the human and mouse genes encoding AFG3L1, a mitochondrial metalloprotease homologous to the human spastic paraplegia protein. Genomics 76:58-65

Krzewska J, Langer T, Liberek K (2001) Mitochondrial Hsp78, a member of the Clp/Hsp100 family in *Saccharomyces cerevisiae*, cooperates with Hsp70 in protein refolding. FEBS Lett 489:92-96

Kutejová E, Durcová G, Surovková E, Kuzela S (1993) Yeast mitochondrial ATP-dependent protease: purification and comparison with the homologous rat enzyme and the bacterial ATP-dependent protease La. FEBS Lett 329:47-50

Kuzmin EV, Karpova OV, Elthon TE, Newton KJ (2004) Mitochondrial respiratory deficiencies signal up-regulation of genes for heat shock proteins. J Biol Chem 279:20672-20677

Laloraya S, Dekker PJT, Voos W, Craig EA, Pfanner N (1995) Mitochondrial GrpE modulates the function of matrix Hsp70 in translocation and maturation of preproteins. Mol Cell Biol 15:7098-7105

Laloraya S, Gambill BD, Craig E (1994) A role for a eukaryotic GrpE-related protein, Mge1p, in protein translocation. Proc Natl Acad Sci USA 91:6481-6485

Langer T (2000) AAA proteases - cellular machines for degrading membrane proteins. Trends Biochem Sci 25:207-256

Lemaire C, Hamel P, Velours J, Dujardin G (2000) Absence of the mitochondrial AAA protease Yme1p restores F_O-ATPase subunit accumulation in an *oxa1* deletion mutant of *Saccharomyces cerevisiae*. J Biol Chem 275:23471-23475

Lenzen CU, Steinmann D, Whiteheart SW, Weis WI (1998) Crystal structure of the hexamerization domain of N-ethylmaleimide-sensitive fusion protein. Cell 94:525-536

Leonhard K, Guiard B, Pellechia G, Tzagoloff A, Neupert W, Langer T (2000) Membrane protein degradation by AAA proteases in mitochondria: extraction of substrates from either membrane surface. Mol Cell 5:629-638

Leonhard K, Herrmann JM, Stuart RA, Mannhaupt G, Neupert W, Langer T (1996) AAA proteases with catalytic sites on opposite membrane surfaces comprise a proteolytic system for the ATP-dependent degradation of inner membrane proteins in mitochondria. EMBO J 15:4218-4229

Leonhard K, Stiegler A, Neupert W, Langer T (1999) Chaperone-like activity of the AAA domain of the yeast Yme1 AAA protease. Nature 398:348-351

Li M, Baumeister P, Roy B, Phan T, Foti D, Luo S, Lee AS (2000) ATF6 as a transcription activator of the endoplasmic reticulum stress element: thapsigargin stress-induced changes and synergistic interactions with NF-Y and YY1. Mol Cell Biol 20:5096-5106

Li Y, Dudek J, Guiard B, Pfanner N, Rehling P, Voos W (2004) The presequence translocase-associated protein import motor of mitochondria. Pam16 functions in an antagonistic manner to Pam18. J Biol Chem 279:38047-38054

Liu T, Lu B, Lee I, Ondrovicova G, Kutejova E, Suzuki CK (2004) DNA and RNA binding by the mitochondrial lon protease is regulated by nucleotide and protein substrate. J Biol Chem 279:13902-13910

Lu B, Liu T, Crosby JA, Thomas-Wohlever J, Lee I, Suzuki CK (2003) The ATP-dependent Lon protease of *Mus musculus* is a DNA-binding protein that is functionally conserved between yeast and mammals. Gene 306:45-55

Lupas AN, Martin J (2002) AAA proteins. Curr Opin Struct Biol 12:746-753

Manning-Krieg UC, Scherer PE, Schatz G (1991) Sequential action of mitochondrial chaperones in protein import into the matrix. EMBO J 10:3273-3280

Martin J (1997) Molecular chaperones and mitochondrial protein folding. J Bioenerg Biomembr 29:35-43

Martinus RD, Garth GP, Webster TL, Cartwright P, Naylor DJ, Hoj PB, Hoogenraad NJ (1996) Selective induction of mitochondrial chaperones in response to loss of the mitochondrial genome. Eur J Biochem 240:98-103

Mokranjac D, Sichting M, Neupert W, Hell K (2003) Tim14, a novel key component of the import motor of the TIM23 protein translocase of mitochondria. EMBO J 22:4945-4956

Mootha VK, Bunkenborg J, Olsen JV, Hjerrild M, Wisniewski JR, Stahl E, Bolouri MS, Ray HN, Sihag S, Kamal M, Patterson N, Lander ES, Mann M (2003) Integrated analysis of protein composition, tissue diversity, and gene regulation in mouse mitochondria. Cell 115:629-640

Nakai M, Kato Y, Ikeda E, Toh-e A, Endo T (1994) Yge1p, a eukaryotic Grp-E homolog, is localized in the mitochondrial matrix and interacts with mitochondrial Hsp70. Biochem Biophys Res Commun 200:435-442

Nakai T, Mera Y, Yasuhara T, Ohashi A (1994) Divalent metal ion-dependent mitochondrial degradation of unassembled subunits 2 and 3 of cytochrome c oxidase. J Biochem (Tokyo) 116:752-758

Nakai T, Yasuhara T, Fujiki Y, Ohashi A (1995) Multiple genes, including a member of the AAA family, are essential for the degradation of unassembled subunit 2 of cytochrome *c* oxidase in yeast mitochondria. Mol Cell Biol 15:4441-4452

Neupert W, Brunner M (2002) The protein import motor of mitochondria. Nature Rev Mol Cell Biol 3:555-565

Neuwald AF, Aravind L, Spouge JL, Koonin EV (1999) AAA+: A class of chaperone-like ATPases associated with the assembly, operation, and disassembly of protein complexes. Genome Res 9:27-43

Nijtmans LGJ, Artal Sanz M, Grivell LA, Coates PJ (2002) The mitochondrial PHB complex: roles in mitochondrial respiratory complex assembly, ageing and degenerative disease. Cell Mol Life Sci 59:143-155

Nijtmans LGJ, de Jong L, Sanz MA, Coates PJ, Berden JA, Back JW, Muijsers AO, Van der Speck H, Grivell LA (2000) Prohibitins act as a membrane-bound chaperone for the stabilization of mitochondrial proteins. EMBO J 19:2444-2451

Niwa H, Tsuchiya D, Makyio H, Yoshida M, Morikawa K (2002) Hexameric ring structure of the ATPase domain of the membrane-integrated metalloprotease FtsH from *Thermus thermophilus* HB8. Structure (Camb) 10:1415-1423

Ogura T, Wilkinson AJ (2001) AAA+ superfamily of ATPases: common structure-diverse function. Genes Cells 6:575-597

Ohlmeier S, Kastaniotis AJ, Hiltunen JK, Bergmann U (2004) The yeast mitochondrial proteome, a study of fermentative and respiratory growth. J Biol Chem 279:3956-3979

Pajic A, Tauer R, Feldmann H, Neupert W, Langer T (1994) Yta10p is required for the ATP-dependent degradation of polypeptides in the inner membrane of mitochondria. FEBS Lett 353:201-206

Paschen SA, Neupert W (2001) Protein import into mitochondria. IUBMB Life 52:101-112

Paul MF, Tzagoloff A (1995) Mutations in *RCA1* and *AFG3* inhibit F_1-ATPase assembly in *Saccharomyces cerevisiae.* FEBS Lett 373:66-70

Pearce DA, Sherman F (1995) Degradation of cytochrome oxidase subunits in mutants of yeast lacking cytochrome *c* and suppression of the degradation by mutation of *yme1.* J Biol Chem 270:1-4

Peltier JB, Ripoll DR, Friso G, Rudella A, Cai Y, Ytterberg J, Giacomelli L, Pillardy J, van Wijk KJ (2004) Clp protease complexes from photosynthetic and non-photosynthetic plastids and mitochondria of plants, their predicted three-dimensional structures, and functional implications. J Biol Chem 279:4768-4781

Pickart CM, Cohen RE (2004) Proteasomes and their kin: proteases in the machine age. Nat Rev Mol Cell Biol 5:177-187

Prokisch H, Scharfe C, Camp DG, 2nd, Xiao W, David L, Andreoli C, Monroe ME, Moore RJ, Gritsenko MA, Kozany C, Hixson KK, Mottaz HM, Zischka H, Ueffing M, Herman ZS, Davis RW, Meitinger T, Oefner PJ, Smith RD, Steinmetz LM (2004) Integrative analysis of the mitochondrial proteome in yeast. PLoS Biol 2:795-804

Rawlings ND, Barrett AJ (1995) Evolutionary families of metallopeptidases. Methods Enzymol 248:183-228

Rehling P, Brandner K, Pfanner N (2004) Mitochondrial import and the twin-pore translocase. Nat Rev Mol Cell Biol 5:519-530

Reichert AS, Neupert W (2004) Mitochondriomics or what makes us breathe. Trends Genet 20:555-562

Rep M, Nooy J, Guélin E, Grivell LA (1996) Three genes for mitochondrial proteins suppress null-mutations in both *AFG3* and *RCA1* when overexpressed. Curr Genet 30:206-211

Rodriguez-Enriquez S, He L, Lemasters JJ (2004) Role of mitochondrial permeability transition pores in mitochondrial autophagy. Int J Biochem Cell Biol 36:2463-2472

Rottgers K, Zufall N, Guiard B, Voos W (2002) The ClpB homolog Hsp78 is required for the efficient degradation of proteins in the mitochondrial matrix. J Biol Chem 277:45829-45837

Rowley N, Prip BC, Westermann B, Brown C, Schwarz E, Barrell B, Neupert W (1994) Mdj1p, a novel chaperone of the DnaJ family, is involved in mitochondrial biogenesis and protein folding. Cell 77:249-259

Russel SM, Burgess RJ, Mayer RJ (1980) Protein degradation in rat liver during post-natal development. Biochem J 192:321-330

Saikawa N, Akiyama Y, Ito K (2004) FtsH exists as an exceptionally large complex containing HflKC in the plasma membrane of *Escherichia coli.* J Struct Biol 146:123-129

Sauer RT, Bolon DN, Burton BM, Burton RE, Flynn JM, Grant RA, Hersch GL, Joshi SA, Kenniston JA, Levchenko I, Neher SB, Oakes ES, Siddiqui SM, Wah DA, Baker TA (2004) Sculpting the proteome with AAA(+) proteases and disassembly machines. Cell 119:9-18

Savel′ev AS, Novikova LA, Kovaleva IE, Luzikov VN, Neupert W, Langer T (1998) ATP-dependent proteolysis in mitochondria: *m*-AAA protease and PIM1 protease exert overlapping substrate specificities and cooperate with the mtHsp70-system. J Biol Chem 273:20596-20602

Schmitt M, Neupert W, Langer T (1995) Hsp78, a Clp homologue within mitochondria, can substitute for chaperone functions of mt-hsp70. EMBO J 14:3434-3444

Schmitt M, Neupert W, Langer T (1996) The molecular chaperone Hsp78 confers compartment-specific thermotolerance to mitochondria. J Cell Biol 134:1375-1386

Schneider HC, Berthold J, Bauer MF, Dietmeier K, Guiard B, Brunner M, Neupert W (1994) Mitochondrial Hsp70/MIM44 complex facilitates protein import. Nature 371:768-774

Sesaki H, Southard SM, Hobbs AE, Jensen RE (2003a) Cells lacking Pcp1p/Ugo2p, a rhomboid-like protease required for Mgm1p processing, lose mtDNA and mitochondrial structure in a Dnm1p-dependent manner, but remain competent for mitochondrial fusion. Biochem Biophys Res Commun 308:276-283

Sesaki H, Southard SM, Yaffe MP, Jensen RE (2003b) Mgm1p, a dynamin-related GTPase, is essential for fusion of the mitochondrial outer membrane. Mol Biol Cell 14:2342-2356

Shah ZH, Hakkaart GAJ, Arku B, DeJong L, Van der Speck H, Grivell L, Jacobs HT (2000) The human homologue of the yeast mitochondrial AAA metalloprotease Yme1p complements a yeast *yme1* disruptant. FEBS Lett 478:267-270

Shah ZH, Migliosi V, Miller SC, Wang A, Friedman TB, Jacobs HT (1998) Chromosomal locations of three human nuclear genes (RPSM12, TUFM, and AFG3L1) specifying putative components of the mitochondrial gene expression apparatus. Genomics 48:384-388

Shaw JM, Nunnari J (2002) Mitochondrial dynamics and division in budding yeast. Trends Cell Biol 12:178-184

Sickmann A, Reinders J, Wagner Y, Joppich C, Zahedi R, Meyer HE, Schonfisch B, Perschil I, Chacinska A, Guiard B, Rehling P, Pfanner N, Meisinger C (2003) The proteome of *Saccharomyces cerevisiae* mitochondria. Proc Natl Acad Sci USA 100:13207-13212

Stahlberg H, Kutejova E, Suda K, Wolpensinger B, Lustig A, Schatz G, Engel A, Suzuki CK (1999) Mitochondrial Lon of *Saccharomyces cerevisiae* is a ring-shaped protease with seven flexible subunits. Proc Natl Acad Sci USA 96:6787-6790

Steglich G, Neupert W, Langer T (1999) Prohibitins regulate membrane protein degradation by the *m*-AAA protease in mitochondria. Mol Cell Biol 19:3435-3442

Suzuki CK, Suda K, Wang N, Schatz G (1994) Requirement for the yeast gene LON in intramitochondrial proteolysis and maintenance of respiration. Science 264:273-276

Takeshige K, Baba M, Tsuboi S, Noda T, Ohsumi Y (1992) Autophagy in yeast demonstrated with proteinase-deficient mutants and conditions for induction. J Cell Biol 119:301-311

Tatsuta T, Model K, Langer T (2005) Formation of membrane-bound ring complexes by prohibitins in mitochondria. Mol Biol Cell 16:248-259

Tauer R, Mannhaupt G, Schnall R, Pajic A, Langer T, Feldmann H (1994) Yta10p, a member of a novel ATPase family in yeast, is essential for mitochondrial function. FEBS Lett 353:197-200

Taylor SW, Fahy E, Zhang B, Glenn GM, Warnock DE, Wiley S, Murphy AN, Gaucher SP, Capaldi RA, Gibson BW, Ghosh SS (2003) Characterization of the human heart mitochondrial proteome. Nat Biotechnol 21:281-286

Teichmann U, van Dyck L, Guiard B, Fischer H, Glockshuber R, Neupert W, Langer T (1996) Substitution of PIM1 protease in mitochondria by *Escherichia coli* Lon protease. J Biol Chem 271:10137-10142

Thorsness PE, Fox TD (1993) Nuclear mutations in *Saccharomyces cerevisiae* that affect the escape of DNA from mitochondria to the nucleus. Genetics 134:21-28

Thorsness PE, White KH, Fox TD (1993) Inactivation of *YME1*, a member of the ftsH-SEC18-PAS1-CDC48 family of putative ATPase-encoding genes, causes increased escape of DNA from mitochondria in *Saccharomyces cerevisiae.* Mol Cell Biol 13:5418-5426

Traven A, Wong JM, Xu D, Sopta M, Ingles CJ (2001) Interorganellar communication. Altered nuclear gene expression profiles in a yeast mitochondrial dna mutant. J Biol Chem 276:4020-4027

Truscott K, Brandner K, Pfanner N (2003) Mechanisms of protein import into mitochondria. Curr Biol 13:R326-337

Tzagoloff A, Yue J, Jang J, Paul MF (1994) A new member of a family of ATPases is essential for assembly of mitochondrial respiratory chain and ATP synthetase complexes in *Saccharomyces cerevisiae.* J Biol Chem 269:26144-26151

Van Dijl JM, Kutejova E, Suda K, Perecko D, Schatz G, Suzuki CK (1998) The ATPase and protease domains of yeast mitochondrial Lon: roles in proteolysis and respiration-dependent growth. Proc Natl Acad Sci USA 95:10584-10589

Van Dyck L, Dembowski M, Neupert W, Langer T (1998) Mcx1p, a ClpX homologue in mitochondria of *Saccharomyces cerevisiae.* FEBS Lett 438:250-254

Van Dyck L, Langer T (1999) ATP-dependent proteases controlling mitochondrial function in the yeast *Saccharomyces cerevisiae.* Cell Mol Life Sci 55:825-842

Van Dyck L, Neupert W, Langer T (1998) The ATP-dependent PIM1 protease is required for the expression of intron-containing genes in mitochondria. Genes Dev 12:1515-1524

Van Dyck L, Pearce DA, Sherman F (1994) *PIM1* encodes a mitochondrial ATP-dependent protease that is required for mitochondrial function in the yeast *Saccharomyces cerevisiae.* J Biol Chem 269:238-242

Voos W, Röttgers K (2002) Molecular chaperones as essential mediators of mitochondrial biogenesis. Biochim Biophys Acta 1592:51-62

Wagner I, Arlt H, van Dyck L, Langer T, Neupert W (1994) Molecular chaperones cooperate with PIM1 protease in the degradation of misfolded proteins in mitochondria. EMBO J 13:5135-5145

Wagner I, Van Dyck L, Savel'ev A, Neupert W, Langer T (1997) Autocatalytic processing of the ATP-dependent PIM1 protease: Crucial function of a pro-region for sorting to mitochondria. EMBO J 16:7317-7325

Wang N, Gottesman S, Willingham MC, Gottesman S, Maurizi MR (1993) A human mitochondrial ATP-dependent protease that is highly homologous to bacterial Lon protease. Proc Natl Acad Sci USA 90:11247-11251

Watabe S, Kimura T (1985) ATP-dependent protease in bovine adrenal cortex. J Biol Chem 260:5511-5517

Weber ER, Hanekamp T, Thorsness PE (1996) Biochemical and functional analysis of the *YME1* gene product, an ATP and zinc-dependent mitochondrial protease from *S. cerevisiae.* Mol Biol Cell 7:307-317

Westermann B, Prip-Buus C, Neupert W, Schwarz E (1995) The role of the GrpE homologue, Mge1p, in mediating protein import and protein folding in mitochondria. EMBO J 14:3452-3460

Ye Y, Meyer HH, Rapoport TA (2001) The AAA ATPase Cdc48/p97 and its partners transport proteins from the ER into the cytosol. Nature 414:652-656

Yoneda T, Benedetti C, Urano F, Clark SG, Harding HP, Ron D (2004) Compartment-specific perturbation of protein handling activates genes encoding mitochondrial chaperones. J Cell Sci 117:4055-4066

Yoshida H, Okada T, Haze K, Yanagi H, Yura T, Negishi M, Mori K (2000) ATF6 activated by proteolysis binds in the presence of NF-Y (CBF) directly to the cis-acting element responsible for the mammalian unfolded protein response. Mol Cell Biol 20:6755-6767

Yu RC, Hanson PI, Jahn R, Brünger AT (1998) Structure of the ATP-dependent oligomerization domain of N-ethylmaleimide sensitive factor complexed with ATP. Nat Struct Biol 5:803-811

Zehnbauer BA, Foley EC, Henderson GW, Markovitz A (1981) Identification and purification of the Lon+ (capR+) gene product, a DNA-binding protein. Proc Natl Acad Sci USA 78:2043-2047

Zhang K, Kaufman RJ (2004) Signaling the unfolded protein response from the endoplasmic reticulum. J Biol Chem 279:25935-25938

Zhang X, Shaw A, Bates PA, Newman RH, Gowen B, Orlova E, Gorman MA, Kondo H, Dokurno P, Lally J, Leonhard G, Meyer H, Van Heel M, Freemont PS (2000) Structure of the AAA ATPase p97. Mol Cell 6:1473-1484

Zhao Q, Wang J, Levichkin IV, Stasinopoulos S, Ryan MT, Hoogenraad NJ (2002) A mitochondrial specific stress response in mammalian cells. EMBO J 21:4411-4419

Graef, Martin
Institut für Genetik, Universität Köln, Zülpicher Str. 47, 50674 Köln, Germany

Kisters-Woike, Brigitte
Institut für Genetik, Universität Köln, Zülpicher Str. 47, 50674 Köln, Germany

Langer, Thomas
Institut für Genetik, Universität Köln, Zülpicher Str. 47, 50674 Köln, Germany
Thomas.Langer@uni-koeln.de.

Nolden, Mark
Institut für Genetik, Universität Köln, Zülpicher Str. 47, 50674 Köln, Germany

Chaperone proteins and peroxisomal protein import

Wim de Jonge, Henk F. Tabak, and Ineke Braakman

Abstract

Peroxisomes are ubiquitous organelles present in most eukaryotic cells. Their role in cellular metabolism is diverse among species. An array of genes involved in the formation and maintenance of peroxisomes has been discovered, and can be categorised into genes important for protein import into the peroxisome and genes involved in the maintenance of the organelles' size and abundance. Thorough cell biological and biochemical studies revealed great detail about the process of peroxisomal protein import. Although involvement of several classes of molecular chaperone proteins in peroxisomal protein import has been demonstrated, details regarding the mechanistic aspects of chaperone involvement in this process are not known yet. This review aims to discuss peroxisomal maintenance, with the emphasis on protein import. A general overview of chaperone proteins and their role in protein import processes will be used as context to discuss the - possible - roles of chaperone proteins in peroxisomal protein import.

1 Introduction

When a protein is synthesized, it faces many difficulties on the pathway to its native, fully folded structure and proper subcellular localisation. Although the roadmap of this pathway is determined by the linear amino-acid sequence of a protein, without help, its maturation will not occur at a high efficiency in a living cell. At this stage, chaperone proteins come into play. Analogous to their human equivalent, they protect the "infant" protein from unwanted inter- and intramolecular interactions in a crowded cellular milieu, ensuring efficient maturation. Chaperones come in various flavours (Hartl and Hayer-Hartl 2002) and similar chaperone systems are found in all kingdoms (Fig. 1).

It is, therefore, easy to envisage that chaperone proteins play an essential role in the living cell. A number of biomolecular processes are supported by chaperone proteins, such as folding and oligomerisation of newly synthesized proteins, protection of proteins from denaturation by environmental stress, protein import into organelles, targeting of proteins for degradation, modulation of protein-protein interactions, and regulation of translation, transcription, and signal transduction pathways.

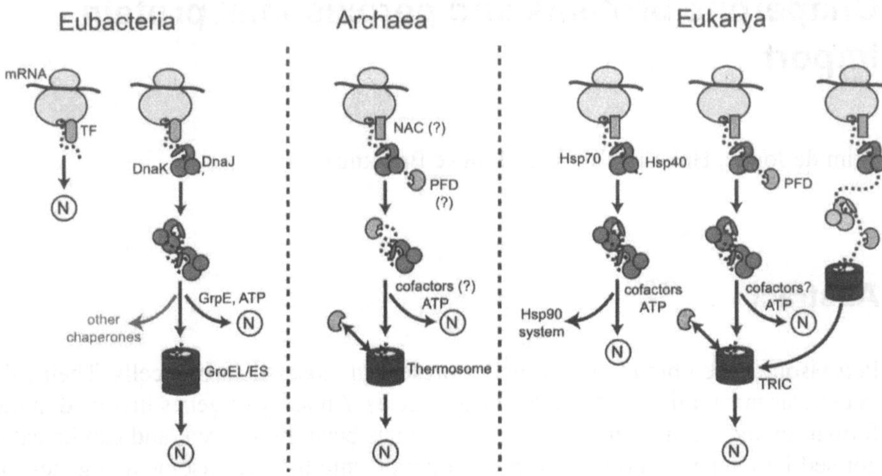

Fig. 1. Variety in chaperone systems. Newly synthesised proteins exit the ribosome, where they first meet ribosome associated chaperones such as trigger factor (TF) in Eubacteria, and nascent chain associated complex (NAC) in Eukarya and possibly in Archaea. Hsp70/40 (DnaJ/K in Eubacteria and Archea) are involved in folding of a part of the pool of newly synthesised proteins, and hand over some proteins to either GroEL/ES in Eubacteria, or similar chaperone systems in Archaea (thermosome) and Eukarya (TRIC). In the latter case, prefoldin serves as a 'targeting factor' for newly synthesised and folding proteins that are client proteins for the TRIC chaperone. Finally, in Eukarya a specific subset of proteins depends on the Hsp90 system for maintaining a proper conformation (Figure adapted and modified after Hartl and Hayer-Hartl 2002).

Chaperones are located in the cytosol and the lumenal cavities of various organelles with the possible exception of peroxisomes. Proteomics studies did not reveal chaperones, except a Lon homolog in mammalian peroxisomes (Kikuchi et al. 2004). Here the chaperones already contribute in the cytosol, which is made possible by the fact that the peroxisomal protein import machinery can accommodate proteins that have attained some conformation. The focus of attention in this review is particularly devoted to the role of chaperones in the formation and maintenance of these remarkable organelles.

2 Peroxisomes

2.1 Formation, maintenance, and function

Peroxisomes are single-membrane bounded organelles present in all eukaryotic kingdoms, with a variety in appearance and metabolic function. Originally, the organelle was defined as containing hydrogen peroxide (H_2O_2) producing oxidases, and the H_2O_2 detoxifying enzyme catalase, hence the name '*peroxi*some'. Examples of peroxisomal metabolic processes are β-oxidation of fatty acids, parts of the

Table 1. Peroxins and their main characteristics Abbreviations used: (species) Mamm - Mammals, Y - Most yeast species, Sc - *Saccharomyces cerevisiae*, Yl - *Yarrowia lipolytica*, put - putative. (localisation) lum - peroxisomal lumen (matrix), im - integral membrane, pm - peripheral membrane, cyt - cytosol, ? - unresolved or conflicting literature reports.

Peroxin	Remarks	Localisation	Species
Pex1p	AAA ATPase	cyt, pm	Y, Mamm
Pex2p	RING protein, translocase?	im	Y, Mamm
Pex3p	Membrane protein stability	im	Y, Mamm
Pex4p	Ubiquitin conjugating enzyme	pm	Y
Pex5p	PTS1 receptor	cyt, pm, im, lum (?)	Y, Mamm
Pex6p	AAA ATPase	cyt, pm	Y, Mamm
Pex7p	PTS2 receptor	cyt, pm, lum (?)	Y, Mamm
Pex8p	PTS1 & PTS2	pm, lum	Y
Pex9p	Unknown function	im	Yl
Pex10p	RING protein, translocase?	im	Y. Mamm
Pex11p	Proliferation	im, pm	Y, Mamm
Pex12p	RING protein, translocase?	im	Y, Mamm
Pex13p	SH3 domain, docking complex	im	Y, Mamm
Pex14p	Docking complex	im, pm	Y, Mamm
Pex15p	Phosphorylated	im	Sc
Pex16p	Membrane protein import	im	Yl, Mamm
Pex17p	Docking complex	im	Sc
Pex18p	Pex7p co-receptor	cyt	Sc
Pex19p	Membrane protein import	pm, cyt	Sc
Pex20p	Thiolase import	cyt, pm	Yl
Pex21p	Pex7p co-receptor	cyt	Sc
Pex22p	Pex4p anchoring	im	Sc
Pex23p	Protein import	im	Yl, Sc (put)
Pex24p	Protein import	im	Yl, Sc (put)
Pex25p	Protein import, proliferation	pm	Sc
Pex26p	Protein import	im	Mamm
Pex27p	Protein import, proliferation	pm	Sc
Pex28p	Proliferation	im	Sc
Pex29p	Proliferation	im	Sc
Pex30p	Proliferation	im	Sc
Pex31p	Proliferation	im	Sc
Pex32p	Proliferation	im	Sc
Djp1p	Hsp70 co-chaperone, protein import	cyt	Sc

glyoxylate cycle or the glycolytic pathway, photorespiration, methanol catabolism, and penicillin synthesis (Hettema 1998). Although the metabolic functions of peroxisomes are diverse among species, the basic mechanism to maintain the organelle in the cell is conserved throughout evolution. This became apparent upon characterisation of the first serendipitously identified peroxisomal targeting signal (PTS1) of luciferase, an enzyme located in the lantern organ of fireflies, which

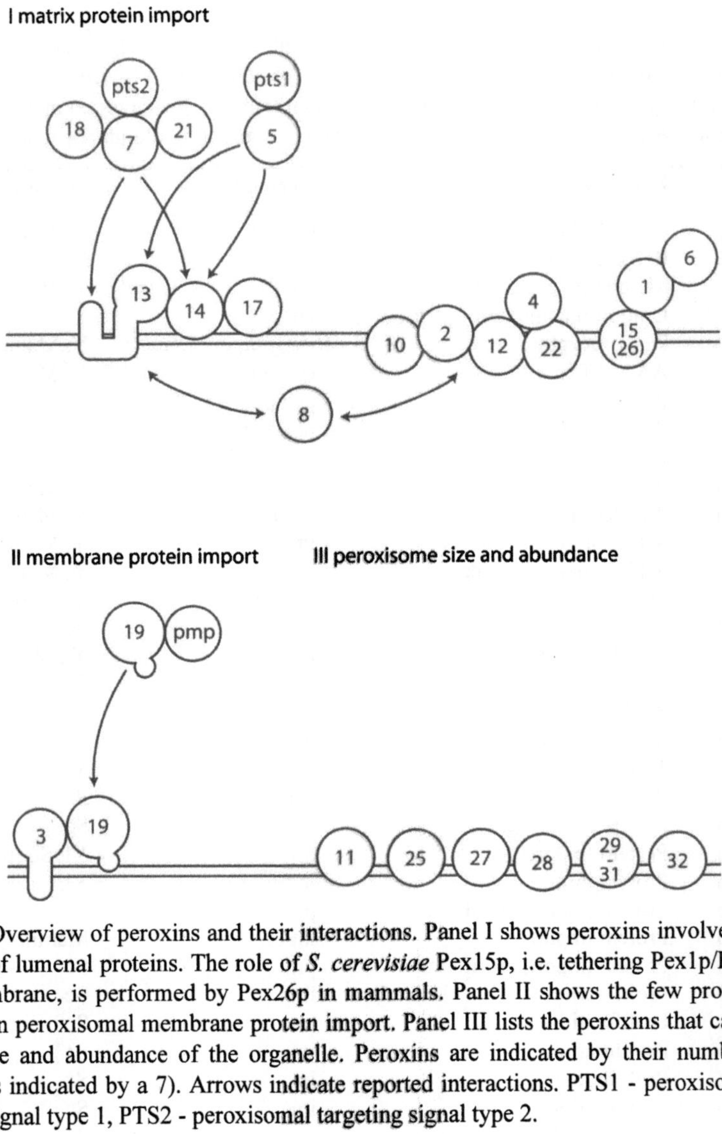

Fig. 2. Overview of peroxins and their interactions. Panel I shows peroxins involved in the import of lumenal proteins. The role of *S. cerevisiae* Pex15p, i.e. tethering Pex1p/Pex6p to the membrane, is performed by Pex26p in mammals. Panel II shows the few proteins involved in peroxisomal membrane protein import. Panel III lists the peroxins that can influence size and abundance of the organelle. Peroxins are indicated by their number (e.g. Pex7p is indicated by a 7). Arrows indicate reported interactions. PTS1 - peroxisomal targeting signal type 1, PTS2 - peroxisomal targeting signal type 2.

was shown to be able to direct a reporter protein to peroxisomes of yeast, mammalian, plant, and insect peroxisomes (Gould et al. 1989, 1990).

Since these initial studies, a series of genetic screens resulted in the characterisation some thirty genes involved in the maintenance of peroxisomes, collectively called peroxins (Distel et al. 1996). A summary of peroxins and their characteristics can be found in Table 1 and Figure 2.

From the early steps in import to the translocation of proteins into the peroxisome, several steps must occur, separated in space and time. In each of these steps, different complexes form and dissociate. In the following paragraphs, an overview of the general properties of peroxisomal protein import is presented. Next, some of the discrete steps in the import process are discussed.

2.2 General properties of protein import

What are the general requirements for peroxisomal protein import? When ATP is depleted import cannot occur in *in vivo*, and in semi *in vitro* experimental systems (Bellion and Goodman 1987; Behari and Baker 1993; Soto et al. 1993; Wendland and Subramani 1993). Because Hsp70, Hsp90, Pex1p, Pex6p, and ANT1 (PMP34) all are ATPases involved in maintenance of, or import into, peroxisomes this is not a surprising finding.

Reports about the intraperoxisomal pH give conflicting results, the intraperoxisomal pH varying from alkaline in human fibroblasts (Dansen et al. 2000), filamentous fungi (van der Lende et al. 2002), and yeast (Van Roermund et al. 2004) to neutral in CHO cells and fibroblasts (Jankowski et al. 2001). It is unclear what causes these differences in pH measurements. Different approaches were used, employing fluorophores coupled to a PTS1, and pH sensitive forms of GFP fused to a PTS1.

In mitochondria, the pH gradient across the outer membrane is essential for the import of mitochondrial precursor proteins. Ionophores did inhibit peroxisomal protein import *in vivo*, suggesting the involvement of a proton gradient in peroxisomal protein import (Bellion and Goodman 1987). Ionophores, however, also diminish the proton gradient across the mitochondrial outer membrane, and thereby, block ATP generation. Therefore, as the authors suggested, the effects observed *in vivo* may be indirect, via ATP depletion. This result is supported by the observation that, in a mammalian peroxisomal import assay using semi-permeabilised cells, ionophores did not inhibit import (Wendland and Subramani 1993). In human fibroblasts, in the absence of import of a subset of peroxisomal proteins, the pH gradient is diminished. In this situation, a peptide containing a peroxisomal targeting signal coupled to a fluorophore could still be imported into peroxisomes (Dansen et al. 2000). Although it is as yet unclear whether a pH gradient across the peroxisomal membrane exists, it is unlikely that it plays a role in the import of peroxisomal proteins.

2.3 Proteins involved in protein import

2.3.1 PTS1 protein import

Genes encoding lumenal and membrane peroxisomal proteins are encoded by the nuclear DNA, implying the requirement for a transport machinery that can specifically recognise these proteins and direct them to the peroxisome. At least two different targeting signals have been identified. The canonical peroxisomal targeting signal type 1 (PTS1) consists of the C-terminal tripeptide Serine-Lysine-Leucine or a conserved variant thereof. PTS1 is able to direct reporter proteins to the peroxisomal compartment of many species, but variants of the signal are not always functional across species (Motley et al. 1995; Elgersma et al. 1996b). Detailed mutagenesis studies on the tolerance of deviations from the generic -SKL sequence revealed a role for amino acids upstream of less conserved variants of PTS1 (Elgersma et al. 1996b; Lametschwandtner et al. 1998; Neuberger et al. 2003).

The PTS1 signal can be recognised by Pex5p, a soluble protein that is mainly present in the cytosol. *In vitro* binding studies show convincingly that Pex5p can bind PTS1 independently of other proteins (Fransen et al. 1995; Terlecky et al. 1995; Wiemer et al. 1995a; Klein et al. 2001). The Pex5p-PTS1 interaction appears to be the first specific step in peroxisomal protein import.

The domain responsible for PTS1 recognition is formed by two sets of three TPR motifs in the COOH-terminal half of Pex5p, connected by a TPR motif-like linker region. TPR motifs are formed by 34 amino acids long antiparallel pairs of alpha-helices, connected by loops, that form a superhelical structure. Mutational analysis of the TPR motifs, combined with homology modelling, revealed a role for residues in the intra-repeat loops of TPR2 and TPR3 in anchoring the lysine at position 2 of SKL, and for nearby residues in recognising the additional residues of the PTS1. In TPR4-7, a set of asparagine residues can be found that could be involved in interactions with the peptide backbone (Klein et al. 2001). These results were in agreement with the structure of the TPR region of human Pex5p, cocrystallized with a PTS1-peptide (Gatto et al. 2000). In the light of more recent results on the importance of residues upstream of PTS1, this is not the complete picture. Co-crystallisation of a PTS1 protein with Pex5p will explain how proteins with a low affinity PTS1 use accessory amino acid residues to ensure proper recognition by Pex5p.

2.3.2 PTS2 protein import

A less common targeting signal is the PTS2, defined as a nonapeptide close to or at the N-terminus with the consensus sequence R/K-L/V/I-X_5-H/Q-L/A (Subramani 1996). PTS2 is necessary and sufficient to direct a reporter protein to peroxisomes (Swinkels et al. 1991; Gietl et al. 1994; Tsukamoto et al. 1994). The few proteins possessing a PTS2 are transported to the peroxisome via the PTS2 receptor Pex7p, characterised by the presence of WD-40 repeats. Pex7p mediated PTS2 import is assisted by either Pex18p/Pex21p (*S. cerevisiae*), Pex20p (*Y. lipolytica*) or by an alternatively spliced form of Pex5p (mammals). For details, see Section 2.4. Mutational analysis of Pex7p was performed in two independent studies, yet in contrast to the PTS1 - Pex5p studies, detailed information on structural aspects of Pex7p – PTS2 recognition is not available yet.

Besides PTS1- and PTS2-mediated import, additional targeting signals have been reported. This concerns rare cases where proteins have found alternative ways to enter peroxisomes. (Interested readers are referred to Klein et al. 2002; Titorenko et al. 2002)

2.3.3 mPTS protein import

Initial research on defining the targeting signal for membrane proteins gave an incomplete picture (Dyer et al. 1996; Jones et al. 2001). Recently, more details about targeting signals for peroxisomal membrane proteins (PMPs) emerged (Rottensteiner et al. 2004). The core sequence was identified as a helical nonapeptide, containing hydrophobic and positively charged amino acids. Although this core sequence is necessary for proper targeting, at least one additional transmembrane region is required for specific insertion into the membrane.

Peroxisomal membrane protein (PMP) insertion depends on Pex3p and Pex19p; in the absence of either one of them, PMPs are mislocalised to the cytosol, and rapidly degraded (Hettema et al. 2000; Jones et al. 2004). Pex3p is an integral membrane protein, and Pex19p is predominantly cytosolic with a small fraction located at the peroxisomal membrane. In cells lacking Pex3 or Pex19p, peroxisomes are absent, in contrast to the peroxisomal 'ghosts' observed when matrix protein import is blocked. When Pex3p, respectively Pex19p, is re-introduced in these cells, peroxisomes arise again (Hettema et al. 2000; Sacksteder et al. 2000). Pex19p has been shown to transiently bind newly synthesized PMPs, and requires Pex3p for membrane localisation (Fang et al. 2004; Jones et al. 2004). The current view is that of Pex19p as a combined targeting receptor – specialised chaperone for PMPs, which uses Pex3p as a docking factor at the acceptor membrane.

2.4 Formation of peroxisomes

The formation of peroxisomes has been a matter of debate since decades, the two extremes being a division model, where peroxisomes multiply by growth and division, and a pathway where the peroxisomal membrane is derived from pre-existing membranes of the endoplasmic reticulum (ER) (Lazarow and Fujiki 1985).

Support for the ER-peroxisome connection was found in morphological studies employing electron microscopy, showing close proximity of peroxisomes to the ER (Novikoff and Novikoff 1972). More recently, direct connections between rough ER and peroxisome precursor structures were detected, shedding new light on the ER-peroxisome connection (Geuze et al. 2003; Tabak et al. 2003).

Attempts to implicate factors involved in classical secretory vesicle transport in the formation of peroxisomes failed (South et al. 2000, 2001; Voorn-Brouwer et al. 2001). In addition, time lapse fluorescence microscopy revealed that, in *S. cerevisiae*, peroxisomes divide and are transported to the daughter cell, dependent on Vps1p (a member of the dynamin family, and the motor protein Myo2p, respectively (Hoepfner et al. 2001). A similar role in peroxisomal fission was found for Dlp1p in mammalian cells (Koch et al. 2003; Li and Gould 2003).

Logically, a model emerges combining both ideas: peroxisomal membranes are derived from the ER via a non-classical secretion route, grow in size, and divide using a machinery dependent on dynamin-like proteins (Vps1p in yeast, Dlp1p in mammals).

2.5 Recycling receptor import model

2.5.1 The Pex5p shuttle

Discussion about functioning of Pex5p has mainly focused on localisation of the protein. Most groups report a predominantly cytosolic localisation for Pex5p, with a small fraction associated with peroxisomes (Van der Leij et al. 1993; Dodt et al. 1995; van der Klei et al. 1995; Wiemer et al. 1995b). The dual localization of Pex5p led to the proposal of the recycling receptor model: Pex5p guides peroxisomal proteins from the cytosol to the peroxisomal membrane, where import can occur. After delivery of the peroxisomal protein, Pex5p recycles to the cytosol, where it can pick up another peroxisomal protein.

ATP depletion and low temperature block peroxisomal protein import *in vivo*. Gould and co-workers found that, when studying import into peroxisomes of fibroblasts, these conditions led to a shift in Pex5p localisation from the cytosol to the peroxisome. Taking away these import-blocking conditions resulted in redistribution of Pex5p to the cytosol (Dodt and Gould 1996). Another condition that caused Pex5p location to the peroxisome is mutation of *PEX2* or *PEX12*.

2.5.2 The Pex5p extended shuttle

Pex5p was also found inside the peroxisome matrix, as shown by electron microscopy (Szilard et al. 1995; van der Klei et al. 1995). These results led to the notion of an extended recycling receptor model, where Pex5p traverses the membrane twice: once with PTS1-containing protein into the organelle, and the second time empty to return to the cytosol (Dodt and Gould 1996). This would be a unique situation since, thus far, only the nuclear pore complex has been shown to allow reversible transport of proteins involved in protein cargo movement.

One study aiming to support the extended recycling receptor model used a protease-recognition site, which is specifically cleaved by an intraperoxisomal protease, combined with an epitope tag fused to the NH_2-terminus of mammalian Pex5p (Dammai and Subramani 2001). Using antibodies that exclusively recognised the cleaved, exposed epitope tag, the authors showed that Pex5p is exposed to the inside of peroxisomes, and recycles back to the cytoplasm. The authors used biochemical criteria to show that Pex5p is present in the peroxisomal matrix, but these results were not conclusive. They did show that the NH_2-terminal region of Pex5p is exposed to the peroxisomal lumen during the import cycle. As an alternative explanation for this result, Pex5p could function similarly to the bacterial SecA protein. SecA "pushes" proteins through the SecYEG translocon at the cytoplasmic membrane of bacteria by a piston-like motion (Economou and Wickner 1994). Perhaps the NH_2-terminus of Pex5p functions as such a piston, explaining why it is at least partially exposed to the peroxisomal lumen. Results from elegant *in vitro* studies, showing the presence of Pex5p inserted into the peroxisomal membrane in different "stages" characterised by differential sensitivity to proteases, might reflect this piston-like motion in the translocation process of peroxisomal proteins (Gouveia et al. 2003).

2.5.3 The Pex7p shuttle

As Pex5p, Pex7p was reported to be present in the cytosol, at the peroxisomal membrane, and inside the peroxisome (Marzioch et al. 1994; Zhang and Lazarow 1994; Elgersma et al. 1998). Upon deletion of *FOX3*, the gene encoding 3-ketoacyl-CoA thiolase, the only PTS2 protein in yeast, Pex7p could not be detected in the peroxisomal fraction anymore, suggesting that binding of Fox3p enhances peroxisomal localisation of Pex7 (Marzioch et al. 1994). As expected, Pex7p and Fox3p could be co-immunoprecipitated from a yeast lysate, indicating that Pex7p and Fox3p interact (Rehling et al. 1996).

Pex7p was co-immunoprecipitated with peroxins that are involved in later steps of the import pathway: Pex14p, Pex17p, and Pex13p. Interestingly, deletion of any of these genes increases the pool of Pex7p-associated Fox3p, strongly suggesting the accumulation of Fox3p in a pre-import stage prior to docking (Girzalsky et al. 1999).

Pex18p and Pex21p are two other proteins involved in the PTS2 pathway. Because they are redundant in function, they were not identified in screens selecting for import defects, but instead by using a two-hybrid approach to identify binding partners of Pex7p. Upon deletion of both *PEX18* and *PEX21*, import of Fox3p is abolished, and Pex7p is not found in the peroxisomal fraction upon subcellular fractionation (Purdue et al. 1998). These results led the authors to suggest a chaperone-like role for Pex18p/Pex21p in guiding Pex7p to the peroxisomal membrane. Because loss of PTS2 protein also abolishes Pex7p association to the peroxisomal fraction, Pex18p and Pex21p, alternatively, may be involved in enhancement of the PTS2-Pex7p interaction, thereby, indirectly enhancing peroxisomal association of Pex7p. Pex18p and Pex21p are not found in higher eukaryotes, where an alternatively spliced, longer form of Pex5p (Pex5L) appears to take over their role (Dodt et al. 2001). In the absence of Pex5L, PTS2 containing proteins are not imported, and Pex5Lp interacts directly with Pex7p (Braverman et al. 1998; Matsumura et al. 2000).

The PTS2 pathway is not conserved throughout all eukaryotes. It is absent in *Caenorhabditis elegans*, and proteins that contain a PTS2 in other organisms have acquired a PTS1 in C. elegans (Motley et al. 2000). Another example of evolutionary divergence in the PTS2 pathway occurs in the yeast *Yarrowia lipolytica*. Here, peroxisomal thiolase is targeted to the peroxisome in the form of a heterotetramer containing two thiolase molecules, and two Pex20p molecules. Pex20p is only found in *Y. lipolytica*, and is the only protein that interacts with thiolase in this particular model organism (Titorenko et al. 1998). Pex20p was suggested to be the ortholog of *S. cerevisiae* Pex18p/Pex21p, and Pex5L in mammals (Einwachter et al. 2001). This is likely because Pex18p and Pex21p were duplicated in *S. cerevisiae* (Dietrich et al. 2004). In addition, *Neurospora crassa* Pex7p acts together with a *Y. lipolytica* Pex20p in PTS2 import (Sichting et al. 2003). A Pex7p homolog has not yet been discovered in *Y. lipolytica,* however, allowing a role of Pex20p as the principal import receptor for PTS2 proteins.

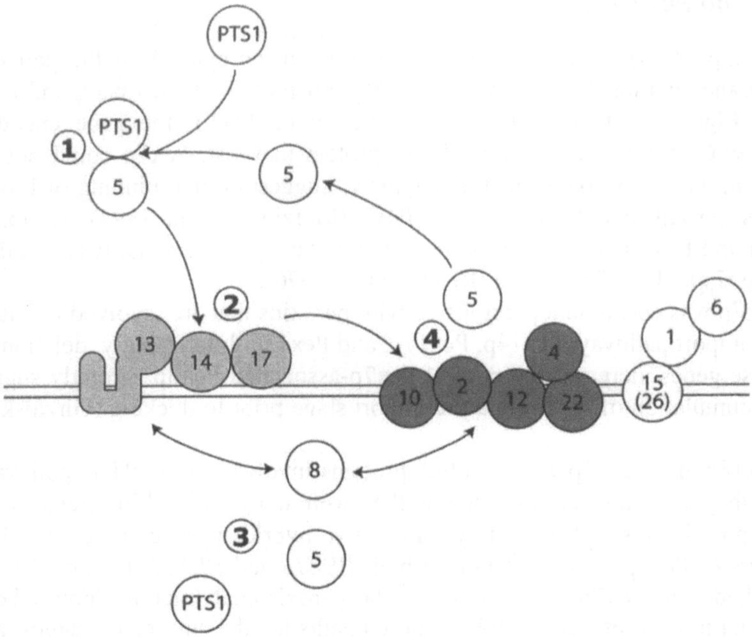

Fig. 3. The peroxisomal protein import cycle. 1. Binding of PTS1 containing cargo to the import receptor Pex5p. 2. Docking of the cargo-receptor complex onto a membrane protein complex formed by Pex13, Pex14p, and Pex17p. 3. Translocation of cargo and receptor across the peroxisomal membrane and subsequent retro-translocation of receptor. Alternatively, translocation of cargo, and 4. recycling of receptor to the cytosol for a new round of import. Arrows indicate movement and formation/dissociation of protein complexes as cargo and receptor progress through the import cycle. Import receptor and the complexes indicated in light grey and dark grey correspond to membrane subcomplexes I and II (See Section 2.6) whose interactions are orchestrated by the intraperoxisomal Pex8p.

2.6 Docking and translocation of peroxisomal proteins

After the initial recognition of the PTS-containing cargo proteins by the cognate receptor, the cargo-receptor complex can interact with proteins at the peroxisomal membrane, a stage in import referred to as docking. Subsequently, the PTS protein will enter the peroxisome, and the receptor will return to the cytosol. This requires two other functionalities: translocation of the cargo, and recycling of the import receptor.

Initial studies suggested that the "docking protein" for Pex5p is Pex13p, an integral peroxisomal membrane protein that contains an SH3 domain (Elgersma et al. 1996a; Erdmann and Blobel 1996; Gould et al. 1996). Later studies revealed that Pex14p is the principal protein on which receptor-cargo complexes dock, and where the PTS1 and PTS2 pathways converge: Pex14p can bind both Pex5p and Pex7p (Albertini et al. 1997). Several interaction studies revealed a complex network of protein-protein interactions, suggesting a cascade of interactions performing the aforementioned functionalities (Agne et al. 2003) (Fig. 3).

A recent investigation using purified components showed that Pex5p loaded with PTS1 peptide preferentially interacts with Pex14p, whereas the empty receptor preferentially interacts with Pex13p (Urquhart et al. 2000). This result suggests that when PTS-containing proteins are "handed over" to the translocation machinery, Pex5p possibly relocates to Pex13p. This study used isolated proteins, and might therefore not represent the events as they occur in protein complexes at the peroxisomal membrane. This caveat becomes more apparent when we consider the complexity of protein assemblies at the peroxisomal membrane, as characterised recently and discussed below (Agne et al. 2003).

Gould and co-workers used the differential instability of Pex5p in some *pex* mutants in an attempt to dissect the sequence of events at the peroxisomal membrane (Collins et al. 2000). The authors classified the import mutants into three major groups: the “docking proteins” (Pex13p, Pex14p), the “translocating” proteins (Pex2p, Pex10p, Pex12p), and the proteins involved in recycling of Pex5p (Pex4p, Pex22p, Pex1p, Pex6p).

Studies on purified protein complexes from rat liver peroxisomes show the existence of two large protein complexes within the peroxisomal membrane. Pex5p and Pex14p form a "core complex", which is more stable than the larger complex found by the same group, which contains Pex2p, Pex12p, Pex5p, and Pex14p (Gouveia et al. 2000; Reguenga et al. 2001). Similar complexes have been found in yeast using affinity purification of subcomplexes from peroxisomal membranes, in an attempt to dissect the sequence of events at the peroxisomal membrane (Girzalsky et al. 1999; Johnson et al. 2001; Hazra et al. 2002). A recent study in *S. cerevisiae* demonstrated a central role for the intraperoxisomal peroxin Pex8p in organising the import-complexes in the peroxisomal membrane (Agne et al. 2003). The authors suggest a working model where the Pex14p/Pex5p/Pex13p/Pex17p complex (I) is the actual translocon and the Pex2p/Pex10p/Pex12p complex (II) likely plays a role in recycling or in exporting Pex5p to the cytosol. Pex8p was suggested to transfer Pex5p from complex I to complex II, while Pex5p is present in the peroxisomal lumen. A model, combining most results discussed above, is shown in Figure 3.

The models are drawn using Pex5p as the cycling receptor. Pex7p likely follows a similar route, depending on the model organism used. In mammals, Pex7p probably moves with Pex5p through the import cycles, since Pex5p and Pex7p physically associate (see previous paragraph). In yeast, at least the early steps in import are similar, with a protein assembly like subcomplex I, where Pex7p replaces Pex5p. The topology of Pex7p in this membrane complex is unknown as yet. It will be informative to know whether Pex7p, like Pex5p, is inserted in a membrane complex, or remains peripherally associated to the import complexes. Both PTS1 and PTS2 are translocated into the peroxisomal lumen, and therefore, Pex5p and Pex7p likely share the machinery for translocation. Within the context of current knowledge about the two import processes, the core translocon is probably formed by (a) component(s) of subcomplex I.

Although many protein interactions have been charted, insight into the real mechanics of protein translocation across the peroxisomal membrane is lacking. A reliable *in vitro* protein import system is much needed to increase our knowledge.

2.7 Folding state and import of peroxisomal proteins

As most peroxisomal lumenal proteins possess a C-terminal targeting signal, import is a posttranslational event. Already in 1973, Lazarow and de Duve showed that most peroxisomal proteins are synthesised on free polyribosomes (As discussed in Lazarow and Fujiki 1985). This result is underlined by *in vivo* pulse-chase experiments: the import substrates are soluble for quite a long time before they get imported, dependent on the protein concerned (Ruigrok et al. in preparation). Although the protein import machinery of peroxisomes can accommodate at least partially folded proteins, it remains an open question to which extent proteins are folded as they cross the peroxisomal membrane.

At least some degree of folding appears to be tolerated by the peroxisomal import machinery *in vivo*. Convincing experiments showed that, in *S. cerevisiae*, the homodimeric enzyme peroxisomal thiolase is imported as a dimer, and that subunits do not mix during the import process. Tagged thiolase lacking a targeting signal could only be recovered as a heterodimer in the peroxisomal fraction, an observation that is highly indicative of "piggy back" translocation of the thiolase lacking the targeting signal (Glover et al. 1994). An often used experimental approach to study the impact of folding on import into mitochondria and the endoplasmic reticulum employs dihydrofolate reductase (DHFR), a protein of which the structure can be stabilised *in vivo* by aminopterin, and *in vitro* by methotrexate. Aminopterin was shown not to prevent import into peroxisomes of DHFR fused to a PTS (Hausler et al. 1996).

Another indication that the translocon of peroxisomes is substantially different from ER and mitochondrial import pores is demonstrated by Danpure's work on alanine:glyoxylate aminotransferase 1 (AGT). AGT is a dimeric protein containing both a peroxisomal and a mitochondrial import signal. Wild type AGT dimerises quickly in the cytoplasm, and is imported into peroxisomes. A naturally occurring mutant, causative for the human disease primary hyperoxaluria 1, prevents dimerisation, resulting in mistargeting of all AGT into mitochondria (Leiper et al. 1996).

An indication of low efficiency of "piggy back" import are experiments performed on peroxisomal malate dehydrogenase (Mdh3p). A massive overexpression of PTS-less tagged Mdh3p was required to force formation of hetero-oligomers and demonstrate "piggy back" import, perhaps because Mdh3p is one of the faster translocating proteins *in vivo* (Elgersma et al. 1996b; Ruigrok et al. in preparation). A more recent report shows that Eci1p, a peroxisomal lumenal protein, stripped of its PTS1 can also be transported in a “piggy-back” fashion by oligomerisation with another peroxisomal protein, Dci1p (Yang et al. 2001). Our own studies on kinetics of protein import into peroxisomes *in vivo* suggest that

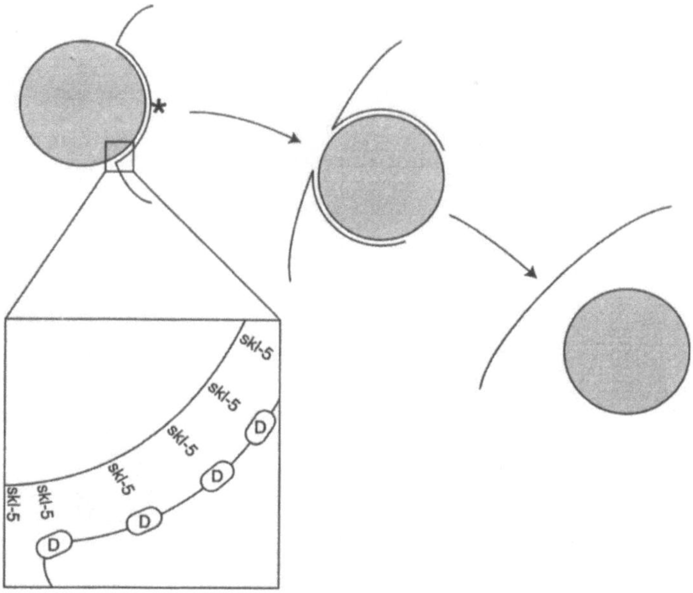

Fig. 4. Putative mechanism for gold particle import. The inset in the first stage of import shows the PTS1-Pex5-Docking complex interaction, which folds the membrane over the gold particle. The asterisk indicates mechanical shearing of the peroxisomal membrane. The membrane slides over the gold particle, and reseals in the last stage, allowing import of the gold particle. Gold particles are depicted in light grey. Abbreviations used: skl-5 - PTS1-Pex5p complex, D - Docking proteins at peroxisomal membrane.

newly synthesised proteins destined for the peroxisome are highly import-competent, as opposed to 'older' import substrate that is not imported yet and gets trapped in an off-pathway state (Ruigrok et al. in preparation).

An example favoured by many authors to support the notion of import of fully folded proteins is the observation that gold particles coated with PTS1-peptides can be imported. Gold particles may not be representative for proteinaceous import substrates; however, it has been reported that hepatocytes take up 17 nm gold particles from their environment, although, these cells do not display phagocytosis (Hardonk et al. 1985). Perhaps such particles traverse the peroxisomal membrane by physical means. When gold particles are coated with PTS1 peptides, Pex5p can interact with both PTS1 and peroxisomal membrane proteins, folding the membrane over the particle, and eventually allowing entry by mechanical shearing (see Fig. 4). The physical properties of gold particles (a rigid, non-deformable structure) might allow this type of import, but these are properties that natural import substrates do not possess.

The idea of the peroxisomal membrane engulfing proteins destined for the peroxisome was already postulated by the endocytosis like hypothesis of McNew and Goodman (McNew and Goodman 1996). A more recent opinion paper on the mechanistics of peroxisomal protein import defined the concept of a "preimplex":

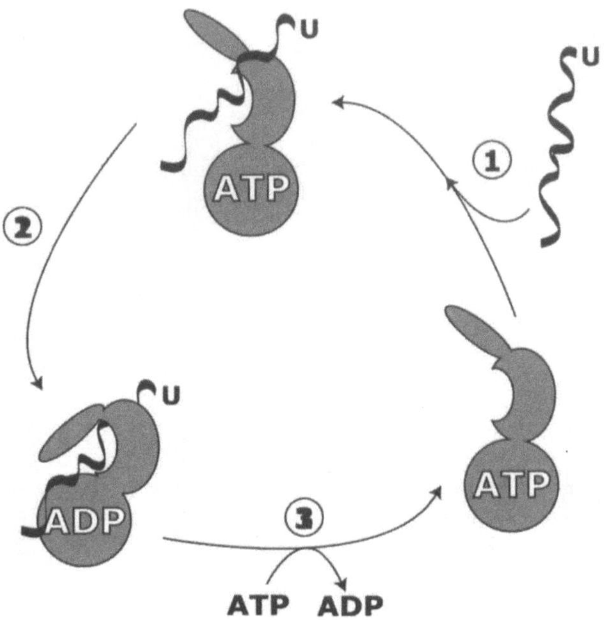

Fig. 5. The chaperone cycle of Hsp70. 1. Binding of substrate to the low-affinity, ATP-bound form of Hsp70. 2. Stimulation of ATPase activity by J-proteins induces the conformational change to a tightly bound, ADP-bound form of Hsp70. The client protein is allowed to fold. 3. Nucleotide exchange stimulates the release of the client protein, and returns Hsp70 to the low-affinity, ATP-bound state. See text for further details. Abbreviations used: U - Unfolded, folding, or misfolded client protein, which exposes regions that can be bound by Hsp70. Hsp70 is depicted in dark grey, and "ATP" or "ADP" indicates the nucleotide state within the schematic representation of Hsp70.

an assembly consisting of multiple receptor-cargo complexes with peroxisomal membrane proteins (Gould et al. 2002). Such large complexes would allow import of discrete 'packages' of peroxisomal proteins via an endocytosis-like process, the 'piggy-back' import of oligomerised peroxisomal proteins, and the occurrence of intra-peroxisomal import receptors.

3 Involvement of Hsp70 in peroxisomal protein import

3.1 Hsp70 family introduction

Hsp70 is an ATPase, and this enzyme activity, located in the N-terminal 44 kDa ATPase domain, is coupled to its chaperone function. When Hsp70 is bound to ATP, it has low affinity for its substrate (folding, or unfolded proteins), whereas the ADP bound state of Hsp70 is characterised by high binding affinity (Schmid et al. 1994). The Hsp70-substrate interaction is mediated by the peptide binding

Box 1. Hsp70 regulatory proteins

After hydrolysis of ATP by Hsp70, the final part of the Hsp70 cycle takes place. Here, the exchange of ADP for ATP, which goes hand in hand with the release of substrate is stimulated by regulatory proteins.

In *E. coli*, chloroplasts, and mitochondria, this function is performed by GrpE (Harrison et al. 1997). In the eukaryotic cytosol and the endoplasmic reticulum, a GrpE-like function is fulfilled by BAG-like proteins like BAG1 (Snl1p in *S. cerevisiae*) in the cytosol, and BAP (Sls1p in *S. cerevisiae*) in the endoplasmic reticulum (Takayama et al. 1997; Kabani et al. 2000; Chung et al. 2002; Sondermann et al. 2002). BAG-like proteins, which contain the Hsp70-regulating "BAG-domain", may be comparable to J-proteins: they use the "BAG-domain" as a module that regulates Hsp70's activity at a specific site in the cell.

In the cytosol, HspBP1 (Fes1p in *S. cerevisiae*) acts as a negative regulator of Hsp70 nucleotide binding (Raynes and Guerriero 1998; Kabani et al. 2002a). It prevents ATP binding to Hsp70 by stimulating the release of this nucleotide before it can be hydrolysed (Kabani et al. 2002b). In mammals, another protein called HIP can interact with the ATPase domain of Hsp70, stabilising the ADP state (Hohfeld et al. 1995).

Finally, two additional classes of proteins, which interact with the COOH-terminus of Hsp70, play a role in the substrate binding by Hsp70. HOP (Sti1p in *S. cerevisiae*) plays an organising role in handing over substrate from Hsp70 to the Hsp90 chaperone system, and thus, promotes substrate release from Hsp70 (Nicolet and Craig 1989; Chen and Smith 1998). CHIP negatively regulates substrate release from Hsp70 (Ballinger et al. 1999), and a homologue of CHIP (small glutamine-rich protein/viral protein U-binding protein (SGT/UBP)) also inhibits substrate release from Hsp70 (Angeletti et al. 2002). The latter protein has a homolog in *S. cerevisiae* (*SGT1*), but the Hsp70-regulating capabilities of this protein have not yet been determined.

domain, which is located C-terminal to the ATPase domain and binds short, mostly hydrophobic, amino acid stretches that are exposed in folding or unfolded proteins (Blond-Elguindi et al. 1993; Rudiger et al. 1997). This on/off cycle of binding prevents unwanted interactions of these hydrophobic segments with the environment (membranes, other proteins), enhancing the chance of productive folding. At the C-terminus of Hsp70 a third domain can be found, often referred to as the 'lid domain'. Upon binding to ATP, Hsp70 is folded such that the peptide-binding domain is freely accessible, allowing the substrate to bind. This interaction is transient in nature, and only upon ATP hydrolysis the conformation of Hsp70 changes such that the peptide is held in the peptide-binding domain, covered by the 'lid domain'. ADP-ATP exchange then changes the conformation of Hsp70, allowing release of the substrate (Buchberger et al. 1995; Mayer et al. 2000) (Fig. 5).

Although Hsp70 can perform this chaperoning cycle on its own, its activity is highly regulated. The best-studied regulators of Hsp70 are the J-proteins, a family of proteins that is characterised by a so-called J-domain of approximately 70 amino acids. These J-proteins, also referred to as Hsp40s, owe their name to DnaJ, the first identified *E. coli* family member. J-domains specifically interact with Hsp70s, and stimulate their ATP hydrolysis activity (Liberek et al. 1991). The peptide-binding domain of Hsp70 shows little substrate specificity (James et al. 1997). This substrate specificity is provided by J-proteins, present at specific subcellular locations or associated to client proteins, recruiting the Hsp70 chaperone to its target (Silver et al. 1993; Cyr et al. 1994). Several other proteins can influence the ATP-ADP state and substrate binding affinity of Hsp70 (Box1).

Hsp70 has duplicated many times in evolution. Whereas the prokaryote *E. coli* only contains two Hsp70s besides DnaK: HscA and HscC (Kawula and Lelivelt 1994; Itoh et al. 1999), *S. cerevisae* contains several Hsp70 families, each consisting of one or more family members. During the ancient genome duplication in *S. cerevisiae*, most Hsp70 families were duplicated (Dietrich et al. 2004) (Table 2). The *SSB* family is ribosome-associated, and presumably acts in early (co-translational) protein folding. The *SSB* family binds to nascent chains, as they emerge from the ribosome, an interaction stimulated by Ssz1p in conjunction with the J-protein Zuo1p. A four-membered family of Ssa proteins forms another group of cytosolic Hsp70. *SSA* comprises an essential gene family, and members of the *SSA* family are functionally redundant, i.e. expression of any of the Ssa proteins can support cell growth in the absence of the other three members (Werner-Washburne et al. 1987). Binding of Hsp70 to substrate is not specific across the *SSA* and *SSB* families, since exchange of the peptide-binding domain of these chaperone families did not affect their function (James et al. 1997). The *SSE* family members are Hsp70-like, but contain an insertion, and are thought to function as co-factors in the Hsp90 chaperone machinery. Finally, one or more intra-organellar Hsp70s can be found in ER and mitochondria (see also Section 3.3).

Less detailed information is known about mammalian Hsp70 subfamilies. The distinction between ribosome-bound and cytosolic Hsp70 is not conserved from yeast to mammals, and stress-inducible Hsp70 as well as constitutively expressed Hsc70 isoforms can be found in the cytosol. Besides differential gene expression, no indication exists of functional differences between Hsc70 and Hsp70. For the purpose of simplicity, Hsc70 and Hsp70 will be referred to as Hsp70. As in yeast, an *SSE*-like family (Hsp105) exists, and intra-organellar Hsp70s can be found in ER, mitochondria, chloroplasts, nucleus, and lysosomes, but not in peroxisomes.

The number of different proteins that can influence the Hsp70 chaperone cycle, Hsp70s' subcellular localisation, and the presence of redundant isoforms of Hsp70 itself reveals an intricate network of regulation, where Hsp70s' activities are regulated in a localised and precise way. A great challenge lies in determining the interplay of these different regulatory networks to ultimately map the way Hsp70s' specificities are achieved in the cell.

Table 2. Duplications in Hsp70s of *S. cerevisiae*

gene	localisation	duplicated (Dietrich et al. 2004)
SSA1	cytosol	no data
SSA2	cytosol	no
SSA3	cytosol	yes (*SSA4*)
SSA4	cytosol	yes (*SSA3*)
SSB1	cytosol	yes (*SSB2*)
SSB2	cytosol	yes (*SSB1*)
SSZ1	cytosol	no
SSE1	cytosol	yes (*SSE2*)
SSE2	cytosol	yes (*SSE1*)
KAR2	ER	no
LHS1	ER	no
SSC1	mitochondria	yes (*ECM10*)
ECM10	mitochondria	yes (*SSC1*)
SSQ1	mitochondria	no

3.2 *In vivo* roles of Hsp70

Hsp70 plays a role in many processes in the cell. Beautiful pulse-chase studies in *E. coli* showed that the *E. coli* Hsp70 (DnaK) transiently associates with newly synthesized proteins, indicating an important role for Hsp70 in *de novo* folding of proteins. In yeast, one class of Hsp70s (Ssb) associates to the ribosome and binds the emerging nascent chain of newly synthesized proteins (Nelson et al. 1992; Pfund et al. 1998). Ssb consists of two almost identical family members. Recently, a ribosome-associated J-protein (Zuo1p) was found in a stoichiometric complex with an Hsp70-like protein (Ssz1p or Pdr13p), which also localises to the ribosome. Ssz1p itself does not bind nascent chains, but the Ssz1p/Zuo1p complex stimulates nascent chain binding to Ssb (Gautschi et al. 2002; Hundley et al. 2002). In mammals, Hsp70 also associates to nascent chains, as was shown in a mammalian translation system (Frydman et al. 1994). Not all proteins, however, depend on Hsp70 for proper cotranslational folding (Nicola et al. 1999).

In yeast, Hsp70s of the cytosolic Ssa family assist at the level of posttranslational folding and perhaps oligomerisation. Maturation of the cytosolic enzyme ornithine transcarbamoylase was followed in a temperature sensitive *ssa1* mutant, in the absence of the other three family members *SSA2*, *SSA3*, and *SSA4*. At the non-permissive temperature, a large decrease of specific activity of the enzyme resulted and a monomeric species accumulated, suggesting delayed oligomerisation or oligomer disassembly (Kim et al. 1998). The tetrameric enzyme peroxisomal catalase (Cta1p) heavily aggregates in the absence of cytosolic Hsp70 (our own unpublished observations). This also implies a role for Hsp70 in folding of cytosolic proteins.

It is clear that Hsp70 also acts on fully folded proteins. One example is the J-protein auxillin, which is associated with clathrin-coated vesicles. Auxillin recruits Hsp70 to depolymerise the clathrin coat into free subunits, releasing the vesicle to join the endosomal system (Ungewickell et al. 1995; Pishvaee et al. 2000). Another example of the capability of Hsp70 to bind folded substrates was demonstrated in elegant studies using surface plasmon resonance by Rapoport and colleagues (Misselwitz et al. 1998). Using purified BiP, an ER Hsp70, and the J-domain of its cognate J-protein Sec63, they showed that BiP can be activated to bind fully folded proteins, provided that the activation of BiP takes place in close proximity to the substrate proteins. Since they could not detect binding to a tightly folded model protein, they assume that BiP binds portions of the protein that are exposed by "thermal breathing".

3.3 Hsp70 and import of proteins into organelles

The role of Hsp70 in protein import processes is diverse and extensively studied. In the next paragraphs, we will highlight some key findings on the role of Hsp70 chaperones in different protein import processes. Our aim is to uncover possible generic mechanisms in Hsp70s functioning in protein import, and to discuss them in the context of peroxisomal protein import. Since peroxisomes do not appear to contain lumenal Hsp70, we will focus our discussion on cytosolic processes prior to import.

Proteins that are destined for import into the ER and mitochondrion traverse the membrane of their target organelle in a partially folded/unfolded state. Two classes of Hsp70 are involved in the import into these organelles: cytosolic and lumenal Hsp70. Cytosolic Hsp70 plays a role in keeping some import substrates in a loosely folded, import competent conformation, whereas lumenal Hsp70 is involved in ensuring unidirectional transport of the polypeptide chain through the translocon.

The role of cytosolic Hsp70 was established in *in vivo* studies in *S. cerevisiae.* Partial depletion of cytosolic Hsp70 by deleting one of the four cytosolic Hsp70s *SSA1*, already resulted in an inability to respire at 37°C (Deshaies et al. 1988). This suggests mitochondrial malfunction, possibly caused by diminished import in the absence of sufficient cytosolic Hsp70. To investigate this, later experiments made use of cells where the complete *SSA* subfamily was deleted, and *SSA1* was expressed under control of a regulated promotor to deplete Hsp70 completely *in vivo* (Deshaies et al. 1988). Pulse-chase experiments revealed precursor accumulation of some but not all mitochondrial and ER precursor proteins. Perhaps some proteins are more easily unfolded, and do not require Hsp70 to keep them loosely folded enough to be translocation competent. Similar observations were made with an *ssa1-45* temperature sensitive (t_S) mutant, also in combination with a t_S mutant of Ydj1p, one of the cytosolic DnaJ proteins. These results again confirmed the involvement of Ssa proteins in mitochondrial and ER import, and showed that Ydj1p works together with Ssa proteins in this process (Becker et al. 1996).

Some *in vitro* evidence also underlines a possible role for Hsp70 to keep proteins in an import-competent state. The precursor to mitochondrial aspartate aminotransferase (pmAAT) was chemically denatured, and refolded in the presence of lysate. Hsp70 was shown to associate to pmAAT, which resulted in a form of pmAAT that could be imported into isolated mitochondria. However, upon depletion of Hsp70, pmAAT was not competent for import anymore (Artigues et al. 2002). Although this situation is different from that in the living cell - a denatured full-length polypeptide chain instead of vectorially translated protein is used as starting material - it does show that Hsp70 in principle can maintain a protein in an import-competent state. Hsp70 appears to act specifically because cytosolic Hsp90, another cytosolic chaperone, also bound to pmAAT and kept it in a partially folded state but did not allow import of pmAAT into mitochondria. Additional evidence for maintenance of import competence by Hsp70 and/or Hsp40 is the observation that Hsp70/Hsp40 stimulate import of prepro α–factor into isolated microsomes by preventing aggregation (Ngosuwan et al. 2002). Using clever experimentation to specifically manipulate levels of ATP outside mitochondria, Asai et al. (2004) showed that cytosolic ATP (and likely cytosolic Hsp70) was important for keeping mitochondrial precursor proteins in an import-competent state.

Although proteins can be presented in a folded state to isolated mitochondria and still be imported efficiently, import occurs more efficiently when the polypeptide chain is (partially) unfolded (Verner and Schatz 1987). A clear example is formed by adenylate cyclase (Adk1p) in yeast. Adk1p is present in the cytosol and mitochondria, a result of competition between Adk1p folding and import (Strobel et al. 2002). Translocation into mitochondria requires unfolded polypeptide, and several factors, including the "pulling" or "trapping" by mitochondrial lumenal Hsp70 and the membrane potential, play a role in actively unfolding the protein at the membrane during translocation (Matouschek et al. 2000; Huang et al. 2002). Studies on the import of a larger set of mitochondrial precursor proteins *in vivo* would provide us with more insight into the specific requirements of Hsp70 in import into mitochondria.

Unlike the above-discussed classical translocation pores, import into several other organelles allows the imported substrate to be folded and even assembled into oligomers. In nuclear import, medium sized proteins can passively diffuse across the nuclear envelope through an aqueous channel of 9 nm diameter formed by the nuclear pore. Even larger structures (at least 20 nm in size) can passage the nuclear pore by active transport. Involvement of Hsp70 in nuclear import was demonstrated using the same conditional mutant in *SSA1* (*ssa1-45*) as used to characterise mitochondrial and ER import.

The lysosome (vacuole in yeast) can accommodate import of folded proteins, and different import mechanisms exist. Lysosomes contain lumenal Hsp70s, and a fraction of cytosolic Hsp70 is associated with the lysosomal membrane (Terlecky and Dice 1993; Agarraberes et al. 1997). When mammalian cells are starved for serum, cytosolic proteins containing a KEFRQ amino-acid sequence, are targeted to the lysosome for degradation. Hsc73 was found to bind to these peptides, and to stimulate lysosomal degradation of a protein containing this sequence in two different *in vitro* assays (Chiang et al. 1989).

Import of a subset of proteins into yeast vacuoles can involve membrane inclusion of oligomeric protein into small vesicles, termed VID vesicles, which are subsequently targeted to the vacuole. When investigating the influence of Hsp70 in an *in vivo* degradation assay, and in an *in vitro* import assay using isolated VID vesicles, a specific requirement for Ssa2p was found (Brown et al. 2000).

3.4 Hsp70 and peroxisomal protein import

Several lines of evidence have implicated Hsp70 in peroxisomal protein import. In a micro-injection based assay, polyclonal antibodies directed against Hsp70 inhibited import of a co-injected PTS1-containing protein. Hsp70 was found to be associated to purified rat liver peroxisomes, and under peroxisome inducing conditions, more Hsp70 associated with peroxisomes (Walton et al. 1994). The PTS2 pathway relies on Hsp70 and Hsp40 for efficient import (Legakis and Terlecky 2001).

Several reports exist on Hsp70s' role in peroxisomal protein import in plants. In cucumber seedlings, two glyoxisomal membrane proteins, PMP73 and PMP61, were shown to be immuno-related to Hsp70 and a DnaJ homologue also present in other plant species (Corpas and RN 1997). Studies in watermelon reveal a Hsp72 gene that could be initiated at two distinct methionines, resulting in a form targeted to plastids, and a form present in the glyoxisomal lumen. The latter localisation was dependent on a PTS2, which was shown to be functional in the yeast *H. polymorpha* (Wimmer et al. 1997). Another group independently reported membrane-associated and lumenal Hsp70 and DnaJ homologs in glyoxisomes (Diefenbach and Kindl 2000). None of the yeast or mammalian Hsp70s carry resemblance to a PTS in their primary structure, suggesting that intraperoxisomal Hsp70 occurs only in the plant kingdom.

In pumpkin seedlings, Hsp70 was found to be peroxisome-associated. The amount of Hsp70 associated with peroxisomes correlated well with the import efficiency: in heat-shocked seedlings more Hsp70 was associated with peroxisomes, and import efficiency increased. As expected, immunodepletion caused a decrease of import into peroxisomes. Hsp70 was suggested to act on the import pathway in the cytosol because prior treatment of peroxisomes with anti-Hsp70 antibody did not affect import. Hsp70 depletion during synthesis of the protein affected import more than Hsp70 depletion after protein synthesis. Taken together with the co-immunoprecipitation of two different import substrates with Hsp70, Hsp70 was concluded to act directly on the import substrate already during its synthesis (Crookes and Olsen 1998).

A recent report directly implicates Hsp70 in peroxisomal protein import via its interaction with Pex5p (Harano et al. 2001). Using the expressed and purified mammalian PTS1 protein acyl-CoA oxidase as bait, both Pex5p and Hsp70 were fished out of mammalian cell lysates. Part of the associated Hsp70 could be released from the complex(es) by ATP, a property indicative of Hsp70-client protein interaction, perhaps reflecting the association of acyl-CoA oxidase with Hsp70.

Other interaction studies, employing PTS1-peptides and purified human Hsp70 and Pex5p, showed that Hsp70 did not have an effect on the binding of Pex5p to PTS1-peptide. The authors concluded that Pex5p-PTS1 interaction is independent

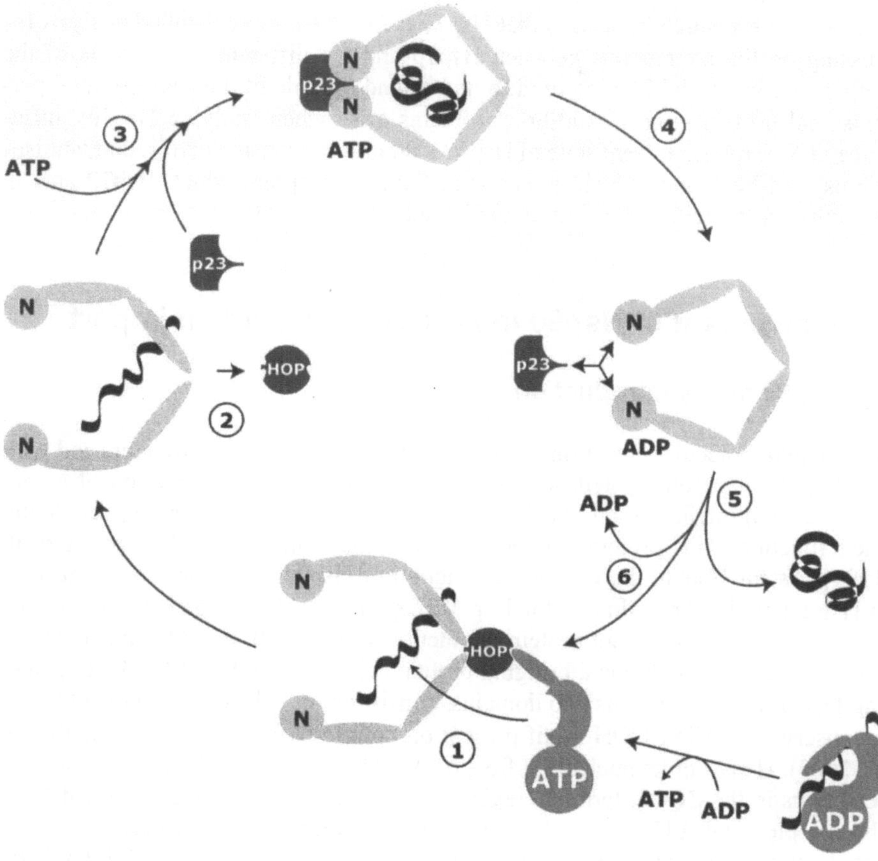

Fig. 6. The chaperone cycle of Hsp90 (After Young et al. 2002). 1. ADP-ATP exchange and substrate release of Hsp70 is stimulated by HOP, which connects Hsp70 to Hsp90. The client protein is next transferred from Hsp70 to Hsp90. 2. HOP dissociates from Hsp90-client protein complex, and ATP binds Hsp90, which induces dimerisation of the NH_2-terminal domains of Hsp90. 3. p23 binds to the dimerised N-termini of Hsp90. 4. Slow ATP hydrolysis by Hsp90 causes the NH_2-termini to monomerise, p23 dissociates from Hsp90. 5. The substrate is released. 6. ADP dissociates, and Hsp90 returns to its nucleotide-free state, ready to accept new client proteins. Abbreviations used: HOP - Hsp90-Hsp70 Organising Protein. Hsp90 dimers are depicted in light grey, cofactors in dark grey, and Hsp70 in medium grey. The NH_2-terminus of Hsp90 is indicated with an N.

of Hsp70, and that Hsp70 may assist folding and oligomerisation of PTS1 proteins. Because Pex5p exists as a tetramer, oligomeric proteins might have higher affinity for Pex5p than a monomeric protein subunit, resulting in more efficient recognition by Pex5p (Harper et al. 2002).

A general problem with Hsp70 immunodepletion is the specificity of the observed effect. All, or at least a subclass, of Hsp70 is removed prior to the start of an import reaction, which can possibly result in aggregation of the import substrate, or misfolding of proteins needed for import. Although the *in vitro* import systems give a rough indication that Hsp70 is involved, more detailed analysis focussing on the interactions between Hsp70 and the different components of the import machinery will be required to understand the role of this chaperone in peroxisomal protein import. Addition problems arise when studying Hsp70s' influence in *S. cerevisiae*. Here several Hsp70 subfamilies, consisting of redundant isoforms, can be found, which makes it difficult to pinpoint which Hsp70 acts in peroxisomal protein import (our unpublished observations).

4 Involvement of Hsp90 in peroxisomal protein import

4.1 Hsp90 family introduction

The Hsp90 molecular chaperone works in concert with a set of co-factors and with the Hsp70 chaperone system to assist in maintaining the conformation of a specific set of proteins, often involved in signalling pathways, thereby, regulating their function. This protein complex is called the "foldosome" (Hutchison et al. 1994). As for Hsp70, the chaperone function of Hsp90 is coupled to its intrinsic ATPase activity. An outline of the Hsp90 chaperone cycle is depicted in Figure 6.

In initial stages of client protein interaction with Hsp90, Hsp70 hands over the client protein to Hsp90 via a bridging protein called HOP (Hsp70-Hsp90 Organising Protein). HOP contains two domains, consisting of TPR repeats, which bind to a conserved EEVD-COOH motif present on both Hsp70 and Hsp90 (Scheufler et al. 2000). Hsp90 in its nucleotide-free, or ADP bound state is dimeric through interactions in the COOH-terminal region of the protein. HOP is then released from the complex and ATP is bound, inducing dimerisation of the NH_2-terminal ATPase domain of Hsp90. This state has a high affinity for p23. Now ATP hydrolysis occurs, which is a slow step in the Hsp90 cycle. p23, substrate, and ADP are released, and Hsp90 is again in the nucleotide-free state, ready to accept a new client protein.

This "minimal Hsp90 chaperone cycle" is a simplified depiction of the actual situation. The p23-bound state of Hsp90, also referred to as the "mature complex", can contain immunophilins, or immunophilin-like proteins, which act in the maturation of specific substrates.

4.2 *In vivo* roles of Hsp90

Unlike Hsp70, Hsp90 does not appear to function in early events of protein folding at the level of the nascent chains (Frydman 2001). Although refolding studies of denatured model proteins showed that Hsp90 can prevent aggregation and can assist in more efficient refolding (Freeman and Morimoto 1996), the same model proteins are not affected in *de novo* folding.

Instead, Hsp90 acts on folded or partially folded proteins to keep them in an active state. This role is exemplified by the well-characterised involvement of Hsp90 in the maturation of signalling proteins such as nuclear hormone receptors, where binding of the receptor to the Hsp90 chaperone machinery ensures that these proteins can be activated by hormones (Picard et al. 1990). Different kinases, such as src- and raf-like kinases and Ste11p in yeast also require Hsp90 for their activity (Xu and Lindquist 1993; Schulte et al. 1995; Louvion et al. 1998). The prominent role of Hsp90 in chaperoning signal transduction pathways explains its being essential in eukaryotes.

4.3 Hsp90 and import of proteins into organelles

Demonstrations of the involvement of Hsp90 in protein import processes are scarce. Hsp90 is involved in import of at least some mitochondrial proteins. Hsp90 interacts with various TPR domain-containing proteins, one of which is Tom70p. Tom70p is an outer membrane protein involved in the import of a subset of mitochondrial proteins (Hines et al. 1990). When the interaction between Tom70p and Hsp90 is disturbed, import of Tom70p dependent mitochondrial preproteins does not occur (Young et al. 2003). The authors concluded that Hsp90 targets mitochondrial preproteins to Tom70p. Tom34p, another protein involved in mitochondrial import also binds Hsp90 (Young et al. 1998). The functional significance of this interaction remains to be established.

Hsp90 is involved in regulating the activity of many signalling proteins. Two examples of signalling proteins, the glucocorticoid receptor (GR), and the dioxin receptor (DR) are targeted to the nucleus in an Hsp90-dependent fashion. Adding geldanamycin, an Hsp90 ATPase inhibitor, to living cells inhibited nuclear import of GR-GFP and DR-GFP fusion proteins. This mechanism appears to rely on immunophilins or immunophilin-like proteins, present in late Hsp90 complexes: FKBP52 and XAP1 for GR and DR, respectively (Silverstein et al. 1999; Kazlauskas et al. 2002). Upon ligand binding to GR, FKBP52 is recruited to the Hsp90-GR complex, and mediates transport of this complex along the cytoskeleton towards the nucleus (Davies et al. 2002). This process is probably mediated via the interaction of FKBP52 with dynein (Galigniana et al. 2002).

As previously mentioned, lamp2a-mediated lysosomal import involves cytosolic Hsp70. Hsp90 is also involved in this import process. Several components of the Hsp90 machinery can associate with import substrates and the lamp2a translocon. Antibodies directed against Hsp70, Hsp40, Hip, or Hop inhibit import into isolated lysosomes. The Hsp90 chaperone machinery might act as a targeting factor for KEFRQ-containing proteins, and/or assist in their translocation (Agarraberes and Dice 2001).

4.4 Hsp90 and peroxisomal protein import

The first report implicating Hsp90 in peroxisomal protein import showed that addition of anti-Hsp90 antibodies to an *in vitro* plant peroxisomal import assay resulted in retarded import of isocitrate lyase (Crookes and Olsen 1998). The role of Hsp70 was different from that of Hsp90: the effect of anti-Hsp90 antibody was pronounced when added both during synthesis and after synthesis, whereas the major effect of anti-Hsp70 antibodies was during synthesis, suggesting a post-translational role for Hsp90. We recently found retarded import of both PTS1 and PTS2 containing proteins in the absence of functional Hsp90 in yeast (our unpublished observations). This general import defect could be informative regarding the function of Hsp90 in peroxisomal protein import. Three possibilities come to mind: 1) Hsp90 acts on the PTS-containing protein, 2) Hsp90 acts at the level of import receptors, 3) Hsp90 acts indirectly through signal transduction.

Arguing against the first scenario is the lack of involvement in *de novo* folding by Hsp90. Hsp90 is found in complex with some nuclear, mitochondrial, and lysosomal proteins and has an active role in their transport. In contrast, in the aforementioned *in vitro* import assay, antibodies directed to Hsp70 precipitated isocitrate lyase, but an interaction between this import substrate and Hsp90 was not observed. This finding suggests that Hsp90 does not act on peroxisomal protein import via interaction with the peroxisomal proteins prior to import, but the same conclusion need not necessarily apply to all peroxisomal proteins.

The second scenario would imply that both the PTS1 and the PTS2 receptor are chaperoned by Hsp90, resulting in a more efficient function of these receptors in the import process. In line with this notion is the observation that Pex5p can be co-immunoprecipitated with Hsp90 using *in vitro* translated proteins (Pratt et al. 2001). Perhaps Hsp90 has a direct role in keeping Pex5p "receptive" for PTS1 recognition. Despite several attempts, we did not find an interaction between Pex5p and Hsp90 in yeast lysates, and therefore, could not confirm this hypothesis (our unpublished observations).

In the final scenario, Hsp90 functions at a later step in peroxisomal protein import, where the PTS1 and PTS2 pathways converge. In this context, it is interesting to note that some peroxins are phosphorylated *in vivo* (Elgersma et al. 1997; Komori et al. 1999). By regulating the activity of kinases, Hsp90 could regulate activity of peroxins.

5 Concluding remarks

Despite the considerable efforts to delineate the contribution of protein folding and chaperones in the process of peroxisomal protein import our insights remain rather limited and fragmentary. This is due to a number of drawbacks:

i. The difficulty to establish the degree of folding *in vivo* of a protein before, during, or after its itinerary from cytosol into the organelle.

ii. The lack of success in establishing a reliable *in vitro* system to study the import of proteins into peroxisomes. This limits experimentation to intact or semi-permeabilised cells, which makes it difficult to distinguish between direct and indirect effects of manipulation. Also, it limits the ability to determine at which step along the import pathway and effect is exerted.

iii. The complexity and redundancy of chaperone families and their regulators makes it difficult to trace the contribution of individual chaperone proteins, even in a genetically tractable organism as *S. cerevisiae*. Most experiments performed in mammalian cells use antibodies, which remove a complete subclass of chaperones, and it is difficult to assess where the chaperone function acts in the import process.

At this moment it is almost impossible to extract a coherent picture about the role of folding and chaperones from the mosaic of various reported observations. Improvement of this impasse awaits the development of *in vitro* techniques to reconstitute partial steps in the peroxisomal protein import pathway.

References

Agarraberes FA, Dice JF (2001) A molecular chaperone complex at the lysosomal membrane is required for protein translocation. J Cell Sci 114:2491-2499

Agarraberes FA, Terlecky SR, Dice JF (1997) An intralysosomal hsp70 is required for a selective pathway of lysosomal protein degradation. J Cell Biol 137:825-834

Agne B, Meindl NM, Niederhoff K, Einwachter H, Rehling P, Sickmann A, Meyer HE, Girzalsky W, Kunau W-H (2003) Pex8p: An intraperoxisomal organizer of the peroxisomal import machinery. Mol Cell 11:635-646

Albertini M, Rehling P, Erdmann R, Girzalsky W, Kiel JAKW, Veenhuis M, Kunau W-H (1997) Pex14p, a peroxisomal membrane protein binding both receptors of the two PTS-dependent import pathways. Cell 89:83-92

Angeletti PC, Walker D, Panganiban AT, Wu Y, Hu Z, Thompson LJ, Yin LY, Patterson C (2002) Small glutamine-rich protein/viral protein U-binding protein is a novel cochaperone that affects heat shock protein 70 activity. Cell Stress Chaperones 7:258-268

Artigues A, Iriarte A, Martinez-Carrion M (2002) Binding to chaperones allows import of a purified mitochondrial precursor into mitochondria. J Biol Chem 277:25047-25055

Asai T, Takahashi T, Esaki M, Nishikawa S, Ohtsuka K, Nakai M, Endo T (2004) Reinvestigation of the requirement of cytosolic ATP for mitochondrial protein import. J Biol Chem 279:19464-19470. Epub 12004 Mar 19464

Ballinger CA, Connell P, Wu Y, Hu Z, Thompson LJ, Yin LY, Patterson C (1999) Identification of CHIP, a novel tetratricopeptide repeat-containing protein that interacts with heat shock proteins and negatively regulates chaperone functions. Mol Cell Biol 19:4535-4545

Becker J, Walter W, Yan W, Craig EA (1996) Functional interaction of cytosolic hsp70 and a DnaJ-related protein, Ydj1p, in protein translocation *in vivo*. Mol Cell Biol 16:4378-4386

Behari R, Baker A (1993) The carboxyl terminus of isocitrate lyase is not essential for import into glyoxysomes in an *in vitro* system. J Biol Chem 268:7315-7322

Bellion E, Goodman JM (1987) Proton ionophores prevent assembly of a peroxisomal protein. Cell 48:165-173

Blond-Elguindi S, Cwirla SE, Dower WJ, Lipshutz RJ, Sprang SR, Sambrook JF, Gething MJ (1993) Affinity panning of a library of peptides displayed on bacteriophages reveals the binding specificity of BiP. Cell 75:717-728

Braverman N, Dodt G, Gould SJ, Valle D (1998) An isoform of pex5p, the human PTS1 receptor, is required for the import of PTS2 proteins into peroxisomes. Hum Mol Genet 7:1195-1205

Brown CR, McCann JA, Chiang HL (2000) The heat shock protein Ssa2p is required for import of fructose-1, 6-bisphosphatase into Vid vesicles. J Cell Biol 150:65-76

Buchberger A, Theyssen H, Schroder H, McCarty JS, Virgallita G, Milkereit P, Reinstein J, Bukau B (1995) Nucleotide-induced conformational changes in the ATPase and substrate binding domains of the DnaK chaperone provide evidence for interdomain communication. J Biol Chem 270:16903-16910

Chen S, Smith DF (1998) Hop as an adaptor in the heat shock protein 70 (Hsp70) and hsp90 chaperone machinery. J Biol Chem 273:35194-35200

Chiang HL, Terlecky SR, Plant CP, Dice JF (1989) A role for a 70-kilodalton heat shock protein in lysosomal degradation of intracellular proteins. Science 246:382-385

Chung KT, Shen Y, Hendershot LM, Siegers K, Moarefi I, Wente SR, Hartl FU, Young JC (2002) BAP, a mammalian BiP-associated protein, is a nucleotide exchange factor that regulates the ATPase activity of BiP. J Biol Chem 277:47557-47563

Collins CS, Kalish JE, Morrell JC, McCaffery JM, Gould SJ (2000) The peroxisome biogenesis factors pex4p, pex22p, pex1p, and pex6p act in the terminal steps of peroxisomal matrix protein import. Mol Cell Biol 20:7516-7526

Corpas F, Trelease RN (1997) The plant 73 kDa peroxisomal membrane protein (PMP73) is immunorelated to molecular chaperones. Eur J Cell Biol 73:49-57

Crookes W, Olsen LJ (1998) The effects of chaperones and the influence of protein assembly on peroxisomal protein import. J Biol Chem 273:17236-17242

Cyr DM, Langer T, Douglas MG (1994) DnaJ-like proteins: molecular chaperones and specific regulators of Hsp70. Trends Biochem Sci 19:176-181

Dammai V, Subramani S (2001) The human peroxisomal targeting signal receptor, Pex5p, is translocated into the peroxisomal matrix and recycled to the cytosol. Cell 105:187-196

Dansen TB, Wirtz KW, Wanders RJ, Pap EH, Behari R, Baker A (2000) Peroxisomes in human fibroblasts have a basic pH. Nat Cell Biol 2:51-53

Davies TH, Ning YM, Sanchez ER (2002) A new first step in activation of steroid receptors: hormone-induced switching of FKBP51 and FKBP52 immunophilins. J Biol Chem 277:4597-4600

Deshaies RJ, Koch BD, Werner-Washburne M, Craig EA, Schekman R (1988) A subfamily of stress proteins facilitates translocation of secretory and mitochondrial precursor polypeptides. Nature 332:800-805

Diefenbach J, H K (2000) The membrane-bound DnaJ protein located at the cytosolic site of glyoxysomes specifically binds the cytosolic isoform 1 of Hsp70 but not other Hsp70 species. Eur J Biochem 267:746-754

Dietrich FS, Voegeli S, Brachat S, Lerch A, Gates K, Steiner S, Mohr C, Pohlmann R, Luedi P, Choi S, Wing RA, Flavier A, Gaffney TD, Philippsen P (2004) The *Ashbya gossypii* genome as a tool for mapping the ancient *Saccharomyces cerevisiae* genome. Science 304:304-307. Epub 2004 Mar 2004

Distel B, Erdmann R, Gould SJ, Blobel G, Crane DI, Cregg JM, Dodt G, Fujiki Y, Goodman JM, Just WW, Kiel JAKW, Kunau W-H, Lazarow PB, Mannaerts GP, Moser HW, Osumi T, Rachubinski RA, Roscher A, Subramani S, Tabak HF, Tsukamoto T, Valle D, van der Klei I, van Veldhoven PP, Veenhuis M (1996) A unified nomenclature for peroxisome biogenesis factors. J Cell Biol 135:1-3

Dodt G, Braverman N, Wong C, Moser A, Moser HW, Watkins P, Valle D, Gould SJ (1995) Mutations in the PTS1 receptor gene, PXR1, define complementation group 2 of the peroxisome biogenesis disorders. Nat Genet 9:115-125

Dodt G, Gould SJ (1996) Multiple PEX genes are required for proper subcellular distribution and stability of Pex5p, the PTS1 receptor: evidence that PTS1 protein import is mediated by a cycling receptor. J Cell Biol 135:1763-1774

Dodt G, Warren D, Becker E, Rehling P, Gould SJ (2001) Domain mapping of human PEX5 reveals functional and structural similarities to *Saccharomyces cerevisiae* Pex18p and Pex21p. J Biol Chem 276:41769-41781

Dyer JM, McNew JA, Goodman JM (1996) The sorting sequence of the peroxisomal integral membrane protein PMP47 is contained within a short hydrophilic loop. J Cell Biol 133:269-280

Economou A, Wickner W (1994) SecA promotes preprotein translocation by undergoing ATP-driven cycles of membrane insertion and deinsertion. Cell 78:835-843

Einwachter H, Sowinski S, Kunau WH, Schliebs W (2001) *Yarrowia lipolytica* Pex20p, *Saccharomyces cerevisiae* Pex18p/Pex21p and mammalian Pex5pL fulfil a common function in the early steps of the peroxisomal PTS2 import pathway. EMBO Rep 2:1035-1039. Epub 2001 Oct 1017

Elgersma Y, Elgersma-Hooisma M, Wenzel T, McCaffery JM, Farquhar MG, Subramani S (1998) A mobile PTS2 receptor for peroxisomal protein import in *Pichia pastoris*. J Cell Biol 140:807-820

Elgersma Y, Kwast L, van den Berg M, Snyder WB, Distel B, Subramani S, Tabak HF (1997) Overexpression of Pex15p, a phosphorylated peroxisomal integral membrane protein required for peroxisome assembly in *S. cerevisiae*, causes proliferation of the endoplasmic reticulum membrane. EMBO J 16:7326-7341

Elgersma Y, Kwast L, Klein A, Voorn-Brouwer T, van den Berg M, Metzig B, America T, Tabak HF, Distel B (1996a) The SH3 domain of the *Saccharomyces cerevisiae* peroxisomal membrane protein Pex13p functions as a docking site for Pex5p, a mobile receptor for the import of PTS1-containing proteins. J Cell Biol 135:97-109

Elgersma Y, Vos A, van den Berg M, Van Roermund CW, van der Sluijs P, Distel B, Tabak HF (1996b) Analysis of the carboxyl-terminal peroxisomal targeting signal 1 in a homologous context in *Saccharomyces cerevisiae*. J Biol Chem 271:26375-26382

Erdmann R, Blobel G (1996) Identification of Pex13p, a peroxisomal membrane receptor for the PTS1 recognition factor. J Cell Biol 135:111-121

Fang Y, Morrell JC, Jones JM, Gould SJ (2004) PEX3 functions as a PEX19 docking factor in the import of class I peroxisomal membrane proteins. J Cell Biol 164:863-875. Epub 2004 Mar 2008

Fransen M, Brees C, Baumgart E, Vanhooren JC, Baes M, Mannaerts GP, van Veldhoven PP (1995) Identification and characterization of the putative human peroxisomal C-terminal targeting signal import receptor. J Biol Chem 270:7731-7736

Freeman BC, Morimoto RI (1996) The human cytosolic molecular chaperones hsp90, hsp70 (hsc70) and hdj-1 have distinct roles in recognition of a non-native protein and protein refolding. EMBO J 15:2969-2979

Frydman J (2001) Folding of newly translated proteins *in vivo*: the role of molecular chaperones. Annu Rev Biochem 70:603-647

Frydman J, Nimmesgern E, Oktsuka K, Hartl FU (1994) Folding of nascent polypeptide chains in a high molecular mass assembly with molecular chaperones. Nature 370:111-117

Galigniana MD, Harrell JM, Murphy PJ, Chinkers M, Radanyi C, Renoir JM, Zhang M, Pratt WB (2002) Binding of hsp90-associated immunophilins to cytoplasmic dynein: direct binding and *in vivo* evidence that the peptidylprolyl isomerase domain is a dynein interaction domain. Biochemistry 41:13602-13610

Gatto GJ Jr, Geisbrecht BV, Gould SJ, Berg JM (2000) Peroxisomal targeting signal-1 recognition by the TPR domains of human PEX5. Nat Struct Biol 7:1091-1095

Gautschi M, Mun A, Ross S, Rospert S, Wu Y, Hu Z, Thompson LJ, Yin LY, Patterson C (2002) A functional chaperone triad on the yeast ribosome. Proc Natl Acad Sci USA 99:4209-4214

Geuze HJ, Murk JL, Stroobants AK, Griffith JM, Kleijmeer MJ, Koster AJ, Verkleij AJ, Distel B, Tabak HF (2003) Involvement of the endoplasmic reticulum in peroxisome formation. Mol Biol Cell 14:2900-2907. Epub 2003 Apr 2904

Gietl C, Faber KN, van der Klei IJ, Veenhuis M (1994) Mutational analysis of the N-terminal topogenic signal of watermelon glyoxysomal malate dehydrogenase using the heterologous host *Hansenula polymorpha*. Proc Natl Acad Sci USA 91:3151-3155

Girzalsky W, Rehling P, Stein K, Kipper J, Blank L, Kunau W-H, Erdmann R (1999) Involvement of Pex13p in Pex14p localization and peroxisomal targeting signal 2-dependent protein import into peroxisomes. J Cell Biol 144:1151-1162

Glover JR, Andrews DW, Rachubinski RA (1994) *Saccharomyces cerevisiae* peroxisomal thiolase is imported as a dimer. Proc Natl Acad Sci USA 91:10541-10545

Gould SJ, Collins CS, Wirtz E, Clayton C (2002) Opinion: peroxisomal-protein import: is it really that complex? Nat Rev Mol Cell Biol 3:382-389

Gould SJ, Kalish JE, Morrell JC, Bjorkman J, Urquhart AJ, Crane DI (1996) Pex13p is an SH3 protein of the peroxisome membrane and a docking factor for the predominantly cytoplasmic PTS1 receptor. J Cell Biol 135:85-95

Gould SJ, Keller GA, Hosken N, Wilkinson J, Subramani S (1989) A conserved tripeptide sorts proteins to peroxisomes. J Cell Biol 108:1657-1664

Gould SJ, Keller GA, Schneider M, Howell SH, Garrard LJ, Goodman JM, Distel B, Tabak H, Subramani S (1990) Peroxisomal protein import is conserved between yeast, plants, insects and mammals. EMBO J 9:85-90

Gouveia AM, Guimaraes CP, Oliveira ME, Reguenga C, Sa-Miranda C, Azevedo JE (2003) Characterization of the peroxisomal cycling receptor, Pex5p, using a cell-free *in vitro* import system. J Biol Chem 278:226-232

Gouveia AM, Reguenga C, Oliveira ME, Sa-Miranda C, Azevedo JE (2000) Characterization of peroxisomal Pex5p from rat liver. Pex5p in the Pex5p-Pex14p membrane complex is a transmembrane protein. J Biol Chem 275:32444-32451

Harano T, Nose S, Uezu R, Shimizu N, Fujiki Y (2001) Hsp70 regulates the interaction between the peroxisome targeting signal type 1 (PTS1)-receptor Pex5p and PTS1. Biochem J 357:157-165

Hardonk MJ, Harms G, Koudstaal J (1985) Zonal heterogeneity of rat hepatocytes in the *in vivo* uptake of 17 nm colloidal gold granules. Histochemistry 83:473-477

Harper CC, Berg JM, Gould SJ, Harano T, Nose S, Uezu R, Shimizu N, Fujiki Y (2002) PEX5 binds the PTS1 independently of Hsp70 and the peroxin PEX12. J Biol Chem 26:26

Harrison CJ, Hayer-Hartl M, Di Liberto M, Hartl F, Kuriyan J (1997) Crystal structure of the nucleotide exchange factor GrpE bound to the ATPase domain of the molecular chaperone DnaK. Science 276:431-435

Hartl FU, Hayer-Hartl M (2002) Molecular chaperones in the cytosol: from nascent chain to folded protein. Science 295:1852-1858

Hausler T, Stierhof YD, Wirtz E, Clayton C (1996) Import of a DHFR hybrid protein into glycosomes *in vivo* is not inhibited by the folate-analogue aminopterin. J Cell Biol 132:311-324

Hazra PP, Suriapranata I, Snyder WB, Subramani S (2002) Peroxisome remnants in pex3delta cells and the requirement of Pex3p for interactions between the peroxisomal docking and translocation subcomplexes. Traffic 3:560-574

Hettema EH (1998), Import of proteins and fatty acids into peroxisomes. PhD thesis University of Amsterdam

Hettema EH, Girzalsky W, Berg Mvd, Erdmann R, Distel B (2000) *Saccharomyces cerevisiae* Pex3p and Pex19p are required for proper localization and stability of peroxisomal membrane proteins. EMBO J 19:223-233

Hines V, Brandt A, Griffiths G, Horstmann H, Brutsch H, Schatz G (1990) Protein import into yeast mitochondria is accelerated by the outer membrane protein MAS70. EMBO J 9:3191-3200

Hoepfner D, van den Berg M, Philippsen P, Tabak HF, Hettema EH (2001) A role for Vps1p, actin, and the Myo2p motor in peroxisome abundance and inheritance in *Saccharomyces cerevisiae.* J Cell Biol 155:979-990. Epub 2001 Dec 2003

Hohfeld J, Minami Y, Hartl FU (1995) Hip, a novel cochaperone involved in the eukaryotic Hsc70/Hsp40 reaction cycle. Cell 83:589-598

Huang S, Ratliff KS, Matouschek A (2002) Protein unfolding by the mitochondrial membrane potential. Nat Struct Biol 9:301-307

Hundley H, Eisenman H, Walter W, Evans T, Hotokezaka Y, Wiedmann M, Craig E (2002) The *in vivo* function of the ribosome-associated Hsp70, Ssz1, does not require its putative peptide-binding domain. Proc Natl Acad Sci USA 99:4203-4208

Hutchison KA, Dittmar KD, Pratt WB (1994) All of the factors required for assembly of the glucocorticoid receptor into a functional heterocomplex with heat shock protein 90 are preassociated in a self-sufficient protein folding structure, a "foldosome". J Biol Chem 269:27894-27899

Itoh T, Matsuda H, Mori H (1999) Phylogenetic analysis of the third hsp70 homolog in *Escherichia coli*; a novel member of the Hsc66 subfamily and its possible co-chaperone. DNA Res 6:299-305

James P, Pfund C, Craig EA (1997) Functional specificity among Hsp70 molecular chaperones. Science 275:387-389

Jankowski A, Kim JH, Collins RF, Daneman R, Walton P, Grinstein S (2001) *In situ* measurements of the pH of mammalian peroxisomes using the fluorescent protein pHluorin. J Biol Chem 276:48748-48753

Johnson MA, Snyder WB, Cereghino JL, Veenhuis M, Subramani S, Cregg JM (2001) *Pichia pastoris* Pex14p, a phosphorylated peroxisomal membrane protein, is part of a PTS-receptor docking complex and interacts with many peroxins. Yeast 18:621-641

Jones JM, Morrell JC, Gould SJ (2001) Multiple distinct targeting signals in integral peroxisomal membrane proteins. J Cell Biol 153:1141-1150

Jones JM, Morrell JC, Gould SJ (2004) PEX19 is a predominantly cytosolic chaperone and import receptor for class 1 peroxisomal membrane proteins. J Cell Biol 164:57-67

Kabani M, Beckerich JM, Brodsky JL (2002a) Nucleotide exchange factor for the yeast Hsp70 molecular chaperone Ssa1p. Mol Cell Biol 22:4677-4689

Kabani M, Beckerich JM, Gaillardin C (2000) Sls1p stimulates Sec63p-mediated activation of Kar2p in a conformation-dependent manner in the yeast endoplasmic reticulum. Mol Cell Biol 20:6923-6934

Kabani M, McLellan C, Raynes DA, Guerriero V, Brodsky JL (2002b) HspBP1, a homologue of the yeast Fes1 and Sls1 proteins, is an Hsc70 nucleotide exchange factor. FEBS Lett 531:339-342

Kawula TH, Lelivelt MJ (1994) Mutations in a gene encoding a new Hsp70 suppress rapid DNA inversion and bgl activation, but not proU derepression, in hns-1 mutant *Escherichia coli.* J Bacteriol 176:610-619

Kazlauskas A, Poellinger L, Pongratz I (2002) Two distinct regions of the immunophilin-like protein XAP2 regulate dioxin receptor function and interaction with hsp90. J Biol Chem 277:11795-11801

Kikuchi M, Hatano N, Yokota S, Shimozawa N, Imanaka T, Taniguchi H (2004) Proteomic analysis of rat liver peroxisome: presence of peroxisome-specific isozyme of Lon protease. J Biol Chem 279:421-428. Epub 2003 Oct 2015

Kim S, Schilke B, Craig EA, Horwich AL (1998) Folding *in vivo* of a newly translated yeast cytosolic enzyme is mediated by the SSA class of cytosolic yeast Hsp70 proteins. Proc Natl Acad Sci USA 95:12860-12865

Klein AT, Barnett P, Bottger G, Konings D, Tabak HF, Distel B (2001) Recognition of peroxisomal targeting signal type 1 by the import receptor Pex5p. J Biol Chem 276:15034-15041

Klein AT, van den Berg M, Bottger G, Tabak HF, Distel B (2002) *Saccharomyces cerevisiae* acyl-CoA oxidase follows a novel, non-PTS1, import pathway into peroxisomes that is dependent on Pex5p. J Biol Chem 277:25011-25019

Koch A, Thiemann M, Grabenbauer M, Yoon Y, McNiven MA, Schrader M (2003) Dynamin-like protein 1 is involved in peroxisomal fission. J Biol Chem 278:8597-8605. Epub 2002 Dec 8523

Komori M, Kiel JA, Veenhuis M (1999) The peroxisomal membrane protein Pex14p of *Hansenula polymorpha* is phosphorylated *in vivo.* FEBS lett 457:397-399

Lametschwandtner G, Brocard C, Fransen M, van Veldhoven P, Berger J, Hartig A (1998) The difference in recognition of terminal tripeptides as peroxisomal targeting signal 1 between yeast and human is due to different affinities of their receptor Pex5p to the cognate signal and to residues adjacent toiIt. J Biol Chem 273:33635-33643

Lazarow PB, Fujiki Y (1985) Biogenesis of peroxisomes. Annu Rev Cell Biol 1:489-530

Legakis J, Terlecky S (2001) PTS2 protein import into mammalian peroxisomes. Traffic 2:252-260

Leiper JM, Oatey PB, Danpure CJ (1996) Inhibition of alanine:glyoxylate aminotransferase 1 dimerization is a prerequisite for its peroxisome-to-mitochondrion mistargeting in Primary Hyperoxaluria Type 1. J Cell Biol 135:939-951

Li X, Gould SJ (2003) The dynamin-like GTPase DLP1 is essential for peroxisome division and is recruited to peroxisomes in part by PEX11. J Biol Chem 278:17012-17020 Epub 12003 Mar 17014

Liberek K, Marszalek J, Ang D, Georgopoulos C, Zylicz M (1991) *Escherichia coli* DnaJ and GrpE heat shock proteins jointly stimulate ATPase activity of DnaK. Proc Natl Acad Sci USA 88:2874-2878

Louvion JF, Abbas-Terki T, Picard D (1998) Hsp90 is required for pheromone signaling in yeast. Mol Biol Cell 9:3071-3083

Marzioch M, Erdmann R, Veenhuis M, Kunau W-H (1994) PAS7 encodes a novel yeast member of the WD-40 protein family essential for import of 3-oxoacyl-CoA thiolase, a PTS2-containing protein, into peroxisomes. EMBO J 13:4908-4918

Matouschek A, Pfanner N, Voos W, Strobel G, Zollner A, Angermayr M, Bandlow W (2000) Protein unfolding by mitochondria. The Hsp70 import motor. EMBO Rep 1:404-410

Matsumura T, Otera H, Fujiki Y (2000) Disruption of the interaction of the longer isoform of Pex5p, Pex5pL, with Pex7p abolishes peroxisome targeting signal type 2 protein import in mammals. Study with a novel Pex5-impared Chinese hamster ovary cell mutant. J Biol Chem 275:21715-21721

Mayer MP, Schroder H, Rudiger S, Paal K, Laufen T, Bukau B (2000) Multistep mechanism of substrate binding determines chaperone activity of Hsp70. Nat Struct Biol 7:586-593

McNew JA, Goodman JM (1996) The targeting and assembly of peroxisomal proteins: some old rules do not apply. Trends Biochem Sci 21:54-58

Misselwitz B, Staeck O, Rapoport TA (1998) J proteins catalytically activate Hsp70 molecules to trap a wide range of peptide sequences. Mol Cell 2:593-603

Motley A, Lumb MJ, Patel PB, Jennings PR, Zoysa PD, Wanders RJA, Tabak HF, Danpure CJ (1995) Identification of the peroxisomal targeting sequence of mammalian alanine: glyoxylate aminotransferase 1. Increased degeneracy and contaxt specificity of the mammalian PTS1 motif and implications for the peroxisome-to-mitochondrion mistargeting of AGT in primary hyperoxaluria type 1. J Cell Biol 131:95-109

Motley AM, Hettema EH, Ketting R, Plasterk R, Tabak HF (2000) *Caenorhabditis elegans* has a single pathway to target matrix proteins to peroxisomes. EMBO reports 1:40-46

Nelson R, T Z, C N, M W-W, EA C (1992) The translation machinery and 70 kd heat shock protein cooperate in protein synthesis. Cell 71:97-105

Neuberger G, Maurer-Stroh S, Eisenhaber B, Hartig A, Eisenhaber F (2003) Motif refinement of the peroxisomal targeting signal 1 and evaluation of taxon-specific differences. J Mol Biol 328:567-579

Ngosuwan J, Wang NM, Fung KL, Chirico WJ (2002) Roles of cytosolic Hsp70 and Hsp40 molecular chaperones in post-translational translocation of presecretory proteins into the endoplasmic reticulum. J Biol Chem 19:19

Nicola A, W C, A H (1999) Co-translational folding of an alphavirus capsid protein in the cytosol of living cells. Nat Cell Biol 1:341-345

Nicolet CM, Craig EA (1989) Isolation and characterization of STI1, a stress-inducible gene from *Saccharomyces cerevisiae*. Mol Cell Biol 9:3638-3646

Novikoff PM, Novikoff AB (1972) Peroxisomes in absorptive cells of mammalian small intestine. J Cell Biol 53:532-560

Pfund C, Lopez-Hoyo N, Ziegelhoffer T, Schilke BA, Lopez-Buesa P, Walter WA, Craig EA (1998) The molecular chaperone Ssb from *Saccharomyces cerevisiae* is a component of the ribosome-nascent chain complex. EMBO J 17:3981-3989

Picard D, Khursheed B, Garabedian MJ, Fortin MG, Lindquist S, Yamamoto KR (1990) Reduced levels of hsp90 compromise steroid receptor action *in vivo*. Nature 348:166-168

Pishvaee B, Costaguta G, Yeung BG, Ryazantsev S, Greener T, Greene LE, Eisenberg E, McCaffery JM, Payne GS (2000) A yeast DNA J protein required for uncoating of clathrin-coated vesicles *in vivo*. Nat Cell Biol 2:958-963

Pratt WB, Krishna P, Olsen LJ (2001) Hsp90-binding immunophilins in plants: the protein movers. Trends Plant Sci 6:54-58

Purdue PE, Yang X, Lazarow PB (1998) Pex18p and Pex21p, a novel pair of related peroxins essential for peroxisomal targeting by the PTS2 pathway. J Cell Biol 143:1859-1869

Raynes DA, Guerriero V Jr (1998) Inhibition of Hsp70 ATPase activity and protein renaturation by a novel Hsp70-binding protein. J Biol Chem 273:32883-32888

Reguenga C, Oliveira ME, Gouveia AM, Sa-Miranda C, Azevedo JE (2001) Characterization of the mammalian peroxisomal import machinery: Pex2p, Pex5p, Pex12p, and Pex14p are subunits of the same protein assembly. J Biol Chem 276:29935-29942

Rehling P, Marzioch M, Niessen F, Wittke E, Veenhuis M, Kunau W-H (1996) The import receptor for the peroxisomal targeting signal 2 (PTS2) in *Saccharomyces cerevisiae* is encoded by the *PAS7* gene. EMBO J 15:2901-2913

Rottensteiner H, Kramer A, Lorenzen S, Stein K, Landgraf C, Volkmer-Engert R, Erdmann R (2004) Peroxisomal membrane proteins contain common Pex19p-binding sites that are an integral part of their targeting signals. Mol Biol Cell 15:3406-3417. Epub 2004 May 3407

Rudiger S, Germeroth L, Schneider-Mergener J, Bukau B (1997) Substrate specificity of the DnaK chaperone determined by screening cellulose-bound peptide libraries. EMBO J 16:1501-1507

Ruigrok CCM, de Jonge W, Tabak HF, Braakman LJ (manuscript in preparation) Efficiency of peroxisomal protein import is determined by competition between translocation and a change into import-incompetence

Sacksteder KA, Jones JM, South ST, Li X, Liu Y, Gould SJ (2000) PEX19 binds multiple peroxisomal membrane proteins, is predominantly cytoplasmic, and is required for peroxisome membrane synthesis. J Cell Biol 148:931-944

Scheufler C, Brinker A, Bourenkov G, Pegoraro S, Moroder L, Bartunik H, Hartl FU, Moarefi I (2000) Structure of TPR domain-peptide complexes: critical elements in the assembly of the Hsp70-Hsp90 multichaperone machine. Cell 101:199-210

Schmid D, Baici A, Gehring H, Christen P, Reinstein J, Bukau B (1994) Kinetics of molecular chaperone action. Science 263:971-973

Schulte TW, Blagosklonny MV, Ingui C, Neckers L (1995) Disruption of the Raf-1-Hsp90 molecular complex results in destabilization of Raf-1 and loss of Raf-1-Ras association. J Biol Chem 270:24585-24588

Sichting M, Schell-Steven A, Prokisch H, Erdmann R, Rottensteiner H (2003) Pex7p and Pex20p of *Neurospora crassa* function together in PTS2-dependent protein import into peroxisomes. Mol Biol Cell 14:810-821

Silver PA, Way JC, James P, Pfund C, Craig EA (1993) Eukaryotic DnaJ homologs and the specificity of Hsp70 activity. Cell 74:5-6

Silverstein AM, Galigniana MD, Kanelakis KC, Radanyi C, Renoir JM, Pratt WB (1999) Different regions of the immunophilin FKBP52 determine its association with the glucocorticoid receptor, hsp90, and cytoplasmic dynein. J Biol Chem 274:36980-36986

Sondermann H, Ho AK, Listenberger LL, Siegers K, Moarefi I, Wente SR, Hartl FU, Young JC (2002) Prediction of novel Bag-1 homologs based on structure/function analysis identifies Snl1p as an Hsp70 co-chaperone in *Saccharomyces cerevisiae*. J Biol Chem 277:33220-33227

Soto U, Pepperkok R, Ansorge W, Just WW (1993) Import of firefly luciferase into mammalian peroxisomes *in vivo* requires nucleoside triphosphates. Exp Cell Res 205:66-75

South ST, Baumgart E, Gould SJ (2001) Inactivation of the endoplasmic reticulum protein translocation factor, Sec61p, or its homolog, Ssh1p, does not affect peroxisome biogenesis. Proc Natl Acad Sci USA 98:12027-12031. Epub 12001 Oct 12022

South ST, Sacksteder KA, Li X, Liu Y, Gould SJ (2000) Inhibitors of COPI and COPII do not block *PEX3*-mediated peroxisome synthesis. J Cell Biol 149:1345-1359

Strobel G, Zollner A, Angermayr M, Bandlow W (2002) Competition of spontaneous protein folding and mitochondrial import causes dual subcellular location of major adenylate kinase. Mol Biol Cell 13:1439-1448

Subramani S (1996) Protein translocation into peroxisomes. J Biol Chem 271:32483-32486

Swinkels BW, Gould SJ, Bodnar AG, Rachubinski RA, Subramani S (1991) A novel, cleavable peroxisomal targeting signal at the amino-terminus of the rat 3-ketoacyl-CoA thiolase. EMBO J 10:3255-3262

Szilard RK, Titorenko VI, Veenhuis M, Rachubinski RA (1995) Pay 32p of the yeast *Yarrowia lipolytica* is an intraperoxisomal component of the matrix protein translocation machinery. J Cell Biol 131:1453-1469

Tabak HF, Murk JL, Braakman I, Geuze HJ (2003) Peroxisomes start their life in the endoplasmic reticulum. Traffic 4:512-518

Takayama S, Bimston DN, Matsuzawa S, Freeman BC, Aime-Sempe C, Xie Z, Morimoto RI, Reed JC (1997) BAG-1 modulates the chaperone activity of Hsp70/Hsc70. EMBO J 16:4887-4896

Terlecky SR, Dice JF (1993) Polypeptide import and degradation by isolated lysosomes. J Biol Chem 268:23490-23495

Terlecky SR, Nuttley WM, McCollum D, Sock E, Subramani S (1995) The *Pichia pastoris* peroxisomal protein PAS8p is the receptor for the C-terminal tripeptide peroxisomal targeting signal. EMBO J 14:3627-3634

Titorenko VI, Nicaud JM, Wang H, Chan H, Rachubinski RA (2002) Acyl-CoA oxidase is imported as a heteropentameric, cofactor-containing complex into peroxisomes of *Yarrowia lipolytica*. J Cell Biol 156:481-494

Titorenko VI, Smith JJ, Szilard RK, Rachubinski RA (1998) Pex20p of the yeast *Yarrowia lipolytica* is required for the oligomerization of thiolase in the cytosol and for its targeting to the peroxisome. J Cell Biol 142:403-420

Tsukamoto T, Hata S, Yokota S, Miura S, Fujiki Y, Hijikata M, Miyazawa S, Hashimoto T, Osumi T (1994) Characterization of the signal peptide at the amino terminus of the rat peroxisomal 3-ketoacyl-CoA thiolase precursor. J Biol Chem 269:6001-6010

Ungewickell E, Ungewickell H, Holstein SE, Lindner R, Prasad K, Barouch W, Martin B, Greene LE, Eisenberg E (1995) Role of auxillin in uncoating clathrin-coated vesicles. Nature 378:632-635

Urquhart AJ, Kennedy D, Gould SJ, Crane DI (2000) Interaction of Pex5p, the type 1 peroxisome targeting signal receptor, with the peroxisomal membrane proteins Pex14p and Pex13p. J Biol Chem 275:4127-4136

van der Klei IJ, Hilbrands RE, Swaving GJ, Waterham HR, Vrieling EG, Titorenko VI, Cregg JM, Harder W, Veenhuis M (1995) The *Hansenula polymorpha* PER3 gene is essential for the import of PTS1 proteins into the peroxisomal matrix. J Biol Chem 270:17229-17236

Van der Leij I, Franse MM, Elgersma Y, Distel B, Tabak HF (1993) PAS10 is a tetratricopeptide-repeat protein that is essential for the import of most matrix proteins into peroxisomes of *Saccharomyces cerevisiae*. Proc Natl Acad Sci USA 90:11782-11786

van der Lende TR, Breeuwer P, Abee T, Konings WN, Driessen AJ (2002) Assessment of the microbody luminal pH in the filamentous fungus *Penicillium chrysogenum*. Biochim Biophys Acta 1589:104-111

Van Roermund CW, De Jong M, L IJ, Van Marle J, Dansen TB, Wanders RJ, Waterham HR (2004) The peroxisomal lumen in *Saccharomyces cerevisiae* is alkaline. J Cell Sci 117:4231-4237

Verner K, Schatz G (1987) Import of an incompletely folded precursor protein into isolated mitochondria requires an energized inner membrane, but no added ATP. EMBO J 6:2449-2456

Voorn-Brouwer T, Kragt A, Tabak HF, Distel B (2001) Peroxisomal membrane proteins are properly targeted to peroxisomes in the absence of COPI- and COPII-mediated vesicular transport. J Cell Sci 114:2199-2204

Walton PA, Wendland M, Subramani S, Rachubinski RA, Welch WJ (1994) Involvement of 70-kD heat-shock proteins in peroxisomal import. J Cell Biol 125:1037-1046

Wendland M, Subramani S (1993) Cytosol-dependent peroxisomal protein import in a permeabilized cell system. J Cell Biol 120:675-685

Werner-Washburne M, Stone DE, Craig EA (1987) Complex interactions among members of an essential subfamily of hsp70 genes in *Saccharomyces cerevisiae*. Mol Cell Biol 7:2568-2577

Wiemer E, Nuttley WM, Bertolaet BL, Li X, Francke U, Wheelock MJ, Anne UK, Johnson KR, Subramani S (1995a) The human PTS1 receptor restores peroxisomal protein import deficiency in cells from patients with fatal peroxisomal disorders. J Cell Biol 130:51-65

Wiemer EA, Terlecky SR, Nuttley WM, Subramani S, van der Klei IJ, Hilbrands RE, Swaving GJ, Waterham HR, Vrieling EG, Titorenko VI, Cregg JM, Harder W, Veenhuis M (1995b) Characterization of the yeast and human receptors for the carboxy-terminal tripeptide peroxisomal targeting signal. Cold Spring Harb Symp Quant Biol 60:637-648. Biotechnology Institute, University of Groningen, Haren, The Netherlands

Wimmer B, Lottspeich F, van der Klei I, Veenhuis M, Gietl C (1997) The glyoxysomal and plastid molecular chaperones (70-kDa heat shock protein) of watermelon cotyledons are encoded by a single gene. Proc Natl Acad Sci USA 94:13624-13629

Xu Y, Lindquist S (1993) Heat-shock protein hsp90 governs the activity of pp60v-src kinase. Proc Natl Acad Sci USA 90:7074-7078

Yang X, Purdue P, Lazarow P (2001) Eci1p uses a PTS1 to enter peroxisomes: either its own or that of a partner, Dci1p. Eur J Cell Biol 80:126-138

Young JC, Hoogenraad NJ, Hartl FU, Obermann WM (2003) Molecular chaperones hsp90 and hsp70 deliver preproteins to the mitochondrial import receptor tom70. Cell 112:41-50

Young JC, Obermann WM, Hartl FU (1998) Specific binding of tetratricopeptide repeat proteins to the C-terminal 12-kDa domain of hsp90. J Biol Chem 273:18007-18010

Zhang JW, Lazarow PB (1994) PEB1 (PAS7) in *Saccharomyces cerevisiae* encodes a hydrophilic, intraperoxisomal protein which is a member of the WD repeat family and is essential for the import of thiolase into peroxisomes. J Cell Biol 129:65-80

Braakman, Ineke

Department of Cellular Protein Chemistry, Utrecht University, Padualaan 8, 3584 CH, Utrecht, The Netherlands
i.braakman@chem.uu.nl

de Jonge, Wim

Department of Cellular Protein Chemistry, Utrecht University, Padualaan 8, 3584 CH, Utrecht, The Netherlands. Current adress: Institute of Information and Computing Sciences, Utrecht University, Padualaan 14, 3584CH Utrecht, The Netherlands.

Tabak, Henk F.

Department of Cellular Protein Chemistry, Utrecht University, Padualaan 8, 3584 CH, Utrecht, The Netherlands, and Laboratory of Cell Biology, University Medical Center Utrecht and Center for Biomedical Genetics, 3584 CX Utrecht, The Netherlands

Proteasomal degradation of misfolded proteins

Robert Gauss, Oliver Neuber, and Thomas Sommer

Abstract

One of the most important functions of cellular quality control systems is to maintain structural fidelity of proteins. Molecular chaperones prevent aggregation and assist folding of newly synthesized proteins in the cytosol and the ER. Furthermore, in concert with ubiquitin-ligases, chaperones detect misfolded or damaged proteins and target them for degradation by the ubiquitin-proteasome system. Some *U-box* ligases link recognition of aberrant cytosolic proteins to degradation. In degradation of malfolded secretory proteins from the ER, recognition by chaperones is separated from the ubiquitin-proteasome system by the ER-membrane. Therefore, dislocation precedes ubiquitination mediated by two *RING*-finger ligases in the ER membrane. Proteins are marked for degradation by the attachment of poly-ubiquitin chains. Poly-ubiquitinated proteins are subsequently recognized by a Cdc48p/p97 complex and delivered to the proteasome where they are degraded. Malfunction of protein degradation by the ubiquitin-proteasome system leads to the generation of severe human diseases.

1 Introduction

Each protein represents a discrete chemical entity with a unique three-dimensional structure essential to its function. Misfolded, partly denatured proteins have a tendency to aggregate and disturb cellular functions. It is, therefore, of prime importance for the cell to ensure that proteins fold correctly into their native conformation and to eliminate any proteins with incorrect structure. Before proteins attain their native folded state, they may expose hydrophobic elements on their surface that are buried in the native conformation. In aqueous phases, hydrophobic molecules tend to aggregate in order to minimize the free enthalpy. Isolated purified proteins will aggregate during folding even at relatively low concentrations. Inside cells, the concentration of folding proteins is significantly higher. Aggregation during the folding process is prevented by molecular chaperones. These ubiquitous and abundant proteins assist folding of both nascent polypeptides still attached to ribosomes, and released, completely synthesised polypeptide chains.

As the native conformation is of such great importance, various 'proof-reading' systems operate to ensure protein integrity. Quality control (QC) of newly synthesized proteins occurs at the level of translation, folding, and maturation. To pass the final checkpoint of the QC system, a protein must have acquired its native conformation. However, correctly folded proteins can also be damaged by aging or cellular stress while performing their cellular functions. Sensor molecules, therefore, constantly monitor the structure of cellular proteins. Surveillance of protein structure is a second function of molecular chaperones. In addition, they play a central role in deciding the fate of a protein. If a protein fails to pass the QC checkpoints or is damaged due to cellular stress, it will either be repaired or destroyed (Fig. 1). The "dual response" to misfolded protein is mediated by two protein systems. Molecular chaperones help damaged proteins to refold and regain their original function. Proteins marked for destruction are passed on to the ubiquitin-proteasome system. The ubiquitin-proteasome system is responsible for the degradation of most cytosolic, nuclear, and ER-resident proteins. The accumulation of unfolded proteins within the cell leads to upregulation of chaperones of the 'heat-shock protein' family that prevent aggregation of aberrant proteins and support refolding. In addition, components of the ubiquitin-proteasome system are expressed at higher level to increase degradation capacity.

In this review, we will give an overview of protein degradation by the ubiquitin-proteasome system and the function of molecular chaperones in detecting aberrant proteins and targeting them for degradation. We will try to outline similarities, but also differences between cytosolic and ER-associated protein degradation and speculate about the function of central components in substrate selection and degradation. Recent publications that indicate a specific role of chaperones in protein degradation will be discussed. Finally, we describe diseases associated with protein degradation and a new approach in the treatment of folding diseases.

2 Recognition and degradation of aberrant proteins in the cytosol

To prevent formation of protein aggregates, newly synthesized proteins that fail to attain their native conformation as well as proteins that have been damaged in response to aging or cellular stress are detected by a cellular quality control system. The QC system recognizes whether a protein is correctly folded or not. In addition, it determines the fate of an aberrant protein. The protein can either be refolded with the assistance of molecular chaperones, or subjected to degradation by the ubiquitin-proteasome system. How are cytosolic proteins recognized as misfolded? Which molecular chaperones are involved in detection of misfolded and damaged proteins in the cytosol? How are proteins marked for degradation and how does the ubiquitin-proteasome system degrade these proteins?

2.1 Hsp70 and Hsp90 chaperones – protein folding and re-folding

Molecular chaperones of the Hsp70 and Hsp90 (heat-shock protein 70 and 90) families are the major components of the cytosolic protein folding and QC system

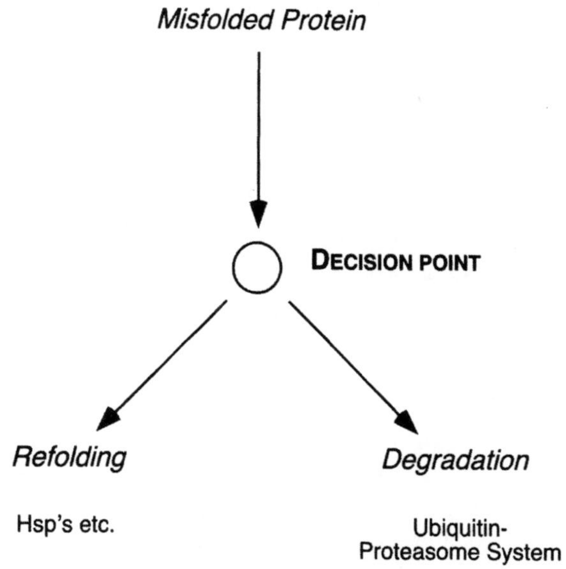

Fig. 1. Dual response to misfolded proteins. Misfolded proteins that are recognized as unfolded can either be refolded or destructed. Molecular chaperones of the Hsp70 and Hsp90 families help proteins to refold and regain their native conformation. The ubiquitin-proteasome system degrades proteins that are destined for destruction.

(Frydman 2001; Wegele et al. 2004). Both chaperone families contribute to protein folding, but they seem to prefer different types of substrates. Hsp70 proteins mainly assist newly synthesized proteins to attain their native structure. In addition, they help to refold partly denatured and even aggregated proteins. The function of Hsp70 is based on its ability to interact transiently with short peptide stretches of substrate proteins in an ATP-dependent manner. The overall architecture of Hsp70s is composed of three distinct domains. An N-terminal ATPase domain, a central peptide-binding domain, and a C-terminal domain (Flaherty et al. 1990; Zhu et al. 1996). The substrate-binding site of Hsp70 is formed by a hydrophobic pocket of two sheets of β-strands (Zhu et al. 1996). It binds patches of hydrophobic amino acids of folding intermediates or misfolded proteins (Mayer et al. 2000; Rudiger et al. 2001). The C-terminal domain functions as a lid that permits entry and release of polypeptide substrates (Bukau and Horwich 1998; Zhu et al. 1996). Opening and closure of the lid seems to be regulated by ATP binding and hydrolysis by the ATPase domain. In the ATP-bound state, the peptide-binding domain possesses a low affinity for substrate proteins. Upon ATP hydrolysis, the chaperone undergoes conformational changes that lead to a strong binding of the substrate molecule (McCarty et al. 1995; Theyssen et al. 1996). ATP hydrolysis is enhanced by Hsp40 co-chaperones. Conversely, the substrate protein is released when the bound ADP is exchanged to ATP by nucleotide exchange factors such as BAG-1 or BAG-1-related proteins.

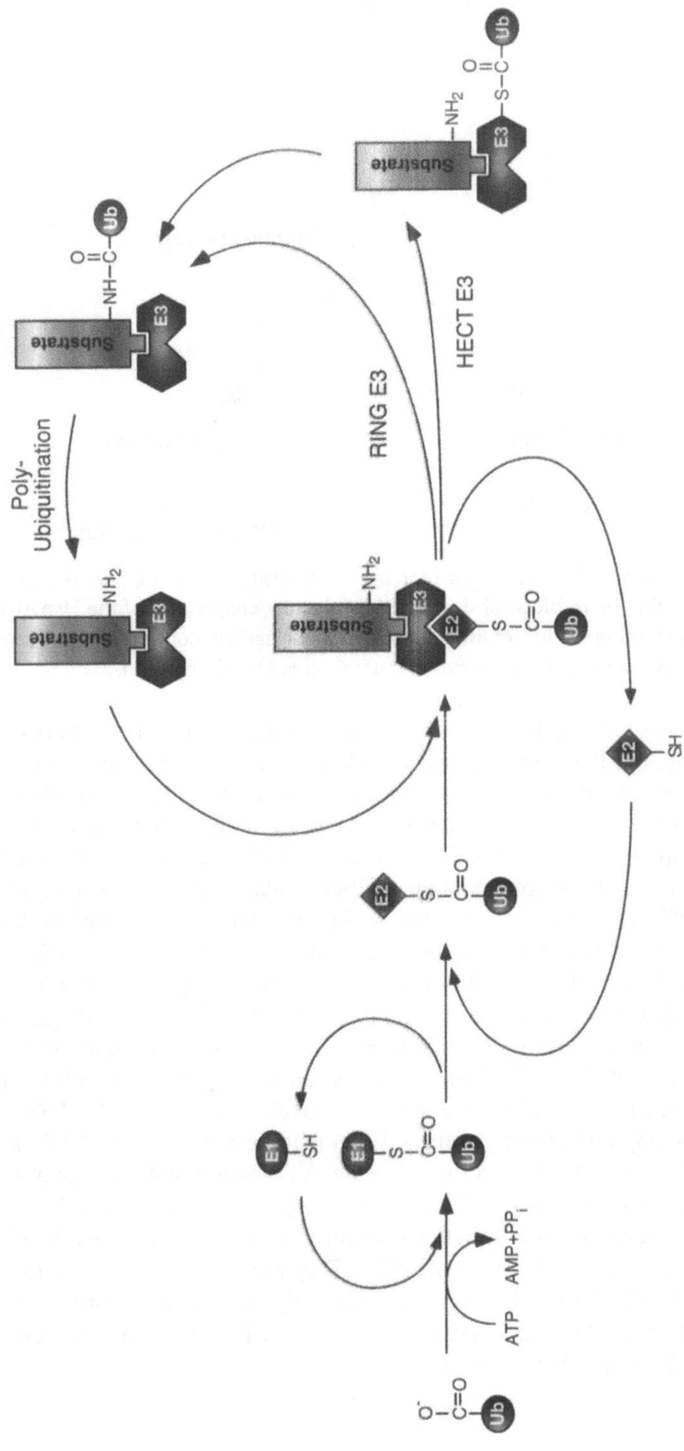
Ub
O⁻
C=O
ATP
AMP+PPi
E1
SH
S
E2
Substrate
E3
NH2
NH-C
O
S-C
Poly-Ubiquitination
RING E3
HECT E3

Fig. 2. (overleaf) The ubiquitination pathway. Ubiquitin is activated by the formation of a thio-ester bond between its C-terminus and the active site of E1 in an ATP-depended manner. Next, ubiquitin is passed on to one of a number of E2 proteins. E2s interact with an even greater number of E3s. Previously E3s might have or might have not recruited a substrate protein. A *RIN*G-finger E3 performs a scaffold building function and brings together substrate and E2 proteins. Ubiquitin is then transferred directly from the E2 to the substrate. In contrast, a *HECT*-domain E3 forms a thio-ester intermediate with ubiquitin. Ubiquitin is transferred to the substrate in a second step.

Hsp90 is one of the most abundant proteins in unstressed cells. It represents at least 1% of the total soluble cytosolic protein (Lai et al. 1984). Hsp90 also promotes protein folding. But in contrast to Hsp70, it prevents unfolding and aggregation of already folded proteins, thus, serves as a "house-keeping" function (Picard 2002). Hsp90 can stabilize meta-stable protein domains and keeps denatured proteins in a folding-competent state (Freeman and Morimoto 1996; Minami et al. 2001). Hsp90 keeps meta-stable signal proteins such as steroid hormone receptors in a conformation that possess a high affinity for their substrates. Binding of a hormone molecule stabilizes the protein.

Hsp70 and Hsp90 function depends on their association with co-factors, so-called co-chaperones. Hsp40 proteins are the main co-chaperones of Hsp70. Isoforms of Hsp40s (like Ydj1p and Sis1p in yeast) regulate the ATPase of Hsp70s (Cyr et al. 1992; Lu and Cyr 1998). In addition, Hsp70/Hsp40 complexes act on nascent polypeptide chains (Gautschi et al. 2001; Hundley et al. 2002). The small acidic protein p23 (Sba1p in yeast) interacts with the N-terminal domain of Hsp90. This interaction seems to stabilize substrate binding in an Hsp90-p23 complex (Dittmar and Pratt 1997). The co-chaperone Hop exhibits a more promiscuous interaction specificity. Hop was shown to bind to Hsp70 and Hsp90 (Perdew and Whitelaw 1991; Wegele et al. 2003) and seems to act as a coupling factor that links Hsp70 and Hsp90. Like most co-chaperones, Hop utilizes *TPR* domains to bind to Hsp70 and Hsp90 chaperones (see Section 2.3.1 for more details about *TPR* motifs).

Although protein folding is mediated and monitored by molecular chaperones such as Hsp70 and Hsp90, chaperones cannot ensure proper folding. Proteins might be kinetically trapped in misfolded conformations or irreversible damaged by oxidation or covalent modification. These proteins have to be destroyed to prevent intracellular aggregates and interference with cellular functions.

2.2 The ubiquitin-proteasome system

Endocytosed proteins and proteins of the late secretory pathway are degraded in lysosomes (in yeast the vacuole). In contrast, cytosolic and nuclear proteins are degraded by the cytosolic ubiquitin-proteasome system. The signal that leads to degradation by the ubiquitin-proteasome system is the covalent attachment of the conserved 76-residue polypeptide ubiquitin to the substrate. Whereas mono-ubiquitination serves as a signal for protein localization, modification, or interaction with partner proteins, formation of poly-ubiquitin chains is necessary for subsequent degradation. Poly-ubiquitination is a multi-step process, which requires at least three classes of enzymes (Hershko and Ciechanover 1998; Pickart 2001): In a first ATP-dependent step, an ubiquitin-activating enzyme (E1) catalyzes the activation of the carboxy-terminus of ubiquitin (Fig. 2). Ubiquitin is conjugated to the E1 by forming a thio-ester-linkage between its C-terminal glycine and an essential cysteine in the E1 active site. Most organisms, including humans and the yeast *S. cerevisiae,* only possess a single E1 enzyme. Activated ubiquitin is subsequently transferred to a cysteine residue in the active site of an ubiquitin-conjugating enzyme (Ubc, E2). Several E2 proteins are known to exist: there are 11 E2s in *S. cerevisiae* and even more in higher organisms, including multiple isoforms of some E2s. All Ubc's share a conserved core domain of about 150 amino acids harbouring the active site with the conserved cysteine residue.

Finally, ubiquitin-protein ligases (E3s) mediate the attachment of ubiquitin residues to the target protein and provide substrate specificity (Fig. 2). According to their structure and the catalytic mechanisms, E3 ligases can be distinguished into two major classes: *HECT* (homology to E6-AP carboxy terminus)-domain E3 ligases receive ubiquitin from an interacting E2 and form a thio-ester intermediate with a cysteine residue in their catalytic domain. Subsequently, ubiquitin is transferred from the *HECT*-domain E3 to the substrate. In contrast, *RING* (really interesting new gene)-finger E3 ligases do not form a covalent intermediate. The *RING*-finger domain performs a scaffold building function. It recruits the substrate and the corresponding Ubc, thus, bringing them in close proximity. Ubiquitin is then transferred directly from the E2 to the substrate.

Successive ubiquitination of the attached ubiquitin molecule leads to the formation of a poly-ubiquitin chain. Each E3 recognizes only a specific set of substrates that share a limited number of ubiquitination signals. In addition, E3s cooperates only with one or a few E2s. Poly-ubiquitin chains can differ in linkage of the lysines of adjacent ubiquitin molecules. Only Lys-48 linked poly-ubiquitin chains are sufficient to mark proteins for proteasomal degradation, whereas Lys11-, Lys29-, or Lys63-linked poly-ubiquitin chains provide signals in different, non-proteolytic processes.

Proteins that are poly-ubiquitinated by a Lys48-linked ubiquitin chain are targeted for degradation by the 26S proteasome. Proteasomes represent a high efficient and tightly regulated proteolytical machinery. They are large multi-subunit proteases, located in the cytosol, and in the nucleus of eucaryotic cells (Baumeister et al. 1998; Bochtler et al. 1999). Proteasomes consist of two subcomplexes: The 20S core particle is a barrel-shaped complex that harbours threonine proteases with their active sites facing the central cavity. The 19S caps flanking the entries of the 20S core particle mediate the recognition of poly-ubiquitinated degradation substrates and exhibit an ATP-dependent unfolding activity. The architecture of the proteasome ensures that only proteins marked for degradation are recognized, transported into the central cavity, and digested.

2.3 Some *U-box* ligases link chaperones to the ubiquitin-proteasome system

How are proteins that are destined for degradation passed on from chaperones to the degradation system? It is reasonable to speculate that chaperones interact with factors that link recognition of aberrant proteins to degradation (Fig. 3). In recent years, several studies identified a new class of ubiquitin ligases, which cooperates with molecular chaperones, in particular Hsp70s and Hsp90s. This class of E3-ligases ubiquitinates chaperone-bound substrates via a 75-amino-acid domain that displays differences to the former known *HECT*- or RING-finger domains. The domain was termed *U-box*. (Hatakeyama and Nakayama 2003; Patterson 2002). Although structurally related to the *RING*-finger domain, it lacks the metal-chelating residues essential for *RING*-finger domain stability (Aravind and Koonin 2000). So far, six mammalian *U-box* proteins have been isolated, and all of them mediate ubiquitination in conjunction with E1 and E2s and in the absence of other E3 components (Hatakeyama et al. 2001).

2.3.1 CHIP – a conductor of protein degradation in the cytosol

The best-characterized *U-box* ligase is the C-terminus of Hsc70 interacting protein (CHIP). It was initially identified in a screen for human proteins containing tricopeptide repeats, *TPR*s (Ballinger et al. 1999). The *TPR* motif consists of an approximately 34 amino acid sequence arranged in two α-helices, which are packed together in a "knobs-in-holes" manner (Sikorski et al. 1990). *TPRs* occur in over 25 proteins of diverse biological function and are conserved from bacteria to humans. Multiple *TPRs* are organized into a right-handed superhelix-like structure with a groove of large surface area available for ligand binding (Das et al. 1998). Proteins with multiple copies of *TPR* motifs function as scaffolding proteins and coordinate the assembly of proteins into multi-subunit complexes. For instance, evolutionary conserved subunits of the anaphase promoting complex (APC), Cdc27p, Cdc16p, and Cdc23p contain multiple *TPRs*. They are thought to mediate the interaction of different APC subunits or might be required to bind APC cofactors or substrates (Peters 1999). In protein phosphatase 5 (PP5), the *TPR* domain mediates the interaction of PP5 and Hsp90 within an Hsp90-glucocorticoid receptor complex (Chinkers 1994; Silverstein et al. 1997).

CHIP consists of a C-terminal *U-box* domain and an N-terminal domain that contains a threefold *TPR* motif. Together with an adjacent charged region this domain binds to a highly conserved EEVD motif in the C-terminus of Hsp70 and Hsp90 (Ballinger et al. 1999; Hatakeyama et al. 2004; Scheufler et al. 2000). Thus, the domain architecture of CHIP link chaperone-mediated recognition of aberrant proteins and degradation. Indeed, it was shown that CHIP participates in degradation of Hsp70 and Hsp90 substrates. In an *in vitro* study, CHIP multi-ubiquitinated thermally denatured firefly luciferase in the presence of an E1 and E2. This occurred only when the unfolded substrate interacted with Hsp90 or Hsp70. Furthermore, CHIP did not mediated ubiquitination of native luciferase

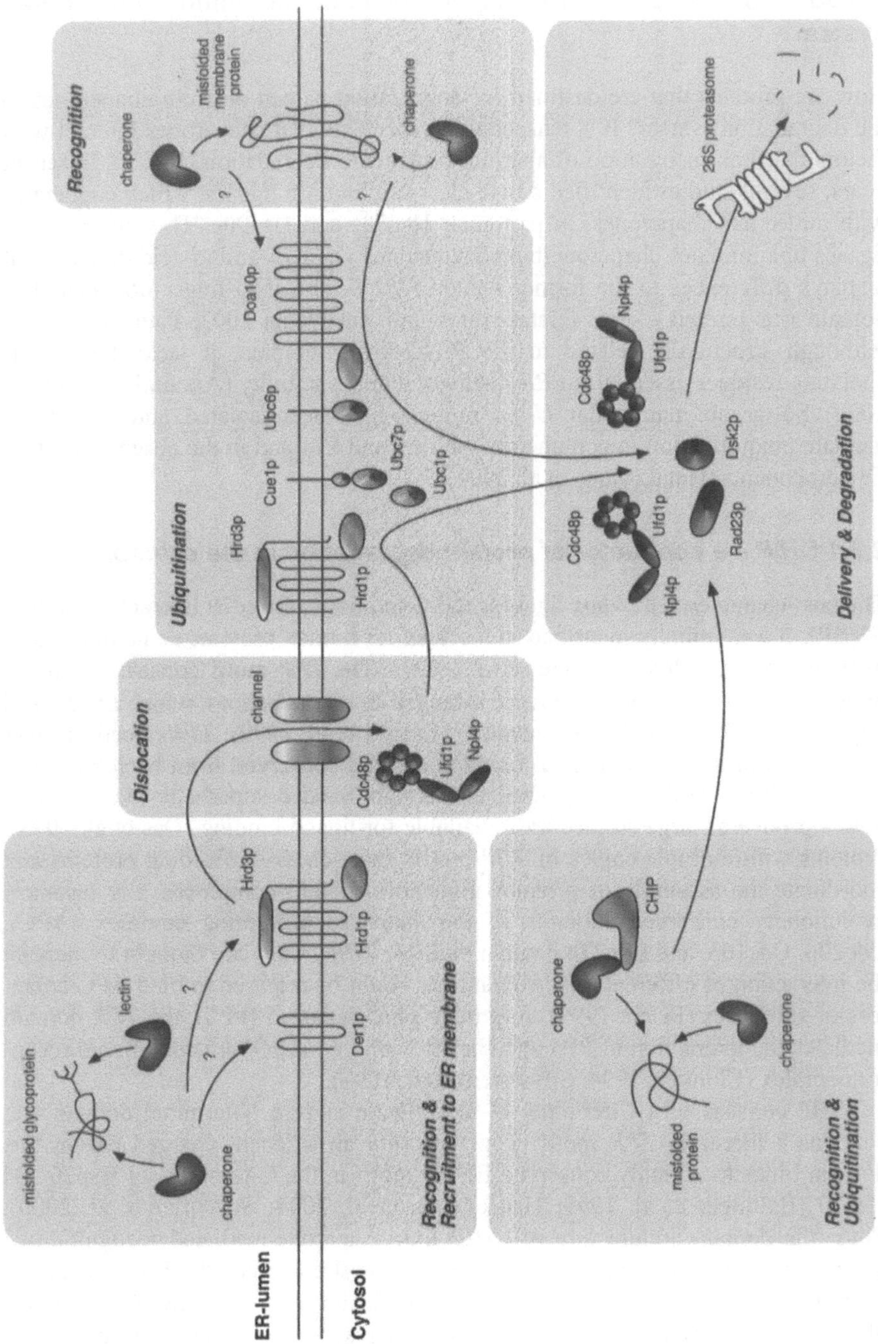

Fig. 3. Hypothetical model of the degradation of misfolded proteins. For details see text.

(Murata et al. 2001). Another study indicated that CHIP abolishes the steroid-binding activity and transactivation potential of the glucocorticoid receptor (GR). GR is known to be an Hsp90 substrate. CHIP dissociates GR from Hsp90 and mediates ubiquitination of GR for degradation by the proteasome (Connell et al. 2001).

CHIP is also involved in the degradation of aberrant proteins from the ER. The cystic fibrosis transmembrane conductance regulator (CFTR) is ubiquitinated by the action of the CHIP *U-box*. Ubiquitination of CFTR requires cooperation of CHIP and Hsc70 in a *TPR*-dependent manner (Meacham et al. 2001). The co-chaperone HspBP1 was recently identified as an inhibitor of the ubiquitin-ligase activity of CHIP when in complex with Hsp70 (Alberti et al. 2004). HspBP1 interferes with CHIP-induced degradation of immature forms of CFTR and stimulates their maturation. Thus, HspBP1 might represent a regulating factor of CHIP in chaperone-mediated degradation.

An additional factor in degradation of CHIP substrates is the Hsp70 co-chaperone BAG-1. It connects the Hsp70/CHIP complex to the 26S proteasome by binding simultaneously to Hsp70 and the proteasome (Demand et al. 2001; Luders et al. 2000). BAG-1 binds to Hsp70 and the Rpn1 subunit of the proteasomal 19S complex via its BAG domain, and its ubiquitin-like domain, respectively. Interestingly, CHIP enhances proteasomal association of BAG-1 by attaching Lys11-linked poly-ubiquitin chains to BAG-1 (Alberti et al. 2002). Recently, EKN1 was identified as an interaction partner of CHIP (Hatakeyama et al. 2004). EKN1 contains a domain that functions in nuclear import, and three *TPR* motifs, which were shown to bind to Hsp70. Thus, substrate recognition of CHIP might depend on direct interaction with Hsp70 or Hsp90, or is mediated by EKN1 interaction with Hsp70.

Chaperone-bound substrates can either be subjected to successive cycles of folding or being degraded because they are terminally misfolded. An assumption is that cofactors associated with the chaperone/substrate complex determine the fate of the bound protein. One set of cofactors promote protein folding, another set targets the protein for destruction. These two classes of cofactors compete for the same binding sites in Hsp70 and Hsp90 (Cyr et al. 2002). One example is the competition of CHIP and Hop for chaperone binding (Connell et al. 2001). This process might also be regulated by additional components like HspBP1. Furthermore, it has been shown that dimerization of human CHIP via a coiled-coil domain is essential for its function (Nikolay et al. 2004).

A molecular clock might provide an alternative determinant of the fate of a protein. In a certain timeframe, a protein is allowed to fold in association with a chaperone complex. Thereafter, factors that belong to the degradation machinery, such as CHIP or probably EKN1, are able to get in contact with the chaperone/substrate complex. This model is supported by the fact, that *U-box* ligases represent a relatively small fraction of cellular ubiquitin ligases. Consequently, the ratio between chaperones and *U-box* ligases has to be well-balanced as it has been shown that only a two-fourfold elevation of the concentration of CHIP is sufficient to route folding intermediates of CFTR or GR for degradation to an exceeding extent (Connell et al. 2001; Meacham et al. 2001).

2.3.2 U-box ligases serve multiple functions

The first identified *U-box* containing protein was Ufd2p from the yeast *S. cerevisiae* (Koegl et al. 1999). Ufd2p was classified as an E4 protein because it functions in elongation of poly-ubiquitin chains: Ufd2p does not cooperate directly with an E2 enzyme, but requires the interaction with an E2/*HECT*-domain E3 complex to perform its function in ubiquitin conjugation. In contrast to Ufd2p, the mammalian UFD2 as well as other *U-box* proteins are able to act as E3 ligases by cooperating with E2 enzymes. Ufd2p was shown to associate with the AAA-ATPase Cdc48p, which, in complex with Ufd1p and Npl4p, is known to function as a delivery-factor of poly-ubiquitinated substrates to the proteasome (described in Chapter 4). In mammals, the Ufd2p homolog UFD2a interacts with p97, the orthologue of yeast Cdc48p (Kaneko et al. 2003). Interestingly, the second mammalian Ufd2p homolog UFD2b displays no interaction with p97. Instead, an association with the chaperone DnaJc7 was demonstrated (Hatakeyama et al. 2004). DnaJc7 contains nine *TPR* motifs which presumably interact with the C-terminus of Hsp70 or Hsp90, and an additionally J-domain, which probably associates with Hsc70 (Pellecchia et al. 1996). Thus, UFD2a and UFD2b might link different chaperone to the ubiquitin-proteasome system.

The human cyclophilin-60, also termed CYC4, consists of an N-terminal *U-box* domain, and a cyclophilin-like domain that possesses peptidyl-prolyl cis/trans isomerase activity (Wang et al. 1996). Members of the cyclophilin family exhibit a co-chaperone-like function by assisting Hsp90 in protecting the cell against stress (Andreeva et al. 1999) and participating in folding of late-stage folding intermediates of hormone receptors and signaling kinases (Caplan 1999; Young et al. 2001). CYC4 is localized to the nucleus. This suggests that it mediates the degradation of nuclear forms of cyclophilin substrates (Wang et al. 1996). The cyclophilin-like domain of CYC4 is also a binding site for the human homolog of the *Drosophila melanogaster* crooked neck gene CRN (Hatakeyama et al. 2004). CRN contains 16 *TPR* motifs, which mediate association with Hsp90 to establish a trimeric CYC4/CRN/Hsp90 complex.

The *U-box* protein UIP5 (KIAA0860) was identified by a yeast two-hybrid screen using the E2 UbcM4 as the bait. In addition to its *U-box* domain, UIP5 contains a *RING*-domain at its C-terminus. Surprisingly, the E3 activity of UIP5 seems to depend only on its *U-box* domain (Pringa et al. 2001). Comparable to Ufd2p in yeast and UFD2a in mammals, UIP5 interacts with Cdc48/p97. PRP19, a human homologue to proteins encoded by the *prp* genes of *S. cerevisiae*, exhibits *U-box* domain mediated ubiquitin ligase activity (Hatakeyama et al. 2001). The *prp* genes in yeast are required for the splicing of nuclear precursor mRNAs (Vijayraghavan et al. 1989). PRP19 contains six WD40 repeats, which are thought to mediate protein-protein interaction. This suggests that PRP19 is involved in regulating RNA splicing by mediating the degradation of spliceosomal components. Whether these processes are linked to chaperones is unknown.

3 Protein degradation from the ER

The ER represents the portal of the secretory pathway where membrane and secretory proteins are synthesized and folded into their native structure. Correctly folded proteins are subsequently recruited into COPII-vesicles and transported to the Golgi. In this compartment, maturation of secretory proteins is completed. Finally, proteins are transported by clathrin-dependent vesicles to endosomes, lysosomes (in yeast the vacuole), and the plasma membrane. Each compartment possesses specific quality control mechanisms, which monitor protein folding and/or maturation, but also sort out misfolded or incorrectly matured proteins. In this chapter, we will concentrate on mechanisms underlying recognition and degradation of misfolded proteins in the ER (Fig. 3).

3.1 Quality control in the ER and the unfolded protein response

Protein folding and post-translational modifications already occur when the polypeptide chain emerges from the amphipathic translocation channel in the ER lumen. Signal peptides are cleaved by the ER-resident signal peptidase and N-linked glycans are transferred onto the polypeptide by oligosaccharyl transferase. To prevent aggregation of newly synthesized proteins due to exposed hydrophobic patches, the ER-resident Hsp70 chaperone BiP/Kar2p binds to polypeptides during translocation. In contrast to the cytosol, the ER lumen maintains an oxidative environment. It supports the formation of disulfide bridges, which stabilize the conformation of proteins. Protein disufide isomerases catalyze disufide interchange reactions, thus, increasing the efficiency of protein folding by unfolding kinetically trapped folding intermediates (Huppa and Ploegh 1998; Misselwitz et al. 1998). The lectins calnexin and calreticulin bind to unfolded glycoproteins. They specifically associate with monoglycosylated trimming intermediates of the N-linked core glycans (Molinari and Helenius 1999). Together with ERp57, a thiol oxidoreductase, they mediate retention of unfolded and misfolded proteins in the ER, and promote proper folding of glycoprotein substrates.

Comparable to the situation in the cytosol, proteins in the process of folding in the ER lumen are continuously monitored by a complex quality control system within the ER (Ellgaard et al. 1999). Molecular chaperones survey the ER to detect misfolded proteins. The accumulation of misfolded proteins within the ER, mainly due to exogenous stress, is detected by the cell and leads to the upregulation of ER-resident chaperons. This mechanism is called unfolded protein response (UPR).

3.2 The ER-associated protein degradation pathway

As accumulation of terminally misfolded proteins results in disturbance of the ER, aberrant proteins are usually subjected to rapid proteolysis. How are these proteins degraded? One possibility is disposal in the lysosomes/vacuole. But normally, proteins are retained in the ER until their maturation is completed. Although some misfolded proteins might be transported to the lysomsomes/vacuole, the majority is degraded by other mechanisms. For a long time it was assumed that specific ER-resident proteases carry out this task. However, it is quiet inconceivable that proteases exist in an environment that supports folding and maturation of proteins, and indeed, such a proteolytical activity has yet to be identified.

Some years ago, studies demonstrated, that the degradation of membrane-bound and ER-lumenal substrates depends on the cytosolic ubiquitin-proteasome pathway (Biederer et al. 1996; Hiller et al. 1996; Jensen et al. 1995; Sommer and Jentsch 1993; Ward et al. 1995). These results provoked a topological problem, as misfolded proteins in the ER lumen and the proteasome are separated by the ER-membrane. This problem could only be resolved by the assumption that degradation of misfolded proteins is preceded by their retrograde translocation back into the cytosol. This process was termed "dislocation". The overall degradation system was called ER-associated protein degradation (Werner et al. 1996). Although specific aspects of protein degradation from the ER are still a matter of intensive research, the process can be divided in general into three steps (Fig. 3):

- recognition of substrates and their subsequent targeting for degradation;
- dislocation back into the cytosol;
- ubiquitinylation and subsequent degradation by the 26S proteasome.

3.3 Selection and recruitment of aberrant proteins

Recognition of misfolded proteins is not as trivial as it may appear at first glance. Aberrant proteins must be distinguished from newly synthesized ones that are still in the process of folding. Since all proteins synthesized in the ER could potentially be trapped irreversibly in a misfolded conformation at different stages of the folding process, degradation cannot depend on a simple and unique signal common to all substrates. What structural features make a protein a substrate of the degradation system? And how are these features recognized?

3.3.1 What makes a protein a substrate of the degradation system?

Studies on AB-toxins suggested a possible determinant that could lead to degradation. AB-toxins are composed of a catalytically active A subunit and a cell binding B subunit. The AB-toxin enters the cell via receptor-mediated endocytosis. It is then retrogradely transported from the plasma membrane to the ER where it exploits the degradation machinery to be dislocated to the cytosol. In the ER, A and B chains are separated, thereby, exposing a patch of hydrophobic amino acids on the surface of the A domain (Deeks et al. 2002; Suhan and Hovde 1998). Absence of this hydrophobic amino acid sequence diminishes export of the toxin into the cytosol and eliminates its toxicity. It seems likely that the hydrophobic sequence is recognized by the ER quality control system and the A subunit is subjected to dislocation to the cytosol.

Likewise, aberrant proteins expose hydrophobic elements on their surface, which could be recognized by chaperones and serve as signal for degradation. Members of virtually all classes of chaperones, except the Hsp60/GroEL family, are found in the ER (Stevens and Argon 1999). The Hsp70 homologue Kar2p/BiP binds hydrophobic polypeptides. Protein folding is mediated by repeated binding and release of Kar2p/BiP to immature polypeptides. Together with its Hsp40 co-chaperones Jem1p and Scj1p, which may enhance the ATPase activity of Kar2p, it retains misfolded proteins in the ER. A protein with some homology to Hsp70, identified as Lhs1p/Cer1p/Ssi1p or GRP170, cannot rescue most mutants of Kar2p and seems to have a distinct function in refolding heat-damaged proteins (Baxter et al. 1996; Craven et al. 1996; Hamilton and Flynn 1996; Saris and Makarow 1998). In addition, a genetic screen led to the identification of Sls1p/Per100p/Sil1p, which interacts with the ATPase domain of Kar2p/BiP and regulate its activity (Kabani et al. 2000; Tyson and Stirling 2000). These chaperones all serve functions in proteins folding, but at least for Kar2p/BiP and its co-chaperones Jem1p and Scj1p, data suggest that they are also involved in disposal of misfolded proteins (see below).

But solvent-exposed hydrophobic patches are also present on newly synthesized proteins. If these signals serve as the only determinants for destruction, terminally misfolded and newly synthesized proteins would be degraded to the same extend. Hydrophobic elements might represent a common determinant, but it is very likely that additional mechanisms exist to select terminally misfolded proteins. These further requirements might be different for glycoproteins and non-glycoproteins, membrane-embedded, or soluble proteins.

A very attractive model invokes a molecular clock that provides a timeframe in which proteins can fold. Calnexin (CNX) and calreticulin (CRT) form the so-called calnexin/calreticulin-cycle, a system that promotes folding of glycoproteins and retains non-native glycoproteins in the ER until they are correctly folded (Ellgaard and Helenius 2003). Initially, a branched 14-subunit oligosaccharide is attached to an asparagine residue of the nascent polypeptide. Two glucose residues are rapidly trimmed by glucosidase I and II. CNX and CRT specifically recognize the resulting $Glc_1Man_{7\text{-}9}GlcNAc_2$-structure (where Glc is glucose, Man mannose, $GlcNAc_2$ *N*-acetyl glucosamine). The remaining glucose residue is removed by glucosidase II, and the protein is released from CNX and CRT. If the protein is correctly folded, it can leave the cycle and exit the ER. If, in contrast, the protein is not correctly folded, it is reglycosylated by UDP-glucose-glycoprotein glucosyl-transferase (GT), and re-enters the cycle. Alternatively, a mannose residue in the middle branch of the oligosaccharide can be removed by α-1,2-mannosidase I. This leads to recognition by Htm1p/EDEM (ER-degradation enhancing 1,2-mannosidase-like protein), the removal of the protein from the CNX-CRT-cycle and degradation (Jakob et al. 2001; Molinari et al. 2003; Nakatsukasa et al. 2001; Oda et al. 2003). The reaction rate of α-1,2-mannosidase I is slow, thereby, providing a timeframe for protein folding. If a protein fails to fold correctly in this time window, or is trapped in a misfolded conformation, it is removed from the folding cycle and subjected to degradation.

Since the CNX-CRT-cycle acts only on glycoproteins, recognition of unglycosylated misfolded proteins requires other mechanisms. Recently, a transmembrane member of the protein disulfide isomerase oxidoreductase family, Eps1p, was shown to mediate the recognition of the model degradation substrate Pma1p-D378N (Wang and Chang 1999, 2003). Pma1p-D378N is a misfolded variant of Pma1p, a yeast plasma membrane [H+]-ATPase. In a Δ*eps1* mutant, Pma1p-D378N escapes the degradation machinery, and is transported to the plasma membrane. In addition, it was shown that Eps1p directly interacts with Pma1p-D378N, but not with native Pma1p. On the other hand, Eps1p is not involved in the degradation of the soluble substrate CPY*, a misfolded variant of the yeast carboxypeptidase Y (Norgaard et al. 2001). These data indicate a role of Eps1p in the recognition of membrane-bound degradation substrates. Additionally, an involvement of a protein disulfide isomerase is reasonable because reduction of disulfide bridges is necessary for unfolding and subsequent degradation by the degradation system. Yet it is not clear, whether Eps1p directly associates with other components of this process.

3.3.2 A single pathway for all substrates?

The degradation system has to accomplish the destruction of a great variety of misfolded proteins. As discussed above, the recognition of different proteins involves different subsets of chaperones and other sensors for aberrant proteins. It is, therefore, very unlikely that a single degradation pathway exists for all aberrant proteins. Instead, one would expect that the degradation of membrane-embedded and soluble proteins, or glycoproteins and non-glycoproteins depends on different components of the degradation system. What different pathways exist, and which subset of proteins does each pathway degrade?

The degradation of misfolded glycoproteins seems to depend on the trimming state of their oligosaccharide moiety (see above). Htm1p/EDEM binds a specific Man_8-oligosaccharide structure generated by ER-mannose I, and removes proteins from the CNX-CRT-cycle (Jakob et al. 2001; Molinari et al. 2003; Nakatsukasa et al. 2001; Oda et al. 2003). Together with factors that recognize the glycoprotein as misfolded, it might be one of the first components of a degradation pathway for glycoproteins. Disposal of proteins with improper disulfide bonds might depend on oxidoreductans such as a protein disulfide isomerase. Apart from its function as disulfide isomerase it also functions in ER quality control and assists aberrant proteins to unfold prior to degradation (Gillece et al. 1999; Tsai et al. 2001). The protein disulfide isomerase Eps1p was shown to recognize a misfolded protein at the ER-membrane (Wang and Chang 2003).

Aberrant soluble and membrane-anchored ER-resident proteins are also degraded depending on distinct factors. In contrast to membrane proteins, misfolded soluble proteins have to be recruited to the ER-membrane subsequent to recognition. It was demonstrated, that the membrane protein Der1p is involved in the degradation of CPY* (Knop et al. 1996). Der1p might interact with lumenal components loaded with misfolded CPY*, thereby, recruiting CPY* to the membrane and the degradation machinery. This is supported by the fact that a membrane-bound CPY* fusion protein is degraded in the absence of Der1p (Taxis et al. 2003). Yet, it is still open whether Der1p interacts directly with ER-resident chaperones or lectins, and other components of the degradation system.

Degradation of misfolded soluble and membrane-embedded ER-proteins also requires the function of different chaperones. Disposal of terminally misfolded proteins in the ER lumen requires Kar2p/BiP and its co-chaperones Jem1p and Scj1p (Nishikawa et al. 2001). In contrast, degradation of membrane proteins such as the cystic fibrosis transmembrane conductance regulator (CFTR), Vph1p (an ATPase in the yeast vacuole), or a mutant variant of Ste6p (the plasma membrane mating factor transporter in yeast) is not dependent on Kar2p/BiP. Instead, the cytosolic Hsp70 chaperone Ssa1p and its Hsp40 co-chaperones Ydj1p and Hlj1p are required for their degradation (Hill and Cooper 2000; Huyer et al. 2004; Zhang et al. 2001).

A recent publication from the Wolf laboratory uncovered substrate-specific requirements for the degradation of different chimeric proteins that contained the same ER-lumenal domain, the substrate CPY* (Taxis et al. 2003). The degradation of CPY* was compared to that of a fusion protein containing CPY* anchored in the ER-membrane by a single transmembrane segment, and to a protein that additionally contained a cytosolic GFP moiety. Wolf and colleagues identified core components required for the degradation of all three substrates. These are the E3 ligase Hrd1p, the two E2s Ubc1p and Ubc7p, the Cdc48p/Ufd1p/Npl4p-complex, and the proteasome. Der1p was dispensable for the degradation of the two membrane-embedded proteins. Unexpectedly, the fusion protein containing the cytosolic GFP moiety was stabilized in mutants of Ssa1p, the cytosolic Hsp40 homologues Hlj1p, Cwc23p, and Jid1p, as well as the Hsp104 chaperone. These data suggest that the properly folded GFP moiety in the cytosol influences the degradation of the fusion protein. One explanation for the observed differences is that the fusion proteins were overexpressed from the constitutive *ADH* promoter. Cell stress caused by elevated levels of CPY* actives degradation pathways that can function independently of the *HRD* and *DER* genes (Spear and Ng 2003). A second model that would account for these data is that cytosolic chaperones involved in the degradation of the GFP-chimera must unfold the GFP moiety after dislocation and prior to degradation by the proteasome.

Taken together, these data indicate that the degradation pathway is chosen with respect to membrane association of the protein. But a recent study proposed that the elimination of a protein depends on its topology and the localization of the misfolded domain with respect to the ER membrane (Vashist and Ng 2004). A protein containing an aberrant ER-lumenal, transmembrane or cytosolic domain might be recognized and degraded by factors in the same compartment as the misfolded domain. Vashist and Ng constructed a number of unstable chimeric substrates. The membrane-anchored substrates contained a defined misfolded domain either on the ER-lumenal or the cytosolic side of the ER-membrane. Upon kinetic and genetic parameters, they defined two distinct degradation pathways. The ER-lumenal pathway degrades soluble substrates and substrates with an aberrant ER-lumenal domain at a turnover rate of 27-35 minutes. Proteins involved in this degradation pathway are Cue1p, Sec12p, Sec18p, Der1p, and Hrd1p. Conversely, the cytosolic pathway degrades membrane proteins with a misfolded domain exposed to the cytosol. In this case, proteins are degraded more rapidly ($t_{1/2}$ 8-12 min). This pathway involved Cue1p and Doa10p, another E3 ligase. Surprisingly, proteins that contained both an aberrant ER-lumenal and cytosolic domain were degraded by the cytosolic pathway. This suggests, that the cytosolic checkpoint precedes the ER-lumenal checkpoint. Taken together, these data support the hypothesis that degradation route taken for a given substrate is determined by the location of the lesion with respect to the ER-membrane.

Disposal of ER-membrane proteins gets even more complex as in some cases the cytosolic degradation system is recruited to the ER-membrane. In cooperation with Hsc70 the *U-box* ligase CHIP senses the folding state of CFTR and targets aberrant forms to destruction by promoting their ubiquitination (Meacham et al. 2001). It remains an interesting question what features of malfolded ER-resident proteins determine the choice of either the cytosolic or the ER-associated degradation system.

In summary, the publications discussed indicate that degradation of terminally misfolded proteins from the ER cannot be described by a single pathway. The degradation system has to eliminate a great variety of aberrant proteins from the ER, which are marked as misfolded by numerous determinates. Therefore multiple pathways exist, each recognizing a specific subset of substrates. These pathways may converge at the stage of ubiquitination or delivery to the proteasome, at the latest.

3.4 Dislocation of terminally misfolded proteins

Misfolded domains exposed to the cytosol are directly accessible to the ubiquitin-proteasome system, and at least these parts of the protein can be destructed. The remaining transmembrane segments might be extracted from the ER membrane by the proteasome. Soluble ER-proteins are separated from the ubiquitin-proteasome system by the ER-membrane and are subjected to dislocation prior to disposal. How are proteins dislocated back into the cytosol, which channel(s) mediates dislocation?

One candidate “dislocon” is the Sec61 complex. This complex has a well-defined function as an import channel during co- or post-translational translocation into the ER (Tsai et al. 2002). The core translocation channel is an assembly of three transmembrane proteins, the α-, β-, and γ-subunit. In yeast, it is composed of Sec61p, which spans the membrane ten times, Sbh1p, with one transmembrane domain, and Sss1p. The pore itself is a passive conduit for polypeptides without directionality. It must associate with additional partners that allow active transport of polypeptides. In co-translational translocation, the ribosome binds to the Sec61 complex and pushes the polypeptides through the channel. For post-translational translocation, the Sec61 complex associates with four additional proteins, Sec62p, Sec63p, Sec71p, and Sec72p. It is suggested, that this “SEC complex” pulls polypeptides through the channel by a ratchet mechanism provided by Kar2p/BiP and its interaction with Sec63p (Panzner et al. 1995).

An involvement of the Sec61 complex in the dislocation reaction was initially proposed by co-immunoprecipitation experiments (Wiertz et al. 1996). In this study, the cytomegalovirus gene product US2 was expressed in mammalian cells, thus, inducing the specific, rapid degradation of MHC (major histocompatibility complex) class I heavy chains in the host cell. Short-lived MHC class I molecules could be co-precipitated with the β-subunit of the Sec61 complex. A second line of evidence came from genetic experiments in yeast. Various *sec61* mutants show defects in the degradation of a number of misfolded substrates (Pilon et al. 1997; Plemper et al. 1997). As the Sec61 channel provides no directionality, reprogramming of the translocon for protein export could be due to association of Sec61 with components of the degradation system. Yet biochemical evidence concerning the function of the Sec61 complex in dislocation is not given.

In contrast to these studies, data obtained from the crystallization of the SecYEG complex from *Methanococcus jannaschii* question an involvement of the translocation channel in protein transport from the ER to the cytosol (Van den Berg et al. 2004). The bacterial SecYEG complex shows a high homology to eukaryotic Sec61 complexes. The structure suggests, that the heterotrimer forms an aqueous pore within the membrane that is shaped like an hourglass, with a ring of hydrophobic residues at its constriction. The pore is plugged by a short helix from the extracellular side/the ER-lumen to prevent unspecific flow through the channel. The structure was determined from an inactive or closed channel. Based on this structure, a prediction was made how the active channel may work. A polypeptide, which is synthesized by the ribosome through the channel, could push the plug aside and be translocated across the membrane into the extracellular space/the ER-lumen. In addition, the constriction of the channel most likely only allows the passage of an unfolded polypeptide. According to this structure, it is hard to imagine how a protein destined for dislocation from the ER lumen into the cytosol might remove the plug to open the channel. In addition, the misfolded protein needs to be completely unfolded to be dislocated through the channel. At least some degradation substrates were detected in an ubiquitinated and glycosylated state (Hiller et al. 1996; Jarosch et al. 2002). Furthermore, data suggest that proteins might be dislocated in a folded state (Fiebiger et al. 2002; Tirosh et al. 2003). Therefore, the specific function of the Sec61 complex in dislocation, if any, remains to be determined.

Derlin-1, a distant mammalian homologue of Der1p contains four transmembrane segments. It is found in association with US11 and with glycosylated and deglycosylated forms of MHC class I molecules, indicating an interaction of Derlin-1 with heavy chains prior and after dislocation. It also appears to interact at least indirectly with VCP/p97. Therefore, it was suggested that Derlin-1 forms a channel for protein dislocation (Ye et al. 2004). However, evidence for this hypothesis is missing. A general role for Derlin-1/Der1p in dislocating proteins from the ER seems very unlikely because disposal of a number of substrates do not rely on Derlin-1/Der1p (Lilley and Ploegh 2004; Taxis et al. 2003).

3.5 Ubiquitination of aberrant ER-resident proteins

Dislocated polypeptides are conjugated with the small polypeptide ubiquitin upon their arrival in the cytosol. The ubiquitin-conjugating enzymes Ubc1p, Ubc6p, and Ubc7p mediate the ubiquitinylation of substrates destined for degradation at the ER membrane. Ubc6p is anchored in the ER membrane by a single transmembrane domain. The catalytic active domain is exposed to the cytosol. In contrast, Ubc1p and Ubc7p are soluble proteins. It was shown that Ubc7p is recruited to the ER membrane by the adaptor protein Cue1p (Biederer et al. 1997). By what mechanism Ubc1p is recruited to the ER membrane is unknown. At least two membrane-bound E3s are specific for the degradation of misfolded proteins from the ER. Both are members of the *RING*-family of ubiquitin ligases. The *RING*-domain is exposed to the cytosol. Hrd1p is required for a number of degradation substrates and seems to cooperate with Ubc1p and Ubc7p (Biederer et al. 1997; Gardner et al. 2000). Hrd1p function depends on its tight interaction with the transmembrane protein Hrd3p. The other yeast E3, Doa10p/Ssm4p, which acts in concert with Ubc7p and Ubc6p, is involved in the turnover of the tail-anchored protein Ubc6p, but also of soluble nuclear and cytosolic proteins (Swanson et al. 2001). Recent studies revealed that Doa10p is involved in degradation of a mutant form of the membrane protein Ste6p (Huyer et al. 2004; Vashist and Ng 2004).

3.6 Do E3 ligases play a central role in the ER degradation system?

Although a number of membrane-bound ubiquitin-ligases exist in the cell, only two E3s, Hrd1p, and Doa10p, are known so far to be involved in the degradation of misfolded proteins from the ER. Both ligases are members of the *RING*-finger ubiquitin-ligase family. *RING*-finger proteins are defined by a pattern of conserved cysteine and histidine residues with a characteristic spacing that allows for the coordination of two zinc ions in a unique cross-brace arrangement. The zinc ion and their ligands are catalytically inert, but essential for folding of the domain. This suggests that *RING*-finger proteins function not as catalysts but as molecular scaffolds that bring other proteins together. Typical cytosolic *RING*-finger class E3s select their substrates and recruit specific Ubc's, which subsequently transfer ubiquitin directly to the substrate. The *RING*-finger domain thus provides a platform for the Ubc's to get in close proximity to their substrates and perform their catalytic function.

The limitation on only two E3 ubiquitin-ligases in the degradation of proteins from the ER evokes a specificity problem. Most cytosolic E3s interact only with a limited number of substrates. In case of protein degradation from the ER, the two E3 ligases mediate recognition and ubiquitination of potentially all ER-resident proteins that might be irreversibly trapped in manifold aberrant conformations. Consequently, substrate selection might not be a primary function of Hrd1p and Doa10p. Presumably these two E3s do not bind their substrates directly, but make use of adaptor proteins. These adaptors should monitor the ER for proteins trapped in irreversibly misfolded conformations, and recruit them to the ubiquitin ligases.

Doa10p was first described to function as an E3 ubiquitin-protein ligase that promotes the degradation of *Deg1*-containing proteins (Swanson et al. 2001). The sequence of this large protein of 151 kD bears several segments with similarity to several known protein motifs. The N-terminus contains a variant of the *RING*-finger motif with high similarity to the human TEB4 protein. Another motif found in Doa10p is a WW domain, which is also present in a number of *HECT*-class E3's. This short conserved domain of approximately 40 amino acids folds as a stable triple stranded sheet (Kanelis et al. 2001; Kasanov et al. 2001). The name *WW* or *WWP* arises from the presence of two tryptophan residues spaced 20-23 amino acids apart, as well as that of a conserved proline. Many *WW* domains bind to proteins with particular proline-motifs (consensus sequence [AP]-P-P-[AP]-Y). Thus, this motif can function as either a co-factor or substrate-binding site. Strikingly, Ubc6p contains a PPxY-motif suggesting that interaction of Doa10p and Ubc6p might be mediated by these two motifs (Chang et al. 2000; Swanson et al. 2001). Furthermore, Doa10p contains 10 to 14 predicted transmembrane segments. As discussed above, Doa10p might be involved in degradation of proteins with a cytosolic degradation signal (Vashist and Ng 2004). If E3 ligases function as scaffold builders, Doa10p might provide a platform for the Ubc's to meet their substrates, which are delivered to this E2-E3 complex by adaptors such as cytosolic chaperones. Additionally, the transmembrane segments of Doa10p might contribute to substrate selection, or be part of a protein dislocation channel that mediates export of transmembrane protein out of the ER.

In contrast to the Doa10p-dependent degradation pathway in which substrate selection and ubiquitination appear at the same side of the ER membrane, the situation for Hrd1p-mediated protein ubiquitination is more complex. Hrd1p mainly functions in degradation of proteins with lumenal lesions (Vashist and Ng 2004). Thus, substrate selection and ubiquitination are separated by the ER membrane. Although Hrd1p was shown to possess an affinity for misfolded proteins and mediate their specific ubiquitination *in vitro* (Bays et al. 2001a), the function of Hrd1p *in vivo* might be limited to ubiquitination of proteins emerging in the cytosol after dislocation. Therefore, Hrd1p must interact with adaptor molecules, which that links its activity to substrate selection. Hrd1p is composed of an N-terminal region with 6 transmembrane segments, and a C-terminal *RING*-domain (Deak and Wolf 2001). Via its ER-lumenal loops, Hrd1p interacts with the adaptor molecule Hrd3p (Gardner et al. 2000).

Hrd3p is a protein of 95,5 kD and is composed of a large ER-lumenal domain, a transmembrane span, and a small cytosol domain. The ER-lumenal domain shows significant similarity to SEL-1 (the *C. elegans* suppressor and/or enhancer of lineage-12) (Ponting 2000). Both proteins contain multiple repeats of a 40-amino-acid sequence, which were named SEL-1-like repeats (*SLP*). *SLPs* were determined as subtypes of tricopeptide repeats (*TPR*), differing from these in possessing a variable number of amino acids between the two *TPR* α-helices (Ponting 2000). The nine *SLR*s present in Hrd3p might function in similar processes as defined for *TPR* motifs. Despite their repeated basic structure, individual *SLR* subunits can be structurally specialized and might serve different function. Providing multiple substrate binding sites, they might contribute to specific recognition of substrates.

Gardener et al. showed that the N-terminal part of the ER-lumenal domain of Hrd3p seems to be essential for its function in degradation of misfolded proteins, whereas the C-terminal part interacts with the ER-lumenal loops of Hrd1p (Gardner et al. 2000). Chaperones such as Kar2p, Jem1p, or Scj1p loaded with terminally misfolded proteins might be recruited to the Hrd1p/Hrd3p complex by interacting with *SLR* motifs of Hrd3p. Consequently, the ER-lumenal domain of Hrd3p might represent a platform where misfolded proteins are delivered to the dislocation/ubiquitination complex.

Summarizing, there is evidence that the two E3 ubiquitin ligases involved in protein degradation from the ER, Doa10p, and Hrd1p/Hrd3p play a central role in this process. They might be the central components of two functionally distinct complexes that perform recognition, dislocation, and ubiquitination of different subsets of misfolded proteins. Their function might be to bring together adaptor molecules loaded with aberrant proteins and Ubc's to mediate ubiquitination. Additionally, the two complexes might be part of a dislocation channel, which transports terminally misfolded proteins out of the ER.

3.7 Driving-force of dislocation

Transport of molecules across a biological membrane requires energy to ensure the vectorial nature of this process. In case of protein translocation into the ER, the driving force is provided by GTP-hydrolysis during protein synthesis at the ribosome, or ATP-hydrolysis by Kar2p/BiP (Wiertz et al. 1996). Dislocating proteins out of the ER probably works in a similar way. One possibility is that ubiquitinylation itself provides directionality. The successive attachment of bulky polyubiquitin moieties to the dislocating protein may prevent backsliding of the substrate into the lumen of the ER. Since this is rather passive dislocation force, it may not suffice to pull partially folded proteins out of the ER. Recently, it was demonstrated that the Cdc48p/Ufd1p/Npl4p complex is also involved in protein degradation at the ER.

4 Cdc48p/p97 – chaperoning poly-ubiquitinated proteins

One of the first activities described for the yeast AAA-ATPase (ATPase associated with various activities) Cdc48p and its mammalian homolog p97 (also termed VCP for valosin-containing protein) was its participation in membrane fusion events in association with the cofactor Shp1p/p47 (Kondo et al. 1997). More recently, numerous studies revealed that Cdc48p/p97 is required for the ubiquitin-dependent proteasomal degradation of cytosolic (Dai et al. 1998; Fu et al. 2003) as well as ER-resident soluble and membrane-bound proteins (Braun et al. 2002; Jarosch et al. 2002; Rabinovich et al. 2002; Ye et al. 2001). To function in the destruction of aberrant proteins, Cdc48p/p97 associates with Ufd1p and Npl4p (Meyer et al. 2000). So far, no influence of the Cdc48p/Ufd1p/Npl4p complex on the ubiquitination of proteins itself was demonstrated. Although the proteasome was shown to be directly involved in pulling poly-ubiquitinated ER-bound proteins out of the ER membrane (Mayer et al. 1998; Walter et al. 2001), the Cdc48p/Ufd1p/Npl4p complex plays a more general role. It might link poly-ubiquitination and subsequent degradation by the proteasome. This suggestion is based upon the analysis of mutations affecting proteasomal function and studies with proteasomal inhibitors that cause accumulation of poly-ubiquitinated proteins in the cytosol. In addition, mutations of Cdc48p/Ufd1p/Npl4p cause an accumulation of degradation substrates in a poly-ubiquitinated form that is still associated with the ER-membrane (Bays et al. 2001b; Jarosch et al. 2002; Ye et al. 2001). Therefore, the Cdc48p/Ufd1p/Npl4p complex acts prior to proteasomal degradation but after poly-ubiquitination.

An essential feature of the Cdc48p/Ufd1p/Npl4p complex is its affinity to poly-ubiquitin chains. Ubiquitin-binding is mediated by several subunits of the complex. Cdc48p/p97 associates with ubiquitinated proteins through direct interaction of its N-terminal domain with poly-ubiquitin chains (Dai and Li 2001). Mammalian Npl4 binds poly-ubiquitin via a putative zinc finger in its NZF (Npl4 zinc finger) domain (Meyer et al. 2002). Moreover, the N-terminal domain of mammalian Ufd1 contains an ubiquitin-binding site that appears to be important for protein degradation and particularly retro-translocation of degradation substrates. The N-terminal domain only recognizes poly-ubiquitin chains that are linked via Lys48 (Ye et al. 2003). This supports the assumption that the function of the Cdc48p/Ufd1p/Npl4p is restricted to protein degradation. Consequently, poly-ubiquitination of proteins may serve as the initial step for recognition of substrates by the Cdc48p/Ufd1p/Npl4p complex.

To prevent dislocated misfolded proteins from moving back into the ER and ensure vectorial transport, dislocation must be coupled to an energy-consuming process. The Cdc48p/Ufd1p/Npl4p complex might provide a driving force due to its properties as an AAA-ATPase. Cdc48p forms a homo-hexameric ring with six-fold symmetry. Upon ATP hydrolysis the complex undergoes strong conformational changes (Rouiller et al. 2000; Zhang et al. 2000a). Successive rounds of ATP hydrolysis coupled to substrate release and re-binding may facilitate pulling polypeptides through an export channel or directly out of the ER membrane into the cytosol.

Alternatively, the Cdc48p/Ufd1p/Npl4p complex may be required for the release of already dislocated substrates from the cytosolic surface of the ER. According to this model, the complex acts as a "segregase". This hypothesis is supported by the fact that poly-ubiquitinated degradation substrates accumulate in non-functional mutants of the Cdc48p/Ufd1p/Npl4p complex and are sensitive to exogenously added protease (Bays et al. 2001b; Jarosch et al. 2002). In addition, the Cdc48p/Ufd1p/Npl4p complex is also involved in the activation of the membrane-bound transcription factor Spt23p (Rape et al. 2001). The complex separates a tightly associated Spt23p homodimer that has been proteolytically processed by the proteasome. Thereby, the active, soluble transcription factor is released into the cytosol. Ubiquitination of the substrate is a prerequisite for this "segregase"-like function by the Cdc48p/Ufd1p/Npl4p complex. Therefore, the

Cdc48p/Ufd1p/Npl4p complex might function in recognizing ubiquitinated substrates and mobilizing them for subsequent processes.

The Cdc48p/Ufd1p/Npl4p complex also participates in delivery of poly-ubiquitinated substrates to the proteasome. Associated p97 and 26S proteasomes were shown to mediate degradation of IκBα (Dai et al. 1998). Likewise, p97 was co-purified with the 19S cap of proteasomes in the absence of ATP (Verma et al. 2000). Substrate delivery to the proteasome and association of the Cdc48p/Ufd1p/Npl4p complex with the proteasome might depend on the cytosolic proteins Rad23p and Dsk2p. These two proteins are involved in the degradation of artificial ubiquitin-fusion degradation (UFD) substrates (Chen and Madura 2002; Rao and Sastry 2002) and CPY* (Medicherla et al. 2004). Deletion mutants of *RAD23* and *DSK2* accumulate CPY* in a poly-ubiquitinated form, which is at least partially attached to the ER-membrane (Medicherla et al. 2004). Both, Rad23p and Dsk2p contain an ubiquitin-associated (*UBA*) domain and an ubiquitin-like (*UBL*) motif. The *UBA* domain was shown to bind poly-ubiquitin chains, the *UBL* motifs seems to associates with the 19S cap of the proteasome (Elsasser et al. 2004; Schauber et al. 1998; Wilkinson et al. 2001). These data suggest that Rad23p and Dsk2p mediate transfer of poly-ubiquitinated substrates from the Cdc48p/Ufd1p/Npl4p complex to the proteasome.

Beside its specific role in protein degradation from the ER, Cdc48p/p97 serves a general function in guiding poly-ubiquitinated substrates to the proteasome. To prevent aggregation, poly-ubiquitinated proteins may always be associated with other proteins, such as Cdc48p/p97 or Rad23p and Disk2p. Cdc48p/p97 has been found in association with mammalian gp78, the homolog of the yeast ubiquitin-protein ligase Hrd1p (Zhong et al. 2004). Moreover, it was shown to interact with the cytosolic *U-box* ligases Ufd2 and UIP5 (see above). Thus, the Cdc48p/Ufd1p/Npl4p complex might directly take over poly-ubiquitinated substrates from ubiquitin ligases. Ubiquitinated substrates might than be passed on to components such as Rad23p and Dsk2p, because *UBA* domains could possess a higher affinity for poly-ubiquitin chains. Taken together, these data suggest that Cdc48p/p97 in cooperation with its cofactors Ufd1p and Npl4p performs a chaperone-like function with specificity for poly-ubiquitinated proteins. It mobilizes substrates, helps to keep them soluble, prevents aggregation, and targets them to the 26S proteasome for degradation.

5 Diseases and toxins – what can go wrong in protein degradation?

Proteolytical pathways of protein degradation keep the cell free of cellular debris. However, malfunction of these systems has major implications for the generation of several human diseases. In addition, some viruses exploit the system to escape their detection by the host organism, and plant and bacterial toxins hijack it to gain their toxicity.

5.1 Diseases associated with protein degradation

Alterations in the structure of a protein account for a number of severe human diseases. The altered conformation can either be inherited or acquired. Mutations that result in a loss-of-function of the mutant gene product cause recessively inherited genetic diseases, such as lysosomal storage diseases, or cystic fibrosis. The mutation of the gene can interfere with the synthesis, maturation, stability, transport, or enzyme activity of the encoded protein. Conversely, conformational rearrangements of a specific protein may result in a tendency to aggregate and become disposited within cellular compartments or tissues. Conformational diseases include α_1-antitrypsin (α_1-AT) deficiency, amyloidoses, the prion disease, and Alzheimer's and Parkinson's disease. Here, we will concentrate on diseases that are associated with protein degradation from the ER.

How can a degradation system be involved in the generation of a disease? The answer arises from the function of the degradation system. It monitors protein conformation and sorts out misfolded or improperly matured proteins. Mutations interfering with protein structure or maturation, thus, may result in degradation of this protein. In some cases, metabolic important proteins become substrates of the degradation pathways when their genetic information is changed. This was demonstrated for the cystic fibrosis transmembrane conductance receptor (CFTR), a chloride channel in the plasma membrane of epithelial cells. Even under normal conditions, folding and maturation of CFTR is very inefficient and 75% of the protein is retained in the ER and degraded. A specific mutation, ΔF508CFTR, alters the conformation of the protein. The mutant protein is fully functional, but it is completely retained in the ER and rapidly degraded (Jensen et al. 1995; Ward et al. 1995). Thus, it does not reach the plasma membrane where it performs its function. The lack of the chloride channel in the plasma membrane leads to cystic fibrosis.

Lysosomal storage disorders (LSD) result from the defective activity of lysosomal enzymes leading to intra-lysosomal accumulation of enzyme substrates. About 50 to 60 soluble (Journet et al. 2002) and at least 7 integral membrane proteins (Eskelinen et al. 2003) reside in the lysosome, and in theory, mutations in each gene that encode for one of these proteins can cause an LSD. One of the first LSD to be clinically described was the sphingolipidose Gaucher disease. The key protein in this most prevalent sphingolipidose is the enzyme glucosylceramide-β-glucosidase (β-glu). Five mutant alleles of β-glu account for the majority of reported cases (Beutler et al. 1992). Common to all these mutation is their destabilizing effect on the conformation of the enzyme. As a consequence, the enzyme does not pass the ER quality control and is degraded from the ER. Deficiency of the enzyme results in the accumulation of its non-degradable substrate glucosylceramide in the lysosome. This accounts for the severe symptoms of Gaucher disease, such as enlargement of the spleen and liver, heart, and kidney injury, involvement of the central nervous system (Futerman and van Meer 2004).

The diseases described result from degradation of mutant proteins. Conversely, a mutation in a component of the degradation system leads to the aberrant expression of proteins. A well-documented example is Parkinson's disease (PD). Patients with PD suffer from rigidity, slowness of movement, tremor, and disturbance of balance. It is characterized by a progressive loss of dopaminergic neurons in the substantia nigra (Dunnett and Bjorklund 1999), and the intra-cellular accumulation of Lewy-bodies, proteinaceous accumulation of eosinophilic material that stain for ubiquitin (Pollanen et al. 1993). Hereditary PD can be caused by mutations in the Parkin gene (Kitada et al. 1998). The wild type parkin gene product was demonstrated to function as a *RING*-finger ubiquitin-protein ligase (Zhang et al. 2000b). The mutant protein is no longer functional. The membrane protein Pael-R was found to be a substrate of Parkin. Accumulation of Pael-R in dopaminergic neurons results in ER-stress induced cell death (Imai et al. 2001). Parkin functions together with the *U-box* ligase CHIP (Imai et al. 2002). CHIP, Hsp70, Parkin, and Pael-R form a complex *in vitro* and *in vivo*. CHIP enhances dissociation of Hsp70 from Parkin and Pael-R, thus, promoting Parkin-mediated ubiquitination of Pael-R.

5.2 Viruses and AB-toxins

Several viruses exploit components of the degradation system to subvert the host immune defense. The elimination of virus-infected cells by $CD8^+$ T lymphocytes is the main defense strategy against viruses. Antigenic peptides derived from virus proteins are presented on the MHC class I molecules on the surface of infected cells. $CD8^+$ T lymphocytes specifically recognize this antigen-MHC class I-complexes and kill the infected cell. MHC class I-dependent antigen presentation requires the loading of virus antigens on MHC class I-molecules in the ER and subsequent transport to the plasma membrane. Viruses have developed a great variety of strategies to interfere with this process, and escape the immune response. HCMV (human cytomegalovirus) infected cells express two proteins, US2 and US11, that target MHC class I molecules for degradation. Each of the proteins can interact with the heavy chains, thereby, converting them into substrates of the degradation pathway. Consequently, the amount of MHC class I molecules on the cell surface is dramatically reduced, and the infected cell is not recognized by cytotoxic T lymphocytes.

Several plant and bacterial toxins enter the target cell by endocytosis. To reach the cytosol, they have to cross the lipid bilayer. In one class of toxins, AB type toxins such as cholera toxin, pertussis toxin, Shiga toxin, and ricin, the task of catalysing the toxin reaction is separated from the task of targeting the toxin into the cell. The A domain is the catalytical active subunit, whereas the B domain is the cell binding component. After receptor binding and endocytosis, the toxins are retrogradely transported via endosomes and the golgi to the ER. Here, A and B chains are separated, thereby, exposing a patch of hydrophobic amino acids on the surface of the A component. This might serve as a recognition motive that targets the A domain for degradation. The toxin is dislocated into the cytosol and a significant portion escapes degradation by the ubiquitin/proteasome pathway to perform its toxic function. The study of AB toxins might reveal useful aspects of the degradation machinery.

These examples demonstrate that the degradation pathway is of great importance not only for cell biologists. If the degradation system is inefficient, disturbed, or misused, the effects for the cell can be fatal. The study of the molecular nature of, and of the interactions between, the components involved should provide the basis for therapeutic tools for the treatment of several severe human diseases.

5.3 Pharmacological chaperones – a new approach in fighting folding diseases

The ER quality control system and the selection of terminally misfolded proteins are based upon conformational criteria and differences in stability of native and non-native conformations. Mutant proteins, such a CFTR, might still be functional, but adopt an altered conformation that is recognized by the degradation machinery. It should be possible to stabilize the conformation and improve maturation of such proteins in order to exit the ER and reach their destination.

Several studies demonstrated that a number of low-molecular-weight compounds are able to support folding of proteins that are defective in patients of inherited diseases. The conformation of the most common mutant CFTR protein, ΔF508CFTR, was stabilized by glycerol. Consequently, the mutant protein could form functional plasma membrane channels (Sato et al. 1996). The term "chemical chaperone" was coined for substances which act unspecific by increasing the fraction of the correct folded variant protein encoded by the mutant gene. Examples are dimethylsulfoxid (DMSO), methyl-β-cyclodextrin, the cellular osmolytes glycerol, and trimethylamine-*N*-oxide as well as some ions. In addition, it was shown that exposure of scrapie-infected cells to chemical chaperones reduced the rate and extent of PrP(Sc) formation (Tatzelt et al. 1996). Chemical chaperones did not affect the existing population of PrP(Sc), but interfered with the conversion of newly synthesized PrP(C) to PrP(Sc). In yeast cells, Hmg2p is the rate-limiting enzyme of sterol synthesis. If the sterol pool of the cell is saturated, Hmg2p attains a conformation accessible for degradation. Chemical chaperones such as glycerol caused a change in Hmg2p structure to the less accessible form (Shearer and Hampton 2004).

At first site, the effect of chemical chaperone is exciting. But the majority of them require relatively high (millimolar) concentrations to be efficient. Similar to molecular chaperones, they exhibit limited specificity and assist the folding of a broad range of proteins. This makes them inapplicable for the treatment of folding diseases. In contrast to the rather unspecific chemical chaperones, other low-molecular-weight compounds specifically stabilize the conformation of certain proteins. These were named "pharmacological chaperones".

The versatility of pharmacological chaperones to assist protein folding, prevent post-secretory misfolding, or stabilize mutant proteins with a preference to misfold is well-documented. Antagonists of vasopressin V_2 receptor (V2R) dramatically increase cell-surface expression and rescue the function of mutant forms of V2R by promoting their proper folding and maturation (Morello et al. 2000).

Membrane-permeable ligands of the δ opoid receptor facilitate maturation and export of the receptor from the ER (Petaja-Repo et al. 2002). Of particular interest are pharmacological chaperones that provide new molecular tools for the treatment of genetic metabolic diseases. First evidence came from a study on Fabry's disease. Fabry's disease is a rare, X-chromosomal linked lysosomal storage disorder, which derives from a defect in lysosomal α-galactosidase A (α-gal A). The absence of the enzyme from lysosomes results in accumulation of enzyme substrates with a terminally α-glycosidically linked galactose residue within vascular epithelial cells. The disposition of substrates accounts for the symptoms of the disease, such as renal failure along with premature myocardial infarction and strokes (Brady et al. 1967). A potent competitive inhibitor of α-gal A, 1-desoxygalactonojirimycin (DGJ), effectively enhanced α-gal A activity when administered at concentrations lower than those usually required for intracellular inhibition of the enzyme (Fan et al. 1999). The conformation of various defective forms of α-gal A was stabilized by DGJ, and enzyme activity was increased about eightfold, up to 45% of normal values. In addition, oral administration of DGJ to transgenic mice substantially elevated the enzyme activity in some organs.

One explanation for the effect that an inhibitor stabilizes enzyme conformation and enhances its activity at low concentration is the fact that enzyme folding and enzyme function take place in different compartments. Lysosomal enzymes have to fold at neutral pH to leave the ER, yet their sequences are optimized for folding and stability at the lysosomal pH of 5. Mutations can further decrease the folding efficiency in the ER, thus, leading to degradation of the mutant proteins. In the acidic environment of the lysosome, these variants would be able to fold into a functional enzyme. Stabilizing the conformation of these mutant proteins in the ER, would restore enzyme function. It has been demonstrated that a glucoconfigurated nojimycin binds 80-fold more strongly to a glucosidase at pH 6.5 than 4.5 (Dale et al. 1985). In the neutral pH of the ER, the inhibitor, for example DGJ, strongly binds to α-gal A. The native conformation of the enzyme is stabilized and even mutant, less stable variants of the enzyme can pass the quality control system. α-Gal A is then transported to the lysosome. In this acidic compartment, the inhibitor is released and the enzyme remains stable. Thus, if a mutant enzyme exhibits catalytical activity but misfolds and, therefore, is degraded, enzyme inhibitors may allow correct maturation and transport without a reduction of enzyme activity.

Over the past years, more and more pharmacological chaperones have been synthesized and tested with respect to influence folding of various mutant enzymes. *N*-(*n*-nonyl)desoxynojirimycin (NN-DGJ) was demonstrated to increase the cellular activity of the N370S mutant form of β-glucosidase (Sawkar et al. 2002). Failure of this enzyme results in Gaucher's disease. Another mutant variant of β-glucosidase, F213I, was effectively transported and matured in the presence of *N*-octyl-β-valienamine (NOV) (Lin et al. 2004). The increase in enzyme activity observed in these studies did not depend on the permanent application of the inhibitor. In case of NN-DGJ, increased enzyme activity persisted at least for six days after withdrawal of the drug.

Current therapeutic strategies include inhibition of substrate production and replacement of the defective enzyme (Schiffmann and Brady 2002). Enzyme replacement for Gaucher disease costs between $100,000 and $750,000 per year. Although the enzyme has been modified to increase uptake into cells, for example, less than 7% are taken up into liver macrophages (Berg-Fussman et al. 1993). Taken the low concentrations needed to achieve an increase in enzyme activity, and the persistence of the effect, pharmacological chaperones offer a valuable therapeutic tool to treat some of the most severe inherited diseases.

6 Conclusions

In recent years, our knowledge of proteasomal degradation of terminally misfolded or damaged proteins became more detailed. The basic mechanisms of protein degradation by the ubiquitin-proteasome system are now well understood. Due to the diversity of substrates the ubiquitin-proteasome system has to eliminate – each protein can putatively be irreversibly trapped in various misfolded conformation or damaged by numerous events – multiple mechanisms exist to recognize proteins as misfolded. Molecular chaperones, initially described to function in protein folding, survey cellular proteins with respect to native structure, and detect unfolded or aberrant proteins. The multiple pathways seem to converge at the point of ubiquitination. Ubiquitin-protein ligases arise as the central components that link recognition of aberrant proteins to the ubiquitin-proteasome system. In the cytosol, *U-box* ligases such as CHIP interact with chaperones loaded with aberrant proteins, ubiquitinate these proteins and target them for degradation. Misfolded ER-lumenal proteins are recognized by ER-resident chaperones. ER-membrane proteins with cytosolic lesions are detected by cytosolic chaperones. Subsequently, two membrane-bound E3 ligases mediate ubiquitination of these substrates. Taken together, the knowledge of how ubiquitin-ligases interact with chaperones loaded with aberrant proteins and how a protein is finally recognized as terminally misfolded will remain an interesting topic for the next years.

Acknowledgement

Part of this work was supported by grants of the "Deutsche Forschungsgemeinschaft". RG receives a fellowship from Boehringer Ingelheim Fonds. We thank Jörg Höhfeld, Ernst Jarosch, and Christian Hirsch for critical reading of the manuscript.

References

Alberti S, Demand J, Esser C, Emmerich N, Schild H, Hohfeld J (2002) Ubiquitylation of BAG-1 suggests a novel regulatory mechanism during the sorting of chaperone substrates to the proteasome. J Biol Chem 277:45920-45927

Alberti S, Bohse K, Arndt V, Schmitz A, Hohfeld J (2004) The cochaperone HspBP1 inhibits the CHIP ubiquitin ligase and stimulates the maturation of the cystic fibrosis transmembrane conductance regulator. Mol Biol Cell 15:4003-4010

Andreeva L, Heads R, Green CJ (1999) Cyclophilins and their possible role in the stress response. Int J Exp Pathol 80:305-315

Aravind L, Koonin EV (2000) The U box is a modified RING finger - a common domain in ubiquitination. Curr Biol 10:R132-R134

Ballinger CA, Connell P, Wu Y, Hu Z, Thompson LJ, Yin LY, Patterson C (1999) Identification of CHIP, a novel tetratricopeptide repeat-containing protein that interacts with heat shock proteins and negatively regulates chaperone functions. Mol Cell Biol 19:4535-4545

Baumeister W, Walz J, Zuhl F, Seemuller E (1998) The proteasome: paradigm of a self-compartmentalizing protease. Cell 92:367-380

Baxter BK, James P, Evans T, Craig EA (1996) SSI1 encodes a novel Hsp70 of the *Saccharomyces cerevisiae* endoplasmic reticulum. Mol Cell Biol 16:6444-6456

Bays NW, Gardner RG, Seelig LP, Joazeiro CA, Hampton RY (2001a) Hrd1p/Der3p is a membrane-anchored ubiquitin ligase required for ER-associated degradation. Nat Cell Biol 3:24-29

Bays NW, Wilhovsky SK, Goradia A, Hodgkiss-Harlow K, Hampton RY (2001b) HRD4/NPL4 is required for the proteasomal processing of ubiquitinated ER proteins. Mol Biol Cell 12:4114-4128

Berg-Fussman A, Grace ME, Ioannou Y, Grabowski GA (1993) Human acid beta-glucosidase. N-glycosylation site occupancy and the effect of glycosylation on enzymatic activity. J Biol Chem 268:14861-14866

Beutler E, Gelbart T, Kuhl W, Zimran A, West C (1992) Mutations in Jewish patients with Gaucher disease. Blood 79:1662-1666

Biederer T, Volkwein C, Sommer T (1996) Degradation of subunits of the Sec61p complex, an integral component of the ER membrane, by the ubiquitin-proteasome pathway. Embo J 15:2069-2076

Biederer T, Volkwein C, Sommer T (1997) Role of Cue1p in ubiquitination and degradation at the ER surface. Science 278:1806-1809

Bochtler M, Ditzel L, Groll M, Hartmann C, Huber R (1999) The proteasome. Annu Rev Biophys Biomol Struct 28:295-317

Brady RO, Gal AE, Bradley RM, Martensson E, Warshaw AL, Laster L (1967) Enzymatic defect in Fabry's disease. Ceramidetrihexosidase deficiency. N Engl J Med 276:1163-1167

Braun S, Matuschewski K, Rape M, Thoms S, Jentsch S (2002) Role of the ubiquitin-selective CDC48(UFD1/NPL4)chaperone (segregase) in ERAD of OLE1 and other substrates. Embo J 21:615-621

Bukau B, Horwich AL (1998) The Hsp70 and Hsp60 chaperone machines. Cell 92:351-366

Caplan AJ (1999) Hsp90's secrets unfold: new insights from structural and functional studies. Trends Cell Biol 9:262-268

Chang A, Cheang S, Espanel X, Sudol M (2000) Rsp5 WW domains interact directly with the carboxyl-terminal domain of RNA polymerase II. J Biol Chem 275:20562-20571

Chen L, Madura K (2002) Rad23 promotes the targeting of proteolytic substrates to the proteasome. Mol Cell Biol 22:4902-4913

Chinkers M (1994) Targeting of a distinctive protein-serine phosphatase to the protein kinase-like domain of the atrial natriuretic peptide receptor. Proc Natl Acad Sci USA 91:11075-11079

Connell P, Ballinger CA, Jiang J, Wu Y, Thompson LJ, Hohfeld J, Patterson C (2001) The co-chaperone CHIP regulates protein triage decisions mediated by heat-shock proteins. Nat Cell Biol 3:93-96

Craven RA, Egerton M, Stirling CJ (1996) A novel Hsp70 of the yeast ER lumen is required for the efficient translocation of a number of protein precursors. Embo J 15:2640-2650

Cyr DM, Lu X, Douglas MG (1992) Regulation of Hsp70 function by a eukaryotic DnaJ homolog. J Biol Chem 267:20927-20931

Cyr DM, Hohfeld J, Patterson C (2002) Protein quality control: U-box-containing E3 ubiquitin ligases join the fold. Trends Biochem Sci 27:368-375

Dai RM, Chen E, Longo DL, Gorbea CM, Li CC (1998) Involvement of valosin-containing protein, an ATPase Co-purified with IkappaBalpha and 26 S proteasome, in ubiquitin-proteasome-mediated degradation of IkappaBalpha. J Biol Chem 273:3562-3573

Dai RM, Li CC (2001) Valosin-containing protein is a multi-ubiquitin chain-targeting factor required in ubiquitin-proteasome degradation. Nat Cell Biol 3:740-744

Dale MP, Ensley HE, Kern K, Sastry KA, Byers LD (1985) Reversible inhibitors of beta-glucosidase. Biochemistry 24:3530-3539

Das AK, Cohen PW, Barford D (1998) The structure of the tetratricopeptide repeats of protein phosphatase 5: implications for TPR-mediated protein-protein interactions. Embo J 17:1192-1199

Deak PM, Wolf DH (2001) Membrane topology and function of Der3/Hrd1p as a ubiquitin-protein ligase (E3) involved in endoplasmic reticulum degradation. J Biol Chem 276:10663-10669

Deeks ED, Cook JP, Day PJ, Smith DC, Roberts LM, Lord JM (2002) The low lysine content of ricin A chain reduces the risk of proteolytic degradation after translocation from the endoplasmic reticulum to the cytosol. Biochemistry 41:3405-3413

Demand J, Alberti S, Patterson C, Hohfeld J (2001) Cooperation of a ubiquitin domain protein and an E3 ubiquitin ligase during chaperone/proteasome coupling. Curr Biol 11:1569-1577

Dittmar KD, Pratt WB (1997) Folding of the glucocorticoid receptor by the reconstituted Hsp90-based chaperone machinery. The initial hsp90.p60.hsp70-dependent step is sufficient for creating the steroid binding conformation. J Biol Chem 272:13047-13054

Dunnett SB, Bjorklund A (1999) Prospects for new restorative and neuroprotective treatments in Parkinson's disease. Nature 399:A32-39

Ellgaard L, Molinari M, Helenius A (1999) Setting the standards: quality control in the secretory pathway. Science 286:1882-1888

Ellgaard L, Helenius A (2003) Quality control in the endoplasmic reticulum. Nat Rev Mol Cell Biol 4:181-191

Elsasser S, Chandler-Militello D, Muller B, Hanna J, Finley D (2004) Rad23 and Rpn10 serve as alternative ubiquitin receptors for the proteasome. J Biol Chem 279:26817-26822

Eskelinen EL, Tanaka Y, Saftig P (2003) At the acidic edge: emerging functions for lysosomal membrane proteins. Trends Cell Biol 13:137-145

Fan JQ, Ishii S, Asano N, Suzuki Y (1999) Accelerated transport and maturation of lysosomal alpha-galactosidase A in Fabry lymphoblasts by an enzyme inhibitor. Nat Med 5:112-115

Fiebiger E, Story C, Ploegh HL, Tortorella D (2002) Visualization of the ER-to-cytosol dislocation reaction of a type I membrane protein. Embo J 21:1041-1053

Flaherty KM, DeLuca-Flaherty C, McKay DB (1990) Three-dimensional structure of the ATPase fragment of a 70K heat-shock cognate protein. Nature 346:623-628

Freeman BC, Morimoto RI (1996) The human cytosolic molecular chaperones hsp90, hsp70 (hsc70) and hdj-1 have distinct roles in recognition of a non-native protein and protein refolding. Embo J 15:2969-2979

Frydman J (2001) Folding of newly translated proteins *in vivo*: the role of molecular chaperones. Annu Rev Biochem 70:603-647

Fu X, Ng C, Feng D, Liang C (2003) Cdc48p is required for the cell cycle commitment point at Start via degradation of the G1-CDK inhibitor Far1p. J Cell Biol 163:21-26

Futerman AH, van Meer G (2004) The cell biology of lysosomal storage disorders. Nat Rev Mol Cell Biol 5:554-565

Gardner RG, Swarbrick GM, Bays NW, Cronin SR, Wilhovsky S, Seelig L, Kim C, Hampton RY (2000) Endoplasmic reticulum degradation requires lumen to cytosol signaling. Transmembrane control of Hrd1p by Hrd3p. J Cell Biol 151:69-82

Gautschi M, Lilie H, Funfschilling U, Mun A, Ross S, Lithgow T, Rucknagel P, Rospert S (2001) RAC, a stable ribosome-associated complex in yeast formed by the DnaK-DnaJ homologs Ssz1p and zuotin. Proc Natl Acad Sci USA 98:3762-3767

Gillece P, Luz JM, Lennarz WJ, de La Cruz FJ, Romisch K (1999) Export of a cysteine-free misfolded secretory protein from the endoplasmic reticulum for degradation requires interaction with protein disulfide isomerase. J Cell Biol 147:1443-1456

Hamilton TG, Flynn GC (1996) Cer1p, a novel Hsp70-related protein required for posttranslational endoplasmic reticulum translocation in yeast. J Biol Chem 271:30610-30613

Hatakeyama S, Yada M, Matsumoto M, Ishida N, Nakayama KI (2001) U box proteins as a new family of ubiquitin-protein ligases. J Biol Chem 276:33111-33120

Hatakeyama S, Nakayama KI (2003) U-box proteins as a new family of ubiquitin ligases. Biochem Biophys Res Commun 302:635-645

Hatakeyama S, Matsumoto M, Yada M, Nakayama KI (2004) Interaction of U-box-type ubiquitin-protein ligases (E3s) with molecular chaperones. Genes Cells 9:533-548

Hershko A, Ciechanover A (1998) The ubiquitin system. Annu Rev Biochem 67:425-479

Hill K, Cooper AA (2000) Degradation of unassembled Vph1p reveals novel aspects of the yeast ER quality control system. Embo J 19:550-561

Hiller MM, Finger A, Schweiger M, Wolf DH (1996) ER degradation of a misfolded luminal protein by the cytosolic ubiquitin-proteasome pathway. Science 273:1725-1728

Hundley H, Eisenman H, Walter W, Evans T, Hotokezaka Y, Wiedmann M, Craig E (2002) The *in vivo* function of the ribosome-associated Hsp70, Ssz1, does not require its putative peptide-binding domain. Proc Natl Acad Sci USA 99:4203-4208

Huppa JB, Ploegh HL (1998) The eS-Sence of -SH in the ER. Cell 92:145-148

Huyer G, Piluek WF, Fansler Z, Kreft SG, Hochstrasser M, Brodsky JL, Michaelis S (2004) Distinct machinery is required in *Saccharomyces cerevisiae* for the endoplasmic reticulum-associated degradation of a multispanning membrane protein and a soluble luminal protein. J Biol Chem 279:38369-38378

Imai Y, Soda M, Inoue H, Hattori N, Mizuno Y, Takahashi R (2001) An unfolded putative transmembrane polypeptide, which can lead to endoplasmic reticulum stress, is a substrate of Parkin. Cell 105:891-902

Imai Y, Soda M, Hatakeyama S, Akagi T, Hashikawa T, Nakayama KI, Takahashi R (2002) CHIP is associated with Parkin, a gene responsible for familial Parkinson's disease, and enhances its ubiquitin ligase activity. Mol Cell 10:55-67

Jakob CA, Bodmer D, Spirig U, Battig P, Marcil A, Dignard D, Bergeron JJ, Thomas DY, Aebi M (2001) Htm1p, a mannosidase-like protein, is involved in glycoprotein degradation in yeast. EMBO Rep 2:423-430

Jarosch E, Taxis C, Volkwein C, Bordallo J, Finley D, Wolf DH, Sommer T (2002) Protein dislocation from the ER requires polyubiquitination and the AAA-ATPase Cdc48. Nat Cell Biol 4:134-139

Jensen TJ, Loo MA, Pind S, Williams DB, Goldberg AL, Riordan JR (1995) Multiple proteolytic systems, including the proteasome, contribute to CFTR processing. Cell 83:129-135

Journet A, Chapel A, Kieffer S, Roux F, Garin J (2002) Proteomic analysis of human lysosomes: application to monocytic and breast cancer cells. Proteomics 2:1026-1040

Kabani M, Beckerich JM, Gaillardin C (2000) Sls1p stimulates Sec63p-mediated activation of Kar2p in a conformation-dependent manner in the yeast endoplasmic reticulum. Mol Cell Biol 20:6923-6934

Kaneko C, Hatakeyama S, Matsumoto M, Yada M, Nakayama K, Nakayama KI (2003) Characterization of the mouse gene for the U-box-type ubiquitin ligase UFD2a. Biochem Biophys Res Commun 300:297-304

Kanelis V, Rotin D, Forman-Kay JD (2001) Solution structure of a Nedd4 WW domain-ENaC peptide complex. Nat Struct Biol 8:407-412

Kasanov J, Pirozzi G, Uveges AJ, Kay BK (2001) Characterizing Class I WW domains defines key specificity determinants and generates mutant domains with novel specificities. Chem Biol 8:231-241

Kitada T, Asakawa S, Hattori N, Matsumine H, Yamamura Y, Minoshima S, Yokochi M, Mizuno Y, Shimizu N (1998) Mutations in the parkin gene cause autosomal recessive juvenile parkinsonism. Nature 392:605-608

Knop M, Finger A, Braun T, Hellmuth K, Wolf DH (1996) Der1, a novel protein specifically required for endoplasmic reticulum degradation in yeast. Embo J 15:753-763

Koegl M, Hoppe T, Schlenker S, Ulrich HD, Mayer TU, Jentsch S (1999) A novel ubiquitination factor, E4, is involved in multiubiquitin chain assembly. Cell 96:635-644

Kondo H, Rabouille C, Newman R, Levine TP, Pappin D, Freemont P, Warren G (1997) p47 is a cofactor for p97-mediated membrane fusion. Nature 388:75-78

Lai BT, Chin NW, Stanek AE, Keh W, Lanks KW (1984) Quantitation and intracellular localization of the 85K heat shock protein by using monoclonal and polyclonal antibodies. Mol Cell Biol 4:2802-2810

Lilley BN, Ploegh HL (2004) A membrane protein required for dislocation of misfolded proteins from the ER. Nature 429:834-840

Lin H, Sugimoto Y, Ohsaki Y, Ninomiya H, Oka A, Taniguchi M, Ida H, Eto Y, Ogawa S, Matsuzaki Y, Sawa M, Inoue T, Higaki K, Nanba E, Ohno K, Suzuki Y (2004) N-octyl-beta-valienamine up-regulates activity of F213I mutant beta-glucosidase in cultured cells: a potential chemical chaperone therapy for Gaucher disease. Biochim Biophys Acta 1689:219-228

Lu Z, Cyr DM (1998) Protein folding activity of Hsp70 is modified differentially by the hsp40 co-chaperones Sis1 and Ydj1. J Biol Chem 273:27824-27830

Luders J, Demand J, Hohfeld J (2000) The ubiquitin-related BAG-1 provides a link between the molecular chaperones Hsc70/Hsp70 and the proteasome. J Biol Chem 275:4613-4617

Mayer MP, Schroder H, Rudiger S, Paal K, Laufen T, Bukau B (2000) Multistep mechanism of substrate binding determines chaperone activity of Hsp70. Nat Struct Biol 7:586-593

Mayer TU, Braun T, Jentsch S (1998) Role of the proteasome in membrane extraction of a short-lived ER-transmembrane protein. Embo J 17:3251-3257

McCarty JS, Buchberger A, Reinstein J, Bukau B (1995) The role of ATP in the functional cycle of the DnaK chaperone system. J Mol Biol 249:126-137

Meacham GC, Patterson C, Zhang W, Younger JM, Cyr DM (2001) The Hsc70 co-chaperone CHIP targets immature CFTR for proteasomal degradation. Nat Cell Biol 3:100-105

Medicherla B, Kostova Z, Schaefer A, Wolf DH (2004) A genomic screen identifies Dsk2p and Rad23p as essential components of ER-associated degradation. EMBO Rep 5:692-697

Meyer HH, Shorter JG, Seemann J, Pappin D, Warren G (2000) A complex of mammalian ufd1 and npl4 links the AAA-ATPase, p97, to ubiquitin and nuclear transport pathways. Embo J 19:2181-2192

Meyer HH, Wang Y, Warren G (2002) Direct binding of ubiquitin conjugates by the mammalian p97 adaptor complexes, p47 and Ufd1-Npl4. Embo J 21:5645-5652

Minami M, Nakamura M, Emori Y, Minami Y (2001) Both the N- and C-terminal chaperone sites of Hsp90 participate in protein refolding. Eur J Biochem 268:2520-2524

Misselwitz B, Staeck O, Rapoport TA (1998) J proteins catalytically activate Hsp70 molecules to trap a wide range of peptide sequences. Mol Cell 2:593-603

Molinari M, Helenius A (1999) Glycoproteins form mixed disulphides with oxidoreductases during folding in living cells. Nature 402:90-93

Molinari M, Calanca V, Galli C, Lucca P, Paganetti P (2003) Role of EDEM in the release of misfolded glycoproteins from the calnexin cycle. Science 299:1397-1400

Morello JP, Salahpour A, Laperriere A, Bernier V, Arthus MF, Lonergan M, Petaja-Repo U, Angers S, Morin D, Bichet DG, Bouvier M (2000) Pharmacological chaperones rescue cell-surface expression and function of misfolded V2 vasopressin receptor mutants. J Clin Invest 105:887-895

Murata S, Minami Y, Minami M, Chiba T, Tanaka K (2001) CHIP is a chaperone-dependent E3 ligase that ubiquitylates unfolded protein. EMBO Rep 2:1133-1138

Nakatsukasa K, Nishikawa S, Hosokawa N, Nagata K, Endo T (2001) Mnl1p, an alpha - mannosidase-like protein in yeast *Saccharomyces cerevisiae,* is required for endoplasmic reticulum-associated degradation of glycoproteins. J Biol Chem 276:8635-8638

Nikolay R, Wiederkehr T, Rist W, Kramer G, Mayer MP, Bukau B (2004) Dimerization of the human E3 ligase CHIP via a coiled-coil domain is essential for its activity. J Biol Chem 279:2673-2678

Nishikawa SI, Fewell SW, Kato Y, Brodsky JL, Endo T (2001) Molecular chaperones in the yeast endoplasmic reticulum maintain the solubility of proteins for retrotranslocation and degradation. J Cell Biol 153:1061-1070

Norgaard P, Westphal V, Tachibana C, Alsoe L, Holst B, Winther JR (2001) Functional differences in yeast protein disulfide isomerases. J Cell Biol 152:553-562

Oda Y, Hosokawa N, Wada I, Nagata K (2003) EDEM as an acceptor of terminally misfolded glycoproteins released from calnexin. Science 299:1394-1397

Panzner S, Dreier L, Hartmann E, Kostka S, Rapoport TA (1995) Posttranslational protein transport in yeast reconstituted with a purified complex of Sec proteins and Kar2p. Cell 81:561-570

Patterson C (2002) A new gun in town: the U box is a ubiquitin ligase domain. Sci STKE 2002:PE4

Pellecchia M, Szyperski T, Wall D, Georgopoulos C, Wuthrich K (1996) NMR structure of the J-domain and the Gly/Phe-rich region of the Escherichia coli DnaJ chaperone. J Mol Biol 260:236-250

Perdew GH, Whitelaw ML (1991) Evidence that the 90-kDa heat shock protein (HSP90) exists in cytosol in heteromeric complexes containing HSP70 and three other proteins with Mr of 63,000, 56,000, and 50,000. J Biol Chem 266:6708-6713

Petaja-Repo UE, Hogue M, Bhalla S, Laperriere A, Morello JP, Bouvier M (2002) Ligands act as pharmacological chaperones and increase the efficiency of delta opioid receptor maturation. Embo J 21:1628-1637

Peters JM (1999) Subunits and substrates of the anaphase-promoting complex. Exp Cell Res 248:339-349

Picard D (2002) Heat-shock protein 90, a chaperone for folding and regulation. Cell Mol Life Sci 59:1640-1648

Pickart CM (2001) Mechanisms underlying ubiquitination. Annu Rev Biochem 70:503-533

Pilon M, Schekman R, Romisch K (1997) Sec61p mediates export of a misfolded secretory protein from the endoplasmic reticulum to the cytosol for degradation. Embo J 16:4540-4548

Plemper RK, Bohmler S, Bordallo J, Sommer T, Wolf DH (1997) Mutant analysis links the translocon and BiP to retrograde protein transport for ER degradation. Nature 388:891-895

Pollanen MS, Dickson DW, Bergeron C (1993) Pathology and biology of the Lewy body. J Neuropathol Exp Neurol 52:183-191

Ponting CP (2000) Proteins of the endoplasmic-reticulum-associated degradation pathway: domain detection and function prediction. Biochem J 351 Pt 2:527-535

Pringa E, Martinez-Noel G, Muller U, Harbers K (2001) Interaction of the ring finger-related U-box motif of a nuclear dot protein with ubiquitin-conjugating enzymes. J Biol Chem 276:19617-19623

Rabinovich E, Kerem A, Frohlich KU, Diamant N, Bar-Nun S (2002) AAA-ATPase p97/Cdc48p, a cytosolic chaperone required for endoplasmic reticulum-associated protein degradation. Mol Cell Biol 22:626-634

Rao H, Sastry A (2002) Recognition of specific ubiquitin conjugates is important for the proteolytic functions of the ubiquitin-associated domain proteins Dsk2 and Rad23. J Biol Chem 277:11691-11695

Rape M, Hoppe T, Gorr I, Kalocay M, Richly H, Jentsch S (2001) Mobilization of processed, membrane-tethered SPT23 transcription factor by CDC48(UFD1/NPL4), a ubiquitin-selective chaperone. Cell 107:667-677

Rouiller I, Butel VM, Latterich M, Milligan RA, Wilson-Kubalek EM (2000) A major conformational change in p97 AAA ATPase upon ATP binding. Mol Cell 6:1485-1490

Rudiger S, Schneider-Mergener J, Bukau B (2001) Its substrate specificity characterizes the DnaJ co-chaperone as a scanning factor for the DnaK chaperone. Embo J 20:1042-1050

Saris N, Makarow M (1998) Transient ER retention as stress response: conformational repair of heat-damaged proteins to secretion-competent structures. J Cell Sci 111 (Pt 11):1575-1582

Sato S, Ward CL, Krouse ME, Wine JJ, Kopito RR (1996) Glycerol reverses the misfolding phenotype of the most common cystic fibrosis mutation. J Biol Chem 271:635-638

Sawkar AR, Cheng WC, Beutler E, Wong CH, Balch WE, Kelly JW (2002) Chemical chaperones increase the cellular activity of N370S beta -glucosidase: a therapeutic strategy for Gaucher disease. Proc Natl Acad Sci USA 99:15428-15433

Schauber C, Chen L, Tongaonkar P, Vega I, Lambertson D, Potts W, Madura K (1998) Rad23 links DNA repair to the ubiquitin/proteasome pathway. Nature 391:715-718

Scheufler C, Brinker A, Bourenkov G, Pegoraro S, Moroder L, Bartunik H, Hartl FU, Moarefi I (2000) Structure of TPR domain-peptide complexes: critical elements in the assembly of the Hsp70-Hsp90 multichaperone machine. Cell 101:199-210

Schiffmann R, Brady RO (2002) New prospects for the treatment of lysosomal storage diseases. Drugs 62:733-742

Shearer AG, Hampton RY (2004) Structural control of endoplasmic reticulum-associated degradation: effect of chemical chaperones on 3-hydroxy-3-methylglutaryl-CoA reductase. J Biol Chem 279:188-196

Sikorski RS, Boguski MS, Goebl M, Hieter P (1990) A repeating amino acid motif in CDC23 defines a family of proteins and a new relationship among genes required for mitosis and RNA synthesis. Cell 60:307-317

Silverstein AM, Galigniana MD, Chen MS, Owens-Grillo JK, Chinkers M, Pratt WB (1997) Protein phosphatase 5 is a major component of glucocorticoid receptor.hsp90 complexes with properties of an FK506-binding immunophilin. J Biol Chem 272:16224-16230

Sommer T, Jentsch S (1993) A protein translocation defect linked to ubiquitin conjugation at the endoplasmic reticulum. Nature 365:176-179

Spear ED, Ng DT (2003) Stress tolerance of misfolded carboxypeptidase Y requires maintenance of protein trafficking and degradative pathways. Mol Biol Cell 14:2756-2767

Stevens FJ, Argon Y (1999) Protein folding in the ER. Semin Cell Dev Biol 10:443-454

Suhan ML, Hovde CJ (1998) Disruption of an internal membrane-spanning region in Shiga toxin 1 reduces cytotoxicity. Infect Immun 66:5252-5259

Swanson R, Locher M, Hochstrasser M (2001) A conserved ubiquitin ligase of the nuclear envelope/endoplasmic reticulum that functions in both ER-associated and Matalpha2 repressor degradation. Genes Dev 15:2660-2674

Tatzelt J, Prusiner SB, Welch WJ (1996) Chemical chaperones interfere with the formation of scrapie prion protein. Embo J 15:6363-6373

Taxis C, Hitt R, Park SH, Deak PM, Kostova Z, Wolf DH (2003) Use of modular substrates demonstrates mechanistic diversity and reveals differences in chaperone requirement of ERAD. J Biol Chem 278:35903-35913

Theyssen H, Schuster HP, Packschies L, Bukau B, Reinstein J (1996) The second step of ATP binding to DnaK induces peptide release. J Mol Biol 263:657-670

Tirosh B, Furman MH, Tortorella D, Ploegh HL (2003) Protein unfolding is not a prerequisite for endoplasmic reticulum-to-cytosol dislocation. J Biol Chem 278:6664-6672

Tsai B, Rodighiero C, Lencer WI, Rapoport TA (2001) Protein disulfide isomerase acts as a redox-dependent chaperone to unfold cholera toxin. Cell 104:937-948

Tsai B, Ye Y, Rapoport TA (2002) Retro-translocation of proteins from the endoplasmic reticulum into the cytosol. Nat Rev Mol Cell Biol 3:246-255

Tyson JR, Stirling CJ (2000) LHS1 and SIL1 provide a lumenal function that is essential for protein translocation into the endoplasmic reticulum. Embo J 19:6440-6452

Van den Berg B, Clemons WM Jr, Collinson I, Modis Y, Hartmann E, Harrison SC, Rapoport TA (2004) X-ray structure of a protein-conducting channel. Nature 427:36-44

Vashist S, Ng DT (2004) Misfolded proteins are sorted by a sequential checkpoint mechanism of ER quality control. J Cell Biol 165:41-52

Verma R, Chen S, Feldman R, Schieltz D, Yates J, Dohmen J, Deshaies RJ (2000) Proteasomal proteomics: identification of nucleotide-sensitive proteasome-interacting proteins by mass spectrometric analysis of affinity-purified proteasomes. Mol Biol Cell 11:3425-3439

Vijayraghavan U, Company M, Abelson J (1989) Isolation and characterization of pre-mRNA splicing mutants of *Saccharomyces cerevisiae*. Genes Dev 3:1206-1216

Walter J, Urban J, Volkwein C, Sommer T (2001) Sec61p-independent degradation of the tail-anchored ER membrane protein Ubc6p. Embo J 20:3124-3131

Wang BB, Hayenga KJ, Payan DG, Fisher JM (1996) Identification of a nuclear-specific cyclophilin which interacts with the proteinase inhibitor eglin c. Biochem J 314 (Pt 1):313-319

Wang Q, Chang A (1999) Eps1, a novel PDI-related protein involved in ER quality control in yeast. Embo J 18:5972-5982

Wang Q, Chang A (2003) Substrate recognition in ER-associated degradation mediated by Eps1, a member of the protein disulfide isomerase family. Embo J 22:3792-3802

Ward CL, Omura S, Kopito RR (1995) Degradation of CFTR by the ubiquitin-proteasome pathway. Cell 83:121-127

Wegele H, Muschler P, Bunck M, Reinstein J, Buchner J (2003) Dissection of the contribution of individual domains to the ATPase mechanism of Hsp90. J Biol Chem 278:39303-39310

Wegele H, Muller L, Buchner J (2004) Hsp70 and Hsp90--a relay team for protein folding. Rev Physiol Biochem Pharmacol 151:1-44

Werner ED, Brodsky JL, McCracken AA (1996) Proteasome-dependent endoplasmic reticulum-associated protein degradation: an unconventional route to a familiar fate. Proc Natl Acad Sci USA 93:13797-13801

Wiertz EJ, Tortorella D, Bogyo M, Yu J, Mothes W, Jones TR, Rapoport TA, Ploegh HL (1996) Sec61-mediated transfer of a membrane protein from the endoplasmic reticulum to the proteasome for destruction. Nature 384:432-438

Wilkinson CR, Seeger M, Hartmann-Petersen R, Stone M, Wallace M, Semple C, Gordon C (2001) Proteins containing the UBA domain are able to bind to multi-ubiquitin chains. Nat Cell Biol 3:939-943

Ye Y, Meyer HH, Rapoport TA (2001) The AAA ATPase Cdc48/p97 and its partners transport proteins from the ER into the cytosol. Nature 414:652-656

Ye Y, Meyer HH, Rapoport TA (2003) Function of the p97-Ufd1-Npl4 complex in retrotranslocation from the ER to the cytosol: dual recognition of nonubiquitinated polypeptide segments and polyubiquitin chains. J Cell Biol 162:71-84

Ye Y, Shibata Y, Yun C, Ron D, Rapoport TA (2004) A membrane protein complex mediates retro-translocation from the ER lumen into the cytosol. Nature 429:841-847

Young JC, Moarefi I, Hartl FU (2001) Hsp90: a specialized but essential protein-folding tool. J Cell Biol 154:267-273

Zhang X, Shaw A, Bates PA, Newman RH, Gowen B, Orlova E, Gorman MA, Kondo H, Dokurno P, Lally J, Leonard G, Meyer H, van Heel M, Freemont PS (2000a) Structure of the AAA ATPase p97. Mol Cell 6:1473-1484

Zhang Y, Gao J, Chung KK, Huang H, Dawson VL, Dawson TM (2000b) Parkin functions as an E2-dependent ubiquitin- protein ligase and promotes the degradation of the synaptic vesicle-associated protein, CDCrel-1. Proc Natl Acad Sci USA 97:13354-13359

Zhang Y, Nijbroek G, Sullivan ML, McCracken AA, Watkins SC, Michaelis S, Brodsky JL (2001) Hsp70 molecular chaperone facilitates endoplasmic reticulum-associated protein degradation of cystic fibrosis transmembrane conductance regulator in yeast. Mol Biol Cell 12:1303-1314

Zhong X, Shen Y, Ballar P, Apostolou A, Agami R, Fang S (2004) AAA ATPase p97/VCP interacts with gp78: a ubiquitin ligase for ER-associated degradation. J Biol Chem 279:45676-45684

Zhu X, Zhao X, Burkholder WF, Gragerov A, Ogata CM, Gottesman ME, Hendrickson WA (1996) Structural analysis of substrate binding by the molecular chaperone DnaK. Science 272:1606-1614

Gauss, Robert
Max-Delbrück-Centrum für Molekulare Medizin, Robert-Rössle-Str. 10, 13025 Berlin, Germany

Neuber, Oliver
Max-Delbrück-Centrum für Molekulare Medizin, Robert-Rössle-Str. 10, 13025 Berlin, Germany

Sommer, Thomas
Max-Delbrück-Centrum für Molekulare Medizin, Robert-Rössle-Str. 10, 13025 Berlin, Germany
tsommer@mdc-berlin.de

Template-induced protein misfolding underlying prion diseases

Luc Bousset, Nicolas Fay, and Ronald Melki

Abstract

Proteins with prion properties are closely associated to a class of fatal neurodegenerative illnesses in mammals and to the emergence and propagation of phenotypic traits in yeast. The structural transition from the correctly folded, native form of a prion protein to a persistent misfolded form that ultimately may cause cell death or the transmission of phenotypic traits are not yet fully understood. The structural and functional properties of mammalian and yeast prions in their soluble and oligomeric forms are presented as are the mechanistic models accounting for this structure-based mode of inheritance. This review highlights a number of unquestioned issues and unanswered questions that may allow a better understanding of the role of prion proteins *in vivo* and their propagation mechanism(s).

1 Prion diseases

Protein misfolding and subsequent aggregation is at the origin of over 20 human diseases termed "conformational" diseases (Sipe and Cohen 2000). A class of fatal neurodegenerative illnesses: Creutzfeldt Jacob's disease (CJD), Gerstmann-Sträussler-Scheinker syndrome (GSS), fatal familial insomnia (FFI), and kuru in humans, Bovine Spongiform Encephalopathies (BSE) in cattle, chronic wasting disease (CWD) in elk and deer, and scrapie in sheep, are peculiar in that they can be transmitted (Prusiner 1998). These diseases are intimately linked to the aggregation of a constitutive protein termed prion protein or PrP in a non-native form. The molecular events at the origin of the emergence and propagation of prion diseases are believed to be due to protein misfolding (Cohen et al. 1994), but are not yet fully understood.

French and English veterinarians first described scrapie in 1732. It is only in 1936 that this disease was proven to be infectious (Cuille and Chelles 1936). A parallel between this disease and the human neurodegenerative disease kuru was first made in 1959 (Hadlow 1959). The size of the infectious particles (Alper et al. 1966) and their considerable resistance toward inactivation by chemical treatments, heat, and irradiation (Stamp et al. 1962; Alper et al. 1966, 1967) very quickly pointed out the unusual nature of this infectious agent.

2 Formulation of the prion hypothesis

The idea that proteins could behave as infectious agents carrying alone the hereditary information that ensures their propagation emerged as early as 1967 when the mathematician Griffith proposed three theoretical pathways allowing an infectious agent made of proteins to replicate (Griffith 1967). This revolutionary idea that contradicts the dogma that only nucleic acids could act as heritable elements remained subject to debate until the scrapie agent was purified (Prusiner et al. 1980, 1981, 1982; Bolton et al. 1982; Diringer et al. 1983) and showed to be mainly composed of a constitutive protein that has a molecular mass of 27-30kDa, termed prion for "infectious protein" (Prusiner 1982) and later on PrP for prion protein.

3 The mammalian prion PrP

3.1 Identification

The identification of the 14 N-terminal amino acid residues of purified PrP paved the way for its identification (Prusiner et al. 1984). PrP is encoded by the chromosomal gene *PRNP* (Chesebro et al. 1985; Oesch et al. 1985). In healthy individuals the highly conserved, constitutive form of the prion protein PrP^c, a glycosylphosphatidylinositol-linked cell surface glycoprotein (Stahl et al. 1987, 1990) is protease-sensitive and is mainly expressed in the central nervous system in lymphatic tissues and at the neuromuscular junction. PrP^c is not essential as PrP-deficient mice are viable and can develop either normally or with minor defects (Raeber et al. 1998). However, mice expressing a truncated version of PrP^c in a null background show neuronal degeneration, suggesting that the protein may be involved in the maintenance and/or regulation of neuronal functions (Shmerling et al. 1998). The strongest evidence for the involvement of PrP^c in scrapie came from the demonstration that mice devoid of PrP are resistant to disease (Bueler et al. 1993).

3.2 Structure

Recombinant PrP^c has a high α-helical content (Pan et al. 1993). The protein is monomeric, has a single disulfide bond (Turk et al. 1988) and two N-glycosylation sites (Endo et al. 1989). The three-dimensional structures of PrP from different vertebrates were determined by NMR spectroscopy (Riek et al. 1996, 1997, 1998; Donne et al. 1997; James et al. 1997; Billeter et al. 1997; Liu et al. 1999; Lopez-Garcia et al. 2000; Calzolai et al. 2000; Zahn et al. 2000; Zhang et al. 2000) as well as by protein crystallography (Knaus et al. 2001). All the structures reveal that the C-terminal residues 121-231 form a globular domain with three α-helices and a short antiparallel β-sheet while the N-terminal segment 23-120 is highly flexible and disordered in solution. While it is reasonable to envisage a change in the flexibility of the N-terminal domain of PrP when the protein is anchored to the cell membrane via the glycosylphosphatidyl-inositol moiety as well as upon N-glycosylation of Asn 181 and 197 (Rudd et al. 1999) no evidence for a major conformational change have been so far revealed upon the covalent attachment of PrP to liposomes (Eberl et al. 2004).

3.3 Folding properties

In individuals developing TSEs, PrP becomes protease-resistant (PrP^{res}) (Prusiner et al. 1982). PrP^{c} and PrP^{res} have the same chemical composition (Stahl et al. 1993). They differ in their secondary structure content. PrP^{c} has a high α-helical content (40%) as measured by FTIR spectroscopy and circular dichroism and very low β-sheets (3%) in agreement with the NMR structure, while the β-sheets and α-helical contents of PrP^{res} are 50% and 20%, respectively (Pan et al. 1993; Caughey et al. 1991; Safar et al. 1993). This together with a number of genetic studies (reviewed in Horwich and Weissman 1997 and in Prusiner 1997) has lead to the view that TSEs are due to a change in the conformation of PrP^{c}. This change leads to the aggregation of PrP into stable high molecular weight oligomers that are resistant to proteinase K treatment and capable of converting PrP^{c} into PrP^{res} (Prusiner 1998; Kocisko et al. 1994). These aggregates when purified by differential centrifugation following detergent solubilization mainly consist of flattened rods (Prusiner et al. 1983). *In vitro* and *in vivo* conditions have been found where the resistance to proteolysis of recombinant PrP increases, presumably following its assembly into high molecular weight aggregates (Caughey 2003; Swietnicki et al. 2000; Saborio et al. 2001; Adler et al. 2003; Deleault et al. 2003; Torrent et al. 2004; Lee and Eisenberg 2003; Ma and Lindquist 1999). Conditions have also been found where the β-sheets content of soluble recombinant PrP increases significantly (Jackson et al. 1999). However, these forms have never been shown to be infectious (Hill et al. 1999) until very recently where *in vitro* assembled fibrils inoculated intracerebrally to transgenic mice overexpressing a truncated PrP (PrP 89-231) were shown to develop a transmissible form of neurologic dysfunction (Legname et al. 2004). If demonstrated to be independent from the overexpression of truncated PrP, in other words reproduced in a natural context, this latter result may constitute the ultimate proof of the key role of the conformation of PrP in the propagation of prion diseases.

3.4 Function

The exact function of PrP^{c} remains unknown. PrP^{c} is clearly nonessential (Raeber et al. 1998). The protein has been reported as being involved in the maintenance and/or regulation of neuronal functions (Shmerling et al. 1998), in oxidative stress (Brown et al. 1999), as a copper-binding protein (Pan et al. 1992; Viles et al. 1999; Brown et al. 1997) involved in the suppression of perturbations related to copper homeostasis (Sakudo et al. 2004), and as a signaling molecule (Mouillet-

Richard et al. 2000). PrP^c has been reported to interact with a variety of proteins (Oesch et al. 1990; Kurschner and Morgan 1996; Yehiely et al. 1997; Jin et al. 2000; Hundt et al. 2001). These important findings did not yet though allow a better understanding of the role of PrP^c. The most important interaction of PrP^c with a partner protein is that with PrP^{res} (Meier et al. 2003). This interaction is at the origin of the "seeded" conversion of PrP^c into PrP^{res}. The efficiency of transmission of prion diseases from one species to another is governed by a phenomenon called "species barrier". It is possible to overcome this barrier by expressing in the recipient animal the *PRNP* gene of the donor animal. A parallel has been made between the ease with which the species barrier is overcome and differences in the amino acid residue composition of PrP^c in different animal species (Scott et al. 1989). Indeed, it is widely believed that in a manner similar to what is observed when proteins are crystallized in solution, the heterogeneity due to amino acid substitutions weakens protein-protein interactions, thus, inhibiting assembly. It is also believed that in a manner similar to the ability of a polypeptide to crystallize in various forms, PrP^c assembles into distinct high molecular weight oligomers that propagate aggregation with different efficiencies. This is at the origin of what is termed prion "strains" that differ in their incubation time preceding the neurological dysfunctions (Scott et al. 1999).

3.5 Cellular processing

Following synthesis in the cytoplasm, PrP^c is translocated into the endoplasmic reticulum where an N-treminal secretory signal peptide of 22 amino acids is cleaved off the polypeptide chain while a glycosylphosphatidylinositol-anchor is added. Mature PrP^c is then directed to the outer surface of the endoplasmic membrane. The half-life of the protein is 6h (Borchelt et al. 1990). It is widely accepted that the conversion of PrP^c into PrP^{res} occurs at the surface of the cell (Enari et al. 2001).

4 The prions in yeast and fungi

In the yeast *Saccharomyces cerevisiae*, two genetic elements, [PSI^+] and [URE3], that fail to behave according to Mendel's laws were discovered nearly 40 years ago (Cox 1965; Lacroute 1971). The efficiency of nonsense suppression is modified in cells exhibiting the [PSI^+] phenotype while [URE3] cells show altered nitrogen metabolism. The non-Mendelian inheritance of [PSI^+] and [URE3] would not be surprising if the genes encoding these traits were not located in the nucleus of yeast cells (Schoun and Lacroute 1969; Cox et al. 1988). Indeed, [PSI^+] is associated with Sup35p an essential component of the translation termination machinery (Hawthorne and Mortimer 1968; Inge-Vechtomov and Andrianova 1970) while [URE3] is associated with Ure2p a polypeptide that acts as a negative regulator of nitrogen metabolism (Mitchell and Magasanik 1984; Courchesne and Magasanik 1988; Coschigano and Magasanik 1991). It was only ten years ago that the prion hypothesis was extended to account for the non-Mendelian behavior of [PSI$^+$] and [URE3] (Wickner 1994).

4.1 Genetic criteria for the prions in yeast and fungi

Three genetic criteria need to be fulfilled in order to establish that an infectious agent is a prion and not a virus, plasmid or other nucleic acid replicon: (i) reversible curability, (ii) induction of prion formation by overexpression of the normal protein, and (iii) a unique phenotypic relationship between the prion state and mutations in the gene of the protein (Wickner 1994).

Some mutations in *SUP35* and *URE2* genes that inactivate Sup35p and Ure2p mimic the [PSI$^+$] and [URE3] phenotypes. However, the wild type phenotypes are not restored when yeast cells are grown in the presence for example of millimolar concentrations of the protein denaturant guanidinium chloride or under high-osmotic strength conditions and cannot reappear without introduction of new DNA in contrast with what happens in the case of authentic [PSI$^+$] and [URE3] phenotypes (Aigle and Lacroute 1975; Singh et al. 1979; Tuite et al. 1981; Lund and Cox 1981). Furthermore, the mutations that lead to the inactivation of Sup35p and Ure2p can reappear with a frequency that is two orders of magnitude lower than the reappearance of the [PSI$^+$] and [URE3] phenotypes in cured cells. Finally, the frequency of *de novo* generation of the [PSI$^+$] and [URE3] phenotypes is highly dependent on the expression level of *SUP35* and *URE2* gene products (Chernoff et al. 1993; Wickner 1994). It increases by 20 to 1000-fold upon overexpression of Ure2p and Sup35p.

4.2 Characteristics

4.2.1 The [PSI$^+$] and [PIN] phenotypes

Sup35p, known also as eRF3, together with Sup45p constitute the translation release factor that recognizes stop codons and releases nascent polypeptides from the last t-RNA. The loss of the function of Sup35p reduces the fidelity with which ribosomes terminate translation at ochre (UAA) stop codons. In *S. cerevisiae*, Sup35p (Swiss-Prot P05453) is a large polypeptide made of 685 amino acid residues that has a calculated molecular mass of 76551Da (Kushnirov et al. 1988). One can distinguish three regions in Sup35p (Fig. 1). The N-terminal region (N) extends from amino acid residues 1 to 122 is rich in Q, N, and G residues (47%). This region is not essential for the role of the protein in translation termination (Ter-Avanesyan et al. 1993, 1994; Derkatch et al. 1996). It is called prion domain (PrD) as it plays a critical role in prion propagation. It includes 5 imperfect repeats of the oligopeptide PQGGYQQYN that resemble to some extent to the octarepeat PHGGGWGQ of mammalian PrP (Tuite 2000). Changes in the number of repeats influences [PSI$^+$] propagation (TerAvanesyan et al.1994; Liu and Lindquist 1999).

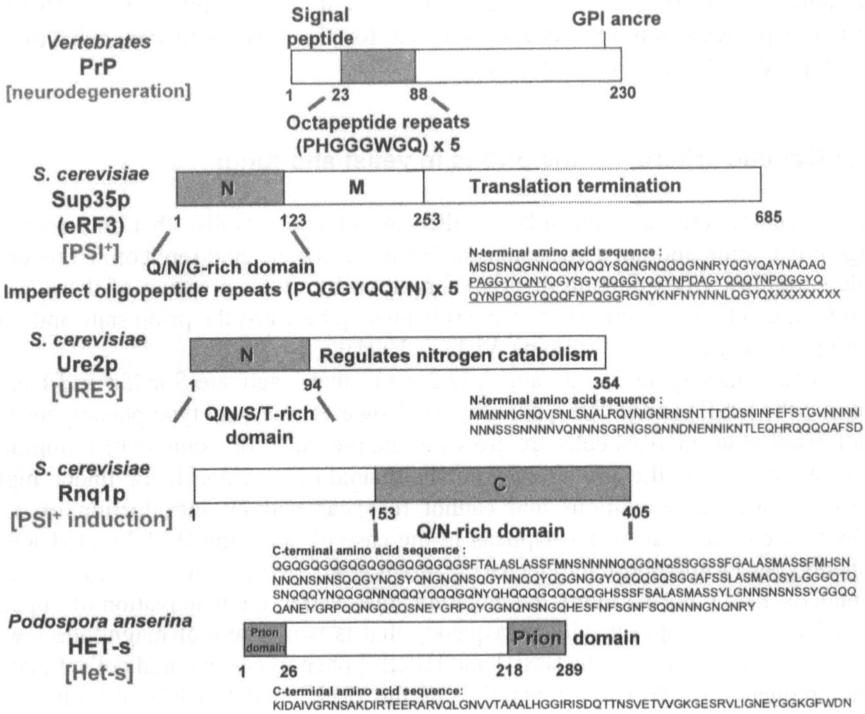

Fig. 1. Comparison of the characteristics of the prions so far identified. The names, function, and phenotypes of each prion so far identified is given. The boundaries between the different domains in each prion are indicated. The prion domains are outlined (in grey) and their amino acid composition or characteristics presented. The consensus sequence of the oligopeptide repeats in PrP and Sup35p are shown. They are also underlined in the N-terminal amino acid sequence of Sup35p.

The middle region (M) has probably a structural function. It extends from amino acid residues 123 to 253. The C-terminal region is the functional domain of the protein providing translation-termination activity (Ter-Avanesyan et al. 1993; Zhouravleva et al. 1995). Sup35p is soluble and functional in [psi$^-$] cells while it is mainly insoluble and nonfunctional in [PSI$^+$] cells (Stansfield et al. 1995; Patino et al. 1996).

The overexpression of Sup35p increases the frequency of [PSI$^+$] appearance. This is also what is observed upon the overexpression of Rnq1p (Swiss-Prot P25367) a 42.5kDa polypeptide rich in Q and N residues in particular in its part extending from amino acid residue 153 to 405 with unknown function (Fig. 1). Rnq1p exists as a soluble protein as well as in insoluble aggregates (Sondheimer and Lindquist 2000). The amount of insoluble Rnq1p increases when the protein is overproduced (Derkatch et al. 2001). In its insoluble state, Rnq1p allows the conversion of [psi$^-$] cells into [PSI$^+$] cells and is at the origin of the phenotype [PIN$^+$] that comes from "[PSI$^+$] inducibility" (Derkatch et al. 1997). The overexpression of many other Q and N-rich polypeptides has [PIN$^+$]-like effects (Derkatch et al. 2001; Osherovich and Weissman 2001). Nevertheless, only [PIN$^+$] is cured when yeast cells are grown on guanidinium chloride and can reappear in cured cells suggesting it is a prion (Derkatch et al. 2000).

4.2.2 The [URE3] phenotype

The exact function of Ure2p is unknown. Functional Ure2p plays a role in nitrogen-dependent repression of ureidosuccinate usage. Indeed, when yeast cells are grown in the presence of nitrogen, the synthesis of enzymes and transporters needed for the assimilation of complex nitrogen sources such as allantoate is turned off. This is believed to be due to the interaction of Ure2p with the transcription factor Gln3p in the cytoplasm, thus, preventing the entry of Gln3p into the nucleus where it allows the transcription of a number of genes among which *DAL5* the gene encoding the allantoate transporter Dal5p, which also transports ureidosuccinate (Courchesne and Magasanik 1988). In *S. cerevisiae*, Ure2p (Swiss-Prot P23202) has a calculated molecular mass of 40271 Da and is made of 354 amino acid residues (Fig. 1). The N-terminal part of Ure2p extending from amino acid residue 1 to 93 (Thual et al. 1999) has an unusual composition. It is very rich in Q, N S and T residues (62%). This region is required for the propagation of the [URE3] phenotype (Masison and Wickner 1995). It is therefore referred to as the prion domain (PrD) of the protein. The C-terminal domain extends from amino acid residues 94 to 354 (Thual et al. 1999). It complements *URE2* gene deletion and constitute, therefore, the functional domain of the protein (Masison and Wickner 1995).

4.2.3 The [Het-s] phenotype

The fusion of two different *Podospora anserina* filaments that contact each other requires among other things that the two filaments are genetically identical at the *het-s* locus (reviewed in Glass et al. 2000). Two common alleles of this locus, called *het-s* and *het-S*, encode cytosolic polypeptides that differ in 13 amino acid residues out of 289 (Turcq et al. 1991). These substitutions are distributed throughout the protein. Cells expressing the *het-s* allele product, the protein HET-s, exist under two states termed [Het-s*] and [Het-s]. Stains in the [Het-s*] state are indifferent to the *het-s/het-S* status of their fusion partner. In contrast, a cell death reaction occurs leading to the death of the heterocaryotic cell upon fusion of a filament in its [Het-s] state with a filament containing the protein HET-S. Indeed, the [Het-s] state meets with the genetic criteria that define a prion (Coustou et al. 1997). The fusion of cells in their [Het-s*] and [Het-s] states, although not involving fusion of the nuclei, yields cells in the [Het-s] state (Beisson-Schecroun 1962). The *het-s* allele is essential for the propagation of the [Het-s] state while the deletion of the gene yields a state similar to the [Het-s*] state. The overexpression of the protein HET-s increases the frequency with which [Het-s] arises *de novo*. Finally, the [Het-s] state is frequently eliminated during sporulation and arises again with a low frequency.

The HET-s protein (TrEMBL Q03689) has a calculated molecular mass of 32kDa. It has no remarkable features (Fig. 1). The C-terminal domain of the protein spanning amino acid residues 218-289 is nowadays presented as essential for prion propagation and is considered as the prion domain of the protein although the N-terminal 26 amino acid residues and point mutations at amino acid residues 23 and 33 were considered up to recently as critical for prion propagation (Deleu et al. 1993; Coustou et al. 1999).

4.2.4 Other potential prions

Any chromosomally encoded protein that undergoes a change into a modified form that is necessary for its own modification is a potential prion. Such proteins are detected if the prion form is toxic but not only as the prion form can also confer selective advantages or disadvantages. If the prion form is toxic, the cell dyes. This is the case of the mammalian prion PrP. If the prion form confers an advantage or a disadvantage a phenotype is observed under selective conditions. This is the case in yeast and fungi where prions do not kill the organism but instead allow cells to read through stop codons, to use poor nitrogen sources, or to forbid fusion between cells, which can be advantageous under particular conditions and disadvantageous under normal conditions.

A number of surveys have been conducted using the unusual content and distribution of Q and N amino acid residues to identify novel yeast prions. Several proteins with Q and N rich domains have been revealed (Michelisch and Weissman 2000; Sondheimer and Lindquist 2000). Novel prions have also been created by fusing the prion domain of Sup35p to polypeptides such as the rat glucocorticoid receptor (Li and Lindquist 2000; True and Lindquist 2000). All the polypeptides with Q and N rich domains do not possess prion properties. This is the case for example of the protein Huntingtin, associated with Huntington disease that exhibits no infectious activity. Furthermore, the mammalian prion PrP and the *P. anserina* prion HET-s are neither rich in Q nor N residues. These facts strongly suggest that the presence of a Q and N rich domains in a polypeptide is not sufficient to make a protein a prion.

4.3 Structural features

Since the presence of Q and N rich domains in a polypeptide is not sufficient to make it of prion nature, what is it that makes a protein a prion? Sup 35p and PrP contain oligopeptide repeats that bear striking resemblance. PrP contains five repeats of the octapeptide PHGGGWGQ while Sup35p contains five imperfect oligopeptide repeat PQGGYQQYN. These oligopeptide repeats play a key role in [PSI$^+$] propagation (Liu and Lindquist 1999) while they appear important but non-essential in mammals (Weissmann et al. 1999). Furthermore, this primary structure features is absent in Ure2p, Rnq1p and HET-s. It cannot therefore be considered as the common feature to all prions.

All prions share a common structural feature. They are all two-domain proteins with a flexible domain that can be either N or C-terminal. The three-dimensional structures of recombinant PrPs from various species have been determined by NMR spectroscopy. While the C-terminal part of all PrP molecules (residues 121-231) is compactly folded, the N-terminal part lacks identifiable secondary structures. No three-dimensional structures are available for the prions from yeast and fungi except a structure for Ure2p lacking the N-terminal domain extending from amino acid residues 1 to 93 (Bousset et al. 2001). This structure reveals, as in the case of the C-terminal domain of PrP, the compact folding of the functional C-terminal domain of Ure2p. Indications on the high exposure to the solvent of the N-terminal parts of Sup35p, Ure2p and the C-terminal parts of Rnq1p and HET-s come from proteolytic and hydrogen/deuterium exchange studies (Thual et al. 1999; Serio et al. 2000; Derkatch et al. 2001; Nazabal et al. 2004).

Purified, full-length prion proteins are soluble and helical (Glover et al. 1997; Jackson et al. 1999; Thual et al. 1999). The soluble forms can convert *in vitro* into high molecular weight oligomers. The conversion reaction is frequently performed at acidic pH in the case of PrP while it occurs at neutral pH in the case of prions from yeast and fungi (Glover et al. 1997; Thual et al. 1999; Derkatch et al. 2001; Dos Reis et al. 2002). The high molecular weight oligomers are either soluble and spherical or insoluble and fibrillar (Jackson et al. 1999; Thual et al. 1999; Serio et al. 2000; Dos Reis et al. 2002). The formation of the fibrils is greatly accelerated by seeding with preformed fibrils (Glover et al. 1997; Thual et al. 1999; Balgerie et al. 2002). Furthermore, there is evidence that *in vitro* assembled fibrils made of recombinant Sup35p and HET-s are infectious since when introduces in their respective hosts they induce the *de novo* appearance of the prion phenotype (Tanaka et al. 2004; King and Diaz-Avalos 2004; Maddelein et al. 2002). This might reveal true also in the case of PrP if the neurologic dysfunction observed upon intracerebral injection of *in vitro* assembled PrP fibrils to transgenic mice overexpressing a truncated PrP (PrP 89-231) (Legname et al. 2004) yield the same results when injected to wild type animals.

5 Properties of the fibrillar forms of prion proteins

The fibrillar forms of prion proteins have an increased resistance to proteolytic treatments, bind the dye Congo red and exhibit yellow green birefringence upon binding of Congo red in polarized light (Thual et al. 2001; Maddelein et al. 2002). These properties are not unique to prion fibrils as biological polymers such as actin filaments and microtubules exhibit increased resistance to proteolysis and yellow-green birefringence in polarized light upon Congo red binding (Bousset et al. 2004). It is claimed that prion fibrils are amyloid as the assembly reaction is accompanied by a change in the circular dichroism spectrum that resemble to an α-helical to β-sheet transition (Glover et al. 1997; Jackson et al. 1999; Schlumpberger et al. 2000). Circular dischroism is not an adequate method to measure conformational changes occurring during the assembly of a polypeptide into protein fibrils. Indeed, although the optical activity in the region 190-230 nm is dominated by the peptide backbone, the CD of an α-helix or a β-sheet is a function of its length. This is not adequately taken into account when homopolymeric structures are used as references to compute the experimental spectra. Suitable proteins with known three-dimensional structures should instead be used to compute the experimental data. In addition, when a protein assembles into highly ordered polymers, individual secondary structure regions that are exposed to the solvent are packed next to each other. It is nearly impossible to adequately compute the contribution to the overall CD spectrum of packed helical and sheet regions even by the use of model proteins with known three-dimensional structures instead of the widely used homopolymeric structures (Cantor and Schimmel 2001).

Amyloids are defined as compact extra-cellular deposits with inherent birefringence that increases intensely upon binding of the dye Congo Red. Amyloids are associated to various pathologies and made of fibrillar materials that exhibit a typical cross-ß structure in X-ray fiber diffraction images (Sipe and Cohen 2000; Westermark et al. 2002). This is a characteristic X-ray diffraction pattern observed with fibrils aligned and oriented with their main axes perpendicular to the X-ray beam. At least two reflections must be present in the X-ray diffraction pattern to define it as that of a cross-β structure: The main chain reflection, a sharp meridional reflection at 4.7Å, corresponding to the separation of two hydrogen-bonded chains and an equatorial reflection at 10Å, also called side chain reflection, corresponding to the packing distance between two juxtaposed β-sheets. The typical anisotropy of the diffraction patterns of amyloid fibrils implies a β-sheet orientation in which hydrogen bonds and the plane of the sheet are parallel to the main fibrils axis, whereas backbone and side chains extend perpendicular to the fibrils axis (Sunde et al. 1997). Fourier Transform infrared (FTIR) spectroscopy is another technique ideally suited for analyzing β-sheet content, the strong hydrogen bonds of amyloid giving rise to specific bands in the IR spectrum between 1623 and 1618 cm^{-1} that can be uniquely assigned as amyloid (Zurdo et al. 2001). Finally, it is widely accepted that the amyloid fibrillogenesis requires the partial unfolding of globular proteins or the partial folding of disordered polypeptides (Rochet and Lansbury 2000).

So far, only HET-s fibrils have been shown to be of amyloid nature (Dos Reis et al. 2002). While fibrils made following dilution from denaturant of Sup35p NM fragment are of amyloid nature (Serio et al. 2000) and seed the aggregation of full-length Sup35p (Glover et al. 1997), the nature of fibrils made of full-length Sup35p has never been accessed. In a manner similar to Sup35p, fibrils made following dilution from denaturant of a synthetic polypeptide reproducing part of the N-terminal domain of Ure2p are of amyloid nature (Taylor et al. 1999). However, fibrils made of native full-length Ure2p under physiologically relevant conditions lack the characteristics of amyloids and are instead made of native-like subunits (Bousset et al. 2002, 2003). Finally, fibrils made of a fragment of PrP (PrP 90-231) that seed efficiently the assembly of the same fragment of the protein have been described recently (Baskakov 2004). However, the amyloid nature of these fibrils is not yet established.

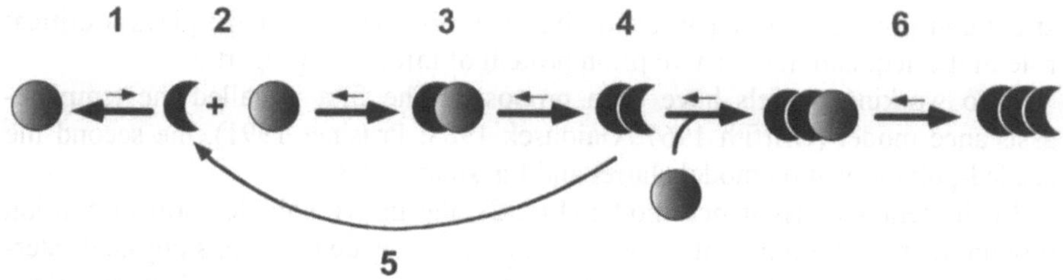

Fig. 2. Template assistance model for prion propagation. 1) The normal form of the prion protein is in equilibrium with the rare abnormal form that is the precursor of the aggregated form. The abnormal form can be of exogenous origin in iatrogenic or dietary infections. 2) The prion protein in its abnormal form interacts with the normal form of the prion protein. 3) Following this interaction, 4) the normal prion protein converts into the abnormal form. 5) The homodimer can then either dissociate, each abnormal monomer converting the cellular pool of native prion protein into the abnormal form through cycles of interaction and dissociation or 6) elongate by interacting with native prion proteins and converting them to the abnormal form.

In summary, the fibrils made *in vitro* of the different prion protein so far identified are not all of amyloid nature. Moreover, a large number of polypeptides, listed in Rochet and Lansbury (2000) that assemble into amyloid fibrils capable of seeding *in vitro* the assembly of the soluble forms of their constituent polypeptides do not bear prion properties. Finally, protein involved in a number of diseases upon assembly into fibrils following subtle conformational changes such as lithostathine (Grégoire et al. 2001) and the members of the serpin superfamily (Lomas and Carrell 2002) do not either have prion properties. It is therefore reasonable to conclude that the nature of the fibrils is not what defines a prion.

6 Soluble oligomeric forms of the prion proteins

Another trait that is common to all proteins with prion properties is their ability to switch from a monomeric form to soluble oligomeric forms that are either spherical or ring shaped (Jackson et al. 1999; Thual et al. 1999; Serio et al. 2000; Bousset et al. 2002). This property is not though specific to prions as numerous proteins assemble into soluble high molecular weight oligomers some of which are toxic and associated to various diseases (Stefani and Dobson 2003). It is therefore not this property either that defines at the molecular level a prion.

7 Mechanistic models for prion propagation

The mechanism of prion propagation is not yet understood at the molecular level. It is widely accepted though that a protein-folding event (partial unfolding of a structured polypeptide or partial folding of a disordered protein) plays a critical role in the acquisition by native prion protein of infectious properties.

Two working models have been proposed. The first is called the template-assistance model (Griffith 1967; Gajdusek 1988; Prusiner 1991), the second the seeded-polymerization model (Jarret and Lansbury 1993).

In the template-assistance model (Fig. 2), the native soluble form of a prion protein interacts with the infectious form. As a consequence of this physical interaction the native form is transformed into the infectious form. The infectious homodimer may then dissociate allowing the interaction of each infectious polypeptide with a native prion protein and the generation of additional infectious forms. In this model, the infectious form acts somehow as a template facilitating the occurrence of an energetically unfavorable protein-folding event. A variant of this scenario has been proposed where the interaction between an unidentified enzyme or chaperone and native prion proteins could favor the protein folding event at the origin of the infectious form of prion proteins (Telling et al. 1995).

In the seeded-polymerization model (Fig. 3), the native monomeric prion molecule P is in equilibrium with a rare and unstable conformational isoform P^*. P^* can be stabilized by complementary association with another P^* molecule. Because P^* is unstable, its concentration must be very low and the formation of low molecular weight oligomers of P^* not favored because the energy gained from intermolecular interactions does not outweigh the entropic cost of binding until a stable nucleus P^n is formed. A number of inherited mutations that destabilize all prion proteins predispose them to convert to their unstable P^* form (Liemann and Glockshuber 1999; Fernandez-Bellot et al. 2000; Uptain et al. 2001). Thus, increasing the concentration of the latter form and favoring their oligomerization. Once stable oligomers are formed, they can grow by incorporation of the P^* form at their ends. Such polymers can break into smaller units, each of which would behave as a seed. The requirement for stable nuclei to form before conversion is stable accounts for the low frequency of occurrence of prion diseases. However, the high efficiency of incorporation of P^* into oligomers and polymers made of P^* accelerates prion propagation justifying a contact-based mode of transmission evoked in iatrogenic or dietary infections. A detailed mathematical model for prion propagation by nucleated polymerization has been developed and its parameters estimated from published data on PrP^{sc} propagation (Masel et al. 1999). The model predicts kinetic aspects of the disease, thus, strengthening the nucleated polymerization hypothesis.

8 Maintenance and inheritance

In mammals, the progression of transmissible prion infections is not fully documented. It is clear however that PrP accumulates in peripheral tissues prior to the invasion of the central nervous system (Hill et al. 1997; Wadsworth et al. 2001; Bosque et al. 2002). The efficiency of transmission of prion disease between species is governed by a phenomenon known as the "species barrier" (Prusiner 1998).

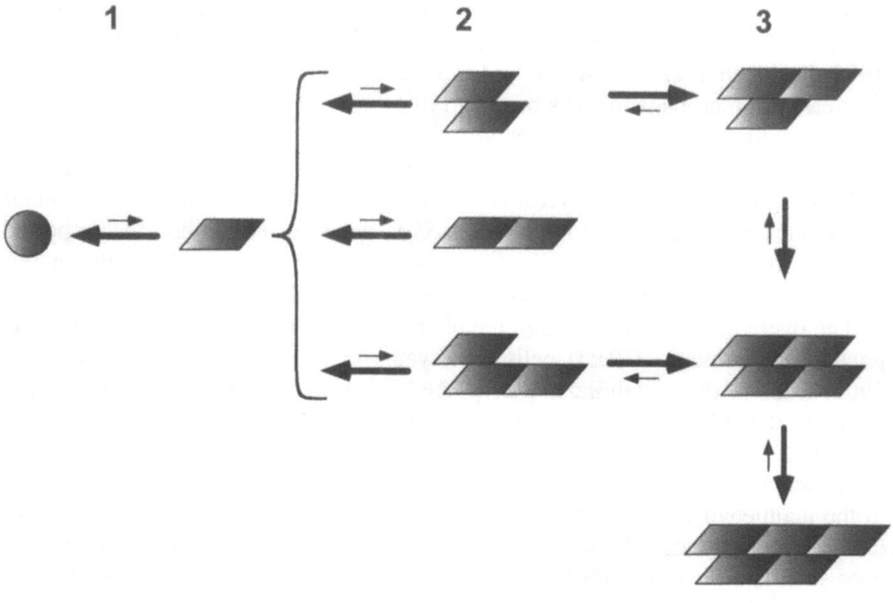

Fig. 3. Seeded polymerization model for prion propagation. 1) The normal form of the prion protein is in equilibrium with the rare abnormal form of the prion protein that is the precursor of the aggregated form. 2) Prion proteins in their abnormal form can interact with each other. The interaction is unstable because the intermolecular interactions are not strong enough to outweigh the entropic cost of binding. Thus, the low molecular weight oligomers that are formed dissociate until a stable nucleus is formed. 3) This nucleus or seed can grow indefinitely from one or both ends depending on the structural properties of the abnormal form of prion protein. It can also break into smaller stable fibrils that can elongate by incorporation of the abnormal form of the prion protein. The seed can be of exogenous origin in iatrogenic or dietary infections.

This barrier is due to differences in primary structure of PrP from different species and to the ability of PrP to adopt a number of different conformational states that stably propagate the same structural forms giving rise to specific neuropathologies and different rates of disease progression that are termed prion "strains" (Bruce 1993; Bessen and March 1994; Caughey et al. 1998).

In yeast, different prion "strains" also called "variants" have been described (Derkatch et al. 1996; Uptain et al. 2001; Schlumpberger et al. 2001; Chien and Weissman 2001). Indeed, a prion phenotype can disappear spontaneously from a yeast strain with a frequency comparable to chromosomal gene mutations (Lund and Cox 1981). The frequency with which a prion phenotype is lost upon cell division defines the stability of a prion "strain". A relationship between prion protein primary structure and prion "strains" has been established (Liu and Lindquist 2000; King 2001; Chien et al. 2003). It is reasonable to envisage that differences in primary structure allow the assembly of prion protein into different fibrillar forms with distinct morphologies, growth rates and polarity (Chien and Weissman 2001).

When yeast prion "strains" were first observed, a difference in the efficiency of curing these strains upon overexpression of Hsp104 was described (Chernoff et al. 1995; Derkatch et al. 1996). The molecular chaperone Hsp104 is strictly required for yeast prion propagation as strains carrying *HSP104* gene deletion are unable to propagate the [PSI^+] (Chernoff et al. 1995) and the [URE3] (Moriyama et al. 2000) phenotypes. HSP104 is not the only cellular factor necessary for the continued propagation of prions in yeast cells. The expression levels of other molecular chaperones are also of great importance. *SSA1* gene expression has been reported to be important for [PSI^+] stability (Jones and Masison 2003) as does other members of the Hsp70 family (Newnam et al. 1999; Chernoff et al. 1999)

The overexpression of Ydj1p (a member of the hsp40 family) and Ssa1p in exponentially growing [URE3] cells (Moriyama et al. 2000; Jones and Masison 2003) leads to the destabilization of [URE3] phenotypes. Finally, Sis1p (another member of the Hsp40 family) is required for [PIN^+] propagation (Sondheimer et al. 2001).

A number of models that account for the role played by molecular chaperones in the maintenance or the destabilization of the prion phenotypes have been published. The first model hypothesizes that the elevated levels of molecular chaperones facilitate the disaggregation of the high molecular weight species of prion proteins that act as seeds (Kushnirov and Ter-Avanesyan 1998; Glover and Lindquist 1998). In the second model, the molecular chaperones sequester either the folding intermediate(s) that assemble into prion aggregates or the cellular factor(s) that are required for the generation of the high molecular weight species of prion proteins that act as seeds (Chernoff et al. 1999). The third model hypothesizes that the faithful transmission of the high molecular weight species of prion proteins that act as seeds from mother to daughter cells is compromised upon the overexpression of molecular chaperones (Cox et al. 2003). Additional scenarios represented in Figure 4 can be proposed. A direct *in vitro* approach where the effect of each purified molecular chaperone on the assembly reaction of soluble Sup35p, Rnq1p, and Ure2p and on *in vitro* assembled fibrils will allow in the future the identification of the exact role of each molecular chaperone in yeast prion propagation.

9 *In vitro* assembly process of prions proteins

While there is a general agreement on the molecular mechanism underlying the formation of the fibrillar form of prion proteins, the nature and function of the conformational changes and the various intermediates involved in the assembly process remain subject to great debate. The first step in the assembly process involves a conformational change whose extent and nature is not yet well established. The formation of PrP^{res} involves the conversion of the soluble monomeric α-helical PrP into a soluble oligomeric β-sheet-rich form of the protein followed by the aggregation with time of these oligomers into fibrils of amyloid nature

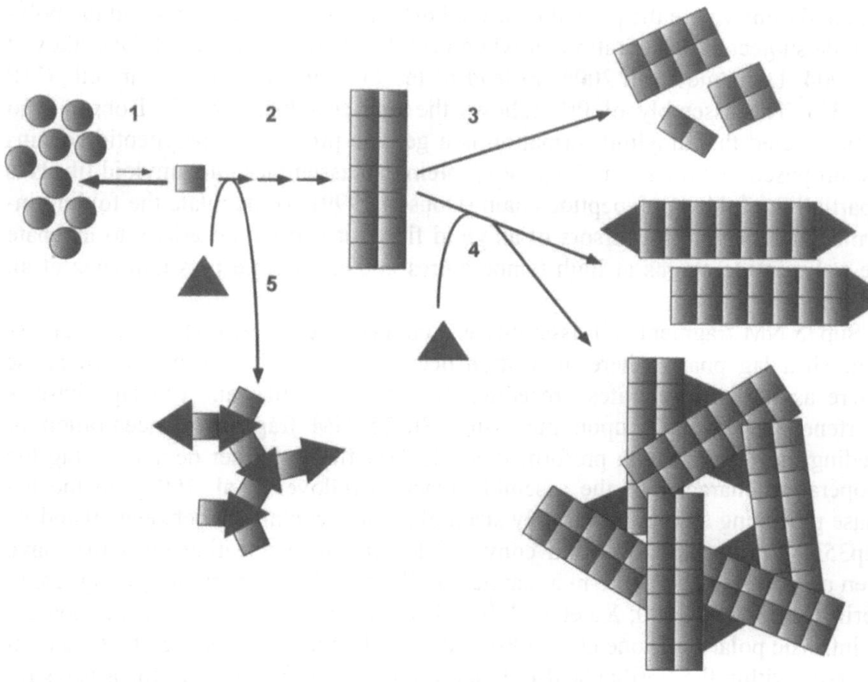

Fig. 4. Molecular chaperones can modulate the aggregation of prion proteins. 1) Molecular chaperones (triangle shaped molecule) can displace the equilibrium between the normal form of the prion protein (molecule with a circular shape) and its abnormal form (molecule with a square shape) either toward the normal form, i.e. refold misfolded prion protein or toward the abnormal form, i.e. unfold native prion protein. In the first case, molecular chaperones will have an unfavorable effect on oligomerization while in the second, molecular chaperones will favor oligomerization. 2) The abnormal prion protein oligomerizes into fibrils that can grow from one or both ends. 3) Molecular chaperones can shear the fibrils, thus, generating additional ends, i.e. incorporation sites for abnormal prion protein. This will favor assembly and decrease the limiting lag phase preceding assembly. 4) Molecular chaperones can cap one or the two ends of the fibrils. This will either inhibit the assembly reaction or have no effect depending on the intrinsic polarity of the fibrillar scaffold and whether the capped end is the end where abnormal prion protein is incorporated. Molecular chaperones can interact with fibrils walls and favor the bundling or the disordered aggregation of the fibrils into very large aggregates. Such an aggregation may lead to the neutralization of prion fibrils, which would reflect in a negative effect on assembly. 5) Finally, molecular chaperones can interact specifically with the abnormal soluble form of prion protein and sequester the precursors of the fibrils into structured or amorphous aggregates. The latter possibility would restrain assembly into fibrils.

(Jackson et al. 1999). This process occurs at pH4.0 or at lower pH values. In addition, the soluble oligomers and the fibrils have neither been shown to have infectious properties nor to seed efficiently the assembly of soluble PrP unless when the seeded assembly reaction is heavily sonicated or mixed (Saborio et al. 2001), i.e. conditions where the pH value of the buffer is subject to changes and the polypeptide subject to denaturation, or when PrP89-231 fragment is used (Baskakov et al. 2004; Legname et al. 2004) instead of the PrP form present in our cells (PrP 23-231). The assembly of PrP follows the rule established by C. Dobson who demonstrated that amyloid formation is a generic property of polypeptide chains and proposed that the form of a given protein that assembles into amyloid fibrils is a partially unfolded polypeptide chain (Dobson 1999). To populate the folding intermediates that are precursors of amyloid fibrils it is often necessary to incubate the polypeptide chains at high temperatures and/or extreme pHs (Jimenez et al. 2000).

Sup35 NM fragment self-assembly is a cooperative process where one can distinguish a lag phase where nucleation occurs followed by an elongation phase where assembly accelerates preceding the onset of a plateau. The lag phase is shortened significantly upon increasing Sup35 NM fragment concentration or seeding the reaction with preformed Sup35 NM fibrils further demonstrating the cooperative character of the assembly reaction (Glover et al. 1997). In the lag phase preceding assembly, variably sized oligomers containing between 20 and 80 Sup35NM molecules form and convert later on into nucleating units that have been proposed to aggregate in a manner similar to colloidal particles giving rise to fibrils (Serio et al. 2000; Xu et al. 2001). These fibrils have been shown to possess an intrinsic polarity (Inoue et al. 2001) that may be the consequence of a structural diversity within the fibrillar scaffold that might account for some of the aspects related to prion strains *in vivo* (DePace and Weissman 2002).

It is widely believed that Sup35p assembles into β-sheet-rich aggregates through the establishment of an array of hydrogen bonds between the side-chains and the main-chains amides of its polyglutamine/asparagine-containing region (Perutz 1999; Perutz et al. 1994, 2002). While this is certainly what occurs in solution when unfolded Sup35NM fragment is diluted from denaturant or when a shorter version of Sup35 N fragment, a peptide reproducing the amino acid stretch 7-13 of Sup35 unrelated to the repeats that modulate the stability of [PSI^+], is used (Balbirnie et al. 2001), no data is available on the assembly reaction of full-length Sup35p. It is therefore not clear whether the assembly reaction of full-length Sup35p is cooperative and the size of the stable nucleus and the structure of full-length Sup35p in its fibrillar form are unknown.

The assembly reaction of full-length Ure2p is also cooperative and follows a lag phase (Thual et al. 1999). The polymerization of Ure2p is greatly accelerated by addition of minute amounts of preformed Ure2p fibrils (Thual et al. 1999), thus, indicating that the formation of nuclei that requires the interaction of many Ure2p molecules is the limiting step in the assembly reaction. The dissociation of the native dimeric form of Ure2p to its constituent monomers is a prerequisite for assembly into fibrils (Bousset et al. 2002). The minimal stable nucleus that is the precursor of Ure2p fibrils is made of 6 Ure2p molecules (Bousset et al. 2002; Fay et al. 2003). Ure2p fibrils grow in a polar manner (Fay et al. 2003) and most interesting, Ure2p remains in a native-like conformation upon assembly into fibrils (Bousset et al. 2002). Finally, it was shown that the N-terminal domain of Ure2p possesses some elements of secondary structure in the soluble form of the protein and the C-terminal domain is tightly involved in the fibrillar scaffold (Bousset et al. 2004) that is not built on the cross-β framework of amyloid (Bousset et al. 2002, 2003). This means that the assembly of authentic Ure2p under physiologically-relevant conditions does not fit well with the general belief that all prion fibrils are of amyloid nature and indicate that subtle conformational changes (Carrell and Gooptu 1998; Huntington et al. 1999; Liu et al. 2001) may govern the assembly reaction of at least some prions.

10 Prions and misfolding diseases, unquestioned issues, and unanswered questions

Proteins that do not remain correctly folded or that have failed to reach their native conformation are removed from the protein pool in the cell and degraded (Schubert et al. 2000; Hershko and Ciechanover 1998). When this quality control process fails to remove the incorrectly folded proteins, small aggregates of misfolded polypeptides form. These aggregates are transported and fused into large intracellular inclusions (Kopito 2000) at the origin of a wide range of diseases (Taylor et al. 2002).

Aggregates of yeast prions are often presented as good models for polyglutamine diseases. A constant feature of the many polyglutamine diseases is the intranuclear localization of the inclusion bodies (Davies et al. 1998) in contrast with the cytosolic inclusion bodies in the case of yeast prions. While the formation of an intra-nuclear inclusion has obvious deleterious consequences, it is not clear how a cytosolic inclusion would influence the growth rate of a cell as observed in *S. cerevisiae.*

Yeast prions do not spread from cell to cell. They are inherited by daughter cells from mothers or passed between partners during mating. The frequencies of appearance and loss of a yeast prion phenotype are 10^{-5} and 10^{-7}, respectively. It is not trivial to imagine a utile function for the loss of function of a protein via an irreversible aggregation process. A plausible function for the behavior of yeast prions has been proposed (Li and Lindquist 2000; True and Lindquist 2000). These epigenetic and metastable traits would provide a selective advantage for yeast cells in fluctuating environments. Whether the prion phenotype mimics the loss or gain of a function, the genes of these proteins are present in an unaltered form in the cell, the loss or gain of function can be considered as reversible provided that the aggregated form of prion protein that perpetuate the structure-based mode of inheritance can be eliminated, counterbalanced or neutralized. Thus, prions might constitute a class of proteins, which activity can be modulated without requirement for genetic mutations.

Protein synthesis is an energy consuming process in the cell. It is therefore very difficult to accept the idea that a cell would have designed a process where the activity of a number of proteins would be regulated by their irreversible aggregation after synthesis, in particular as the transcription and translation of the genes encoding these proteins, i.e. their synthesis, is not turned off. In addition, while the regulation of inducible genes is very efficient, immediate and costless, the regulation of the activity of a protein through its irreversible aggregation is inefficient, slow and energy consuming. The efficiency of the prion inactivation depends on the amount of soluble form of the prion protein left in yeast cells. In cells exhibiting the [PSI$^+$] phenotype for example, the essential protein Sup35 is in part functional as witnessed by translation termination. This is probably due to the fact that the critical concentration for aggregation of prion proteins is not zero but in the nano molar range. *In vitro* prion proteins assemble into fibrils within a timescale (hours when the solutions are not shaken) incompatible with shutting off an activity following changes in the environment. Finally, the synthesis of prion proteins in cells exhibiting prion phenotypes are not shut off, i.e. neo-synthesized prion proteins are destined to aggregate, which is pretty illogical and energy consuming.

The only case where the inactivation of a protein via its aggregation may have indirect beneficial aspects is the case of Sup35p. The inactivation of Sup35p leads to an enhanced suppression of non-sense codons and non-sense codon mutations, i.e. a reduction in translational fidelity. The [PSI$^+$]-mediated read through naturally occurring stop codons or open reading frames (ORFs) that have acquired inactivating stop codons mutations surely alter the function and/or the stability of encoded proteins. It may though allow the expression of proteins that are not naturally expressed, thus, contributing to phenotypic changes. In the case the new phenotype is advantageous, the population will grow and mutations will arise to fix the trait by eliminating stop codons that are relevant to the phenotype and require [PSI$^+$] for read through and the cell will occupy new niches. If this scenario occurs, the protein profiles of wild type and yeast cells exhibiting the prion phenotype should differ significantly when compared using a proteomic tools.

The vast majority of Ure2p and Sup35p homologues that are expressed in various yeast strains possess asparagine/glutamine-rich N-terminal extensions. Apart from a unique exception, none induce [URE3] or [PSI$^+$] phenotypes in their respective yeast strains. In contrast a number of these homologues induce the prion phenotype when expressed in *S. cerevisiae*. The reason for that is unclear and suggests either that the prion behavior of these proteins is limited to *S. cerevisiae* or that the observed phenotype is not due to Ure2p or Sup35p aggregation but to indirect physiological events occurring within *S. cerevisiae* and only in this yeast strain. A careful analysis of the physiological changes that accompany the propagation of the prion phenotype by comparing the proteomes of wild type and cells exhibiting the prion phenotypes should allow a better comprehension of the molecular events at the origin of these phenotypes in *S. cerevisiae*.

The assembly process of yeast prions is often considered as being solely due to their polyglutamine/asparagine extensions (see the review of Tuite and Cox 2003). This simplistic view needs to be revised. Indeed, the assembly process of a variant Ure2p truncated of its amino acids 15-42 was recently described but not discussed, as it would have deserved. While the Q and N proportion in the N-terminal domain of full-length Ure2p is 46% (43 residues out of 93) this proportion increases to 54% (35 residues out of 65) in the Ure2pΔ15-42 variant. In marked contrast with what one would expect if the assembly of Ure2p was driven by the polyglutamine/asparagine extensions the Ure2pΔ15-42 variant does not assemble under conditions where full-length Ure2p does (Jiang et al. 2004). This clearly indicates either that assembly is not driven by the polyglutamine/asparagine extensions of yeast prions or that the structure of Ure2p N-terminal domain is lost upon removal of amino acid residues 15 to 42, strongly suggesting that the N-terminal domain of Ure2p is structured unlike the widely accepted view of a highly unstructured and flexible domain.

11 Conclusions and perspectives

Important steps have been made in understanding the biology of mammalian and yeast prions during the last decade. This includes the characterization of the structure and folding dynamics of mammalian and yeast prions. Although a number of folding intermediates have been identified, the folding intermediates that assemble into the low and high molecular weight oligomers involved in prion propagation are still unidentified. Their identification is crucial for developing tools that inhibit prion assembly. A better characterization of the structural properties of prion fibrils is a prerequisite for developing tools that allow either fibrillar deposits disassembly or lowering their toxicity. The identification of the partners of the different prion proteins will also allow a better comprehension of prion disease/phenotype genesis. Finally, although mammalian and yeast prions and their respective propagation processes share similarities, they differ in a number of features, in particular the involvement of molecular chaperones. Thus, caution must be taken in directly extrapolating findings with yeast prions to mammalian prions or even to mammalian proteins rich in glutamine and asparagine residues.

References

Adler V, Zeiler B, Kryukov V, Kascsak R, Rubenstein R, Grossman A (2003) Small, highly structured RNAs participate in the conversion of human recombinant PrP(Sen) to PrP(Res) *in vitro*. J Mol Biol 332:47-57

Aigle M, Lacroute F (1975) Genetic aspects of [URE3] a non-Mendelian cytoplasmically inherited mutation in yeast. Mol Gen Genet 136:327-335

Alper T, Haig DA, Clarke MC (1966) The exceptionally small size of the scrapie agent. Biochem Biophys Res Commun 22:278-284

Alper T, Cramp WA, Haig DA, Clarke MC (1967) Does the agent of scrapie replicate without nucleic acid? Nature 214:764-766

Balbirnie M, Grothe R, Eisenberg DS (2001) An amyloid-forming peptide from the yeast prion Sup35 reveals a dehydrated b-sheet structure for amyloid. Proc Natl Acad Sci USA 98:2375-2380

Balgerie A, Dos Reis S, Ritter C, Chaignepain S, Coulary-Salin B, Forge V, Bathany, K, Lascu I, Schmitter JM, Riek R, Saupe SJ (2003) Domain organization and structure-function relationship of the HET-s prion protein of *Podospora anserina*. EMBO J 22:2071-2081

Baskakov IV (2004) Autocatalytic conversion of recombinant prion proteins displays a species barrier. J Biol Chem 279:7671-7677

Bessen RA, March RF (1994) Distinct PrP properties suggest the molecular basis of strain variation in transmissible mink encephalopathy. J Virol 68:7859-7868

Beisson-Schecroun J (1962) Incompatibilité cellulaire et intéractions nucléo-cytoplasmiques dans les phénomènes de barrage chez *Podospora anserina*. Ann Genet 4:3-50

Billeter M, Riek R, Wider G, Hornemann S, Glockshuber R, Wuthrich K (1997) Prion protein NMR structure and species barrier for prion diseases. Proc Natl Acad Sci USA 94:7281-7285

Bolton DC, McKinley MP, Prusiner SB(1982) Identification of a protein that purifies with scrapie prion. Science 218:1309-1311

Borchelt DR, Scott M, Taraboulos A, Stahl N, Prusiner SB (1990) Scrapie and cellular prion proteins differ in their kinetics of synthesis and topology in cultured cells. J Cell Biol 110:743-752

Bosque PJ, Ryou C, Telling G, Peretz D, Legname G, DeArmond SJ, Prusiner SB (2002) Prion in skeletal muscle. Pro Natl Acad Sci USA 99:3812-3817

Bousset L, Belrhali H, Janin J, Melki R, Morera S (2001) Structure of the globular region of the prion protein Ure2 from the yeast *Saccharomyces cerevisiae*. Structure 9:39–46

Bousset L, Thomson NH, Radford SE, Melki R (2002) The yeast prion Ure2p retains its native alpha-helical conformation upon assembly into protein fibrils *in vitro*. EMBO J 21:2903-2911

Bousset L, Briki F, Doucet J, Melki R (2003) The native-like conformation of Ure2p in fibrils assembled under physiologically relevant conditions switches to an amyloid-like conformation upon heat-treatment of the fibrils. J Struct Biol 141:132-142

Bousset L, Redeker V, Decottignies P, Dubois S, Le Marechal P, Melki R (2004) Structural characterization of the fibrillar form of the yeast *Saccharomyces cerevisiae* prion Ure2p. Biochemistry 43:5022-5032

Brown DR, Wong BS, Hafiz F, Clive C, Haswell SJ, Jones IM (1999) Normal prion protein has an activity like that of superoxide dismutase. Biochem J 344:1-5

Brown DR, Qin K, Herms JW, Madlung A, Manson J, Strome R, Fraser PE, Kruck T, von Bohlen A, Schulz-Schaeffer W, Giese A, Westaway D, Kretzschmar H (1997) The cellular prion protein binds copper *in vivo*. Nature 390:684-687

Bruce ME (1993) Scrapie strain variation and mutation. Brit Med Bull 49:822-838

Bueler H, Aguzzi A, Sailer A, Greiner RA, Autenried P, Aguet M, Weissmann C (1993) Mice devoid of PrP are resistant to scrapie. Cell 73:1339-1347

Calzolai L, Lysek DA, Guntert P, von Schroetter C, Riek R, Zahn R, Wuthrich K (2000) NMR structures of three single-residue variants of the human prion protein. Proc Natl Acad Sci USA 97:8340-8345

Cantor CR, Schimmel PR (2001) Biophysical Chemistry, twelfth printing. WH Freeman and Co. New York pp 409-431

Carrell RW, Gooptu B (1998) Conformational changes and disease--serpins, prions, and Alzheimer's. Curr Opin Struct Biol 8:799-809

Caughey B (2003) Prion protein conversions: insight into mechanisms, TSE transmission barriers and strains. Br Med Bull 66:109-120

Caughey BW, Dong A, Bhat KS, Ernst D, Hayes SF, Caughey WS (1991) Secondary structure analysis of the scrapie-associated protein PrP 27-30 in water by infrared spectroscopy. Biochemistry 30:7672-7680

Caughey B, Raymond GJ, Bessen RA (1998) Strain dependent differences in beta-sheet conformations of abnormal prion protein. J Biol Chem 273:32230-32235

Chernoff YO, Derkatch IL, Inge-Vechtomov SG (1993) Multicopy SUP35 gene induces *de novo* appearance of psi-like factors in the yeast *Saccharomyces cerevisiae*. Curr Genet 24:268-270

Chernoff YO, Lindquist SL, Ono B, Inge-Vechtomov SG, Liebman SW (1995) Role of the chaperone protein Hsp104 in propagation of the yeast prion-like factor [psi+]. Science 268:880-884

Chernoff YO, Newnam GP, Kumar J, Allen K, Zink AD (1999) Evidence for a protein mutator in yeast: Role of Hsp70-related chaperone Ssb in formation, stability and toxicity of the [PSI+] prion. Mol Cell Biol 19:8103-8112

Chernoff YO, Newnam GP, Kumar J, Allen K, Zink AD (1999) Evidence for a protein mutator in yeast: Role of Hsp70-related chaperone Ssb in formation, stability and toxicity of the [PSI+] prion. Mol Cell Biol 19:8103-8112

Chesebro B, Race R, Wehrly K, Nishio J, Bloom M, Lechner D, Bergstrom S, Robbins K, Mayer L, Keith JM, et al. (1985) Identification of scrapie prion protein-specific mRNA in scrapie-infected and uninfected brain. Nature 315:331-333

Chien P, DePace AH, Collins SR, Weissman JS (2003) Generation of prion transmission barriers by mutational control of amyloid conformations. Nature 424:948-951

Chien P, Weissman JS (2001) Conformational diversity in a yeast prion dictates its seeding specificity. Nature 410:223-227

Cohen FE, Pan KM, Huang Z, Baldwin M, Fletterick RJ, Prusiner SB (1994) Structural clues to prion replication. Science 264:530-531

Coschigano PM, Magasanik B (1991) The URE2 gene product of *Saccharomyces cerevisiae* plays an important role in the cellular response to the nitrogen source and has homology to glutathione-S-transferases. Mol Cell Biol 11:822-832

Courchesne WE, Magasanik B (1988) Regulation of nitrogen assimilation in *Saccharomyces cerevisiae*: Roles of the URE2 and GLN3 genes. J Bacteriol 170:708-713

Coustou V, Deleu C, Saupe S, Begueret J (1997) The protein product of the het-s heterokaryon incompatibility gene of the fungus *Podospora anserina* behaves as a prion analog. Proc Natl Acad Sci USA 94:9773-9778

Cox BS (1965) PSI, a cytoplasmic suppressor of super-suppressor in yeast. Heredity 20:505-521

Cox BS, Ness F, Tuite MF (2003) Analysis of the generation and segregation of propagons: entities that propagate the [PSI+] prion in yeast. Genetics 165:23-33

Cox BS, Tuite MF, McLaughlin CS (1988) The Psi factor of yeast: A problem in inheritance. Yeast 4:159-179

Cuille J, Chelle PL (1936) Pathologie animale. La maladie dite tremblante du mouton est-elle inoculable? C R Acad Sci (Paris) 203:1552-1554

Davies SW, Beardsall K, Turmaine M, DiFiglia M, Aronin N, Bates GP (1998) Are neuronal intranuclear inclusions the common neuropathology of triplet-repeat disorders with polyglutamine-repeat expansions? Lancet 351:131-133

DePace AH, Weissman JS (2002) Origins and kinetic consequences of diversity in Sup35 yeast prion fibers. Nat Struct Biol 9:389-396

Deleault NR, Lucassen RW, Supattapone S (2003) RNA molecules stimulate prion protein conversion. Nature 425:717-720

Deleu C, Clavé C, Bégueret J (1993) A single amino acid difference is sufficient to elicit vegetative incompatibility in the fungus *Podospora anserina*. Genetics 135:45-52

Derkatch IL, Chernoff YO, Kushnirov VV, Inge-Vechtomov SG, Liebman SW (1996) Genesis and variability of [PSI] prion factors in *Saccharomyces cerevisiae*. Genetics 144:1375-1386

Derkatch IL, Bradley ME, Zhou P, Chernoff YO, Liebman S (1997) Genetic and environmental factors affecting the *de novo* appearance of the [PSI+] prion in *Saccharomyces cerevisiae*. Genetics 147:507-519

Derkatch IL, Bradley ME, Masse SV, Zadorsky SP, Polozkov GV, Inge-Vechtomov SG, Liebman SW (2000) Dependence and independence of [PSI(+)] and [PIN(+)]: a two-prion system in yeast? EMBO J 19:1942-1952

Derkatch IL, Bradley ME, Hong JY, Liebman SW (2001) Prions affect the appearance of other prions: the story of [PIN(+)]. Cell 106:171-182

Diringer H, Gelderblom H, Hilmert H, Ozel M, Edelbluth C, Kimberlin RH (1983) Scrapie infectivity, fibrils and low molecular weight protein. Nature 306:476-478

Dobson CM (1999) Protein misfolding, evolution and disease. Trends Biochem Sci 24:329-332

Donne DG, Viles JH, Groth D, Mehlhorn I, James TL, Cohen FE, Prusiner SB, Wright PE, Dyson HJ (1997) Structure of the recombinant full-length hamster prion protein PrP(29-231): the N terminus is highly flexible. Proc Natl Acad Sci USA 94:13452-13457

Dos Reis S, Coulary-Salin B, Forge V, Lascu I, Begueret J, Saupe SJ (2002) The HET-s prion protein of the filamentous fungus *Podospora anserina* aggregates *in vitro* into amyloid-like fibrils. J Biol Chem 277:5703-5706

Eberl H, Tittmann P, Glockshuber R (2004) Characterization of recombinant, membrane-attached full-length prion protein. J Biol Chem 279:25058-25065

Enari M, Flechsig E, Weissmann C (2001) Scrapie prion protein accumulation by scrapie-infected neuroblastoma cells abrogated by exposure to a prion protein antibody. Proc Natl Acad Sci USA 98:9295-9299

Endo T, Groth D, Prusiner SB, Kobata A (1989) Diversity of oligosaccharide structures linked to asparagines of the scrapie prion protein. Biochemistry 28:8380-8388

Fay N, Inoue Y, Bousset L, Tagich H, Melki R (2003) Assembly of the yeast prion Ure2p into protein fibrils: Thermodynamic and kinetic characterization. J Biol Chem 278:30199-30205

Fernandez-Bellot E, Guillemet E, Cullin C (2000) The yeast prion [URE3] can be greatly induced by a functional mutated URE2 allele. EMBO J 19:3215-3222

Gajdusek DC (1988) Transmissible and non-transmissible amyloidoses: Autocatalyticv post-translational conversion of host precursor proteins to ß-pleated conformations. J Neuroimmunol 20:95-110

Glass NL, Jacobson DJ, Shiu PK (2000) The genetics of hyphal fusion and vegetative incompatibility in filamentous ascomycete fungi. Annu Rev Genet 34:165-186

Glover JR, Kowal AS, Schirmer EC, Patino MM, Liu JJ, Lindquist S (1997) Self-seeded fibers formed by Sup35, the protein determinant of [PSI+], a heritable prion-like factor of *S. cerevisiae*. Cell 89:811-819

Glover JR, Lindquist S (1998) Hsp104, Hsp70, and Hsp40: a novel chaperone system that rescues previously aggregated proteins. Cell 94:73-82

Gregoire C, Marco S, Thimonier J, Duplan L, Laurine E, Chauvin JP, Michel B, Peyrot V, Verdier JM (2001) Three-dimensional structure of the lithostathine protofibril, a protein involved in Alzheimer's disease. EMBO J 20:3313-3321

Griffith JS (1967) Self-replication and scrapie. Nature 215:1043-1044

Hadlow WJ (1959) Scrapie and Kuru. Lancet 2:289-290

Hawthorne DC, Mortimer RK (1968) Genetic mapping of nonsense suppressors in yeast. Genetics 60:735-742

Hershko A, Ciechanover A (1998) The ubiquitin system for protein degradation. Annu Rev Biochem 61:761-807

Hill AF, Antoniou M, Collinge J (1999) Protease-resistant prion protein produced *in vitro* lacks detectable infectivity. J Gen Virol 80:11-14

Hill AF, Zeidler, M, Ironside J, Collinge J (1997) Diagnosis of new variant Creutzfeldt-Jakob disease by tonsil biopsy. Lancet 349:99-100

Horwich AL, Weissman JS (1997) Deadly conformations Protein misfolding in prion disease. Cell 89:499-510

Hundt C, Peyrin JM, Haik S, Gauczynski S, Leucht C, Rieger R, Riley ML, Deslys JP, Dormont D, Lasmezas CI, Weiss S (2001) Identification of interaction domains of the prion protein with its 37-kDa/67-kDa laminin receptor. EMBO J 20:5876-5886

Huntington JA, Pannu NS, Hazes B, Read RJ, Lomas DA, Carrell RW (1999) A 2.6Å structure of a serpin polymer and implications for conformational disease. J Mol Biol 293:449-455

Inge-Vechtomov SG, Andrianova VM (1970) Recessive super-suppressors in yeast. Genetika 6:103-115

Inoue Y, Kishimoto A, Hirao J, Yoshida M, Taguchi H (2001) Strong growth polarity of yeast prion fiber revealed by single fiber imaging. J Biol Chem 276:35227-35230

Jackson GS, Hosszu LL, Power A, Hill AF, Kenney J, Saibil H, Craven CJ, Waltho JP, Clarke AR, Collinge J (1999) Reversible conversion of monomeric human prion protein between native and fibrilogenic conformations. Science 283:1935-1937

James TL, Liu H, Ulyanov NB, Farr-Jones S, Zhang H, Donne DG, Kaneko K, Groth D, Mehlhorn I, Prusiner SB, Cohen FE (1997) Solution structure of a 142-residue recombinant prion protein corresponding to the infectious fragment of the scrapie isoform. Proc Natl Acad Sci USA 94:10086-10091

Jarret JT, Lansbury PT (1993) Seeding "one-dimensional-crystallization" of amyloid: a pathogenic mechanism in Alzheimer's disease and scrapie? Cell 73:1055-1058

Jiang Y, Li H, Zhu L, Zhou JM, Perrett S (2004) Amyloid nucleation and hierarchical assembly of Ure2p fibrils: role of asparagine/glutamine repeat and nonrepeat regions of the prion domain. J Biol Chem 279:3361-3369

Jimenez JL, Guijarro JI, Orlova E, Zurdo J, Dobson CM, Sunde M, Saibil HR (1999) Cryo-electron microscopy structure of an SH3 amyloid fibril and model of the molecular packing. EMBO J 18:815-821

Jin T, Gu Y, Zanusso G, Sy M, Kumar A, Cohen M, Gambetti P, Singh N (2000) The chaperone protein BiP binds to a mutant prion protein and mediates its degradation by the proteasome. J Biol Chem 275:38699-38704

Jones GW, Masison DC (2003) *Saccharomyces cerevisiae* Hsp70 mutations affect [PSI+] prion propagation and cell growth differently and implicate Hsp40 and tetratricopeptide repeat cochaperones in impairment of [PSI+]. Genetics 163:495-506

King CY (2001) Supporting the structural basis of prion strains: Induction and identification of [PSI] variants. J Mol Biol 307:1247-1260

King CY and Diaz-Avalos R (2004) Protein-only transmission of three yeast prion strains. Nature 428:319-323

Knaus KJ, Morillas M, Swietnicki W, Malone M, Surewicz WK, Yee VC (2001) Crystal structure of the human prion protein reveals a mechanism for oligomerization. Nat Struct Biol 8:770-774

Kocisko DA, Come JH, Priola SA, Chesebro B, Raymond GJ, Lansbury PT, Caughey B (1994) Cell-free formation of protease-resistant prion protein. Nature 370:471-474

Kopito RR (2000) Aggresomes, inclusion bodies and protein aggregation. Trends Cell Biol 10:524-530

Kurschner C, Morgan JI (1996) Analysis of interaction sites in homoand heteromeric complexes containing Bcl-2 family members and the cellular prion protein. Brain Res Mol Brain Res 37:249-258

Kushnirov VV, Ter-Avanesyan MD (1998) Structure and replication of yeast prions. Cell 94:13-16

Kushnirov VV, Ter-Avanesyan MD, Telckov MV, Surguchov AP, Smirnov VN, Inge-Vechtomov SG (1988) Nucleotide sequence of the SUP2 (SUP35) gene of *Saccharomyces cerevisiae.* Gene 66:45-54

Lacroute F (1971) Non-Mendelian mutation allowing ureidosuccinic acid uptake in yeast. J Bacteriol 106:519-522

Lee S, Eisenberg D (2003) Seeded conversion of recombinant prion protein to a disulfide-bonded oligomer by a reduction-oxidation process. Nat Struct Biol 10:725-730

Legname G, Baskakov IV, Nguyen HO, Riesner D, Cohen FE, DeArmond SJ, Prusiner SB (2004) Synthetic mammalian prions. Science 305:673-676

Li L, Lindquist S (2000) Creating a protein-based element of inheritance. Science 287:661-664

Liemann S, Glockshuber R (1999) Influence of amino acid substitutions related to inherited human prion diseases on the thermodynamic stability of the cellular prion protein. Biochemistry 38:3258-3267

Liu H, Farr-Jones S, Ulyanov NB, Llinas M, Marqusee S, Groth D, Cohen FE, Prusiner SB, James TL (1999) Solution structure of Syrian hamster prion protein rPrP(90-231). Biochemistry 38:5362-5377

Liu Y, Gotte G, Libonati M, Eisenberg D (2001) A domain-swapped RNase A dimer with implications for amyloid formation. Nat Struct Biol 8:211-214

Liu JJ, Lindquist S (1999) Oligopeptide-repeat expansions modulate 'protein-only' inheritance in yeast. Nature 400:573-576

Lomas DA, Carrell RW (2002) Serpinopathies and the conformational dementias. Nat Rev Genet 3:759-768

Lopez Garcia F, Zahn R, Riek R, Wuthrich K (2000) NMR structure of the bovine prion protein. Proc Natl Acad Sci USA 97:8334-8339

Lund PM, Cox BS (1981) Reversion analysis of [psi] mutations in *Saccharomyces cerevisiae.* Genet Res 37:173-182

Ma J, Lindquist S (1999) *De novo* generation of a PrPSc-like conformation in living cells. Nat Cell Biol 1:358-361

Maddelein ML, Dos Reis S, Duvezin-Caubet S, Coulary-Salin B, Saupe SJ (2002) Amyloid aggregates of the HET-s prion protein are infectious. Proc Natl Acad Sci USA 99:7402-7407

Masel J, Jansen VAA, Nowak MA (1999) Quantifying the kinetic parameters of prion replication. Biophys Chem 77:139-152

Masison DC, Wickner RB (1995) Prion-inducing domain of yeast Ure2p and protease resistance of Ure2p in prion-containing cells. Science 270:93-95

Meier P, Genoud N, Prinz M, Maissen M, Rulicke T, Zurbriggen A, Raeber AJ, Aguzzi A (2003) Soluble dimeric prion protein binds PrP(Sc) *in vivo* and antagonizes prion disease. Cell 113:49-60

Michelitsch MD, Weissman, JS (2000) A census of glutamine/asparagine-rich regions: Implications for their conserved function and the prediction of novel prions. Proc Natl Acad Sci USA 97:11910-11915

Mitchell AP, Magasanik B (1984) Regulation of glutamine-repressible gene products by GLN3 function in *Saccharomyces cerevisiae.* Mol Cell Biol 4:2758-2766

Moriyama H, Edskes HK, Wickner RB (2000) [URE3] prion propagation in *Saccharomyces cerevisiae*: requirement for chaperone Hsp104 and curing by overexpressed chaperone Ydj1p. Mol Cell Biol 20:8916-8922

Mouillet-Richard S, Ermonval M, Chebassier C, Laplanche JL, Lehmann S, Launay JM, Kellermann O (2000) Signal transduction through prion protein. Science 289:1925-1928

Nazabal A, Dos Reis S, Bonneu M, Saupe SJ, Schmitter JM (2004) Conformational transition occuring upon amyloid aggregation of the HET-s prion protein of *Podospora anserina* analyzed by Hydrogen/Deuterium exchange and mass spectrometry. Biochemistry 42:8852-8861

Newnam GP, Wegrzyn RD, Lindquist SL, Chernoff YO (1999) Antagonistic interaction between yeast chaperones Hsp104 and Hsp70 in prion curing. Mol Cell Biol 19:1325-1333

Oesch B, Teplow DB, Stahl N, Serban D, Hood LE, Prusiner SB (1990) Identification of cellular proteins binding to the scrapie prion protein. Biochemistry 29:5848-5855

Oesch B, Westaway D, Walchli M, McKinley MP, Kent SB, Aebersold R, Barry RA, Tempst P, Teplow DB, Hood LE, et al.(1985) A cellular gene encodes scrapie PrP 27-30 protein. Cell 40:735-746

Osherovich LZ, Weissman JS (2001) Multiple Gln/Asn-rich prion domains confer susceptibility to induction of the yeast [PSI(+)] prion. Cell106:183-194

Pan KM, Stahl N, Prusiner SB (1992) Purification and properties of the cellular prion protein from Syrian hamster brain. Protein Sci 1:1343-1352

Pan KM, Baldwin M, Nguyen J, Gasset M, Serban A, Groth D, Mehlhorn I, Huang Z, Fletterick RJ, Cohen FE, Prusiner SB (1993) Conversion of alpha-helices into beta-sheets features in the formation of the scrapie prion proteins. Proc Natl Acad Sci USA 90:10962-10966

Patino MM, Liu JJ, Glover JR, Lindquist S (1996) Support for the prion hypothesis for inheritance of a phenotypic trait in yeast. Science 273:622-626

Perutz MF (1999) Glutamine repeats and neurodegenerative diseases: molecular aspects. Trends Biochem Sci 24:58-63

Perutz MF, Johnson T, Suzuki M, Finch JT (1994) Glutamine repeats as polar zippers: their possible role in inherited neurodegenerative diseases. Proc Natl Acad Sci USA 91:5355-5358

Perutz MF, Pope BJ, Owen D, Wanker EE, Scherzinger E (2002) Aggregation of proteins with expanded glutamine and alanine repeats of the glutamine-rich and asparagine-rich domains of Sup35 and of the amyloid b-peptide of amyloid plaques. Proc Natl Acad Sci USA 99:5596-5600

Prusiner SB(1982) Novel proteinaceous infectious particles cause scrapie. Science 216:136-144

Prusiner SB (1991) Molecular biology of prion diseases. Science 252:1515-1522

Prusiner SB (1997) Prion diseases and the BSE crisis. Science 278:245-251

Prusiner SB (1998) Prions. Proc Natl Acad Sci USA 95:13363-13383

Prusiner SB, Groth DF, Cochran SP, Masiarz FR, McKinley MP, Martinez HM (1980) Molecular properties, partial purification, and assay by incubation period measurements of the hamster scrapie agent. Biochemistry 19:4883-4891

Prusiner SB, McKinley MP, Groth DF, Bowman KA, Mock NI, Cochran SP, Masiarz FR (1981) Scrapie agent contains a hydrophobic protein. Proc Natl Acad Sci USA 78:6675-6679

Prusiner SB, Bolton DC, Groth DF, Bowman KA, Cochran SP, McKinley MP (1982) Further purification and characterization of scrapie prions. Biochemistry 26:6942-6950

Prusiner SB, McKinley MP, Bowman KA, Bolton DC, Bendheim PE, Groth DF, Glenner GG (1983) Scrapie prions aggregate to form amyloid-like birefringent rods. Cell 35:349-358

Prusiner SB, Groth DF, Bolton DC, Kent SB, Hood LE (1984) Purification and structural studies of a major scrapie prion protein. Cell 38:127-34

Raeber AJ, Brandner S, Klein MA, Benninger Y, Musahl C, Frigg R, Roeckl C, Fischer MB, Weissmann C, Aguzzi A (1998) Transgenic and knockout mice in research on prion diseases. Brain Pathol 8:715-733

Riek R, Hornemann S, Wider G, Billeter M, Glockshuber R, Wuthrich K (1996) NMR structure of the mouse prion protein domain PrP(121-321). Nature 382:180-182

Riek R, Hornemann S, Wider G, Glockshuber R, Wuthrich K (1997) NMR characterization of the full-length recombinant murine prion protein, mPrP(23-231). FEBS Lett 413:282-288

Riek R, Wider G, Billeter M, Hornemann S, Glockshuber R, Wuthrich K (1998) Prion protein NMR structure and familial human spongiform encephalopathies. Proc Natl Acad Sci USA 95:11667-11672

Rochet JC, Lansbury PT Jr (2000) Amyloid fibrillogenesis: themes and variations. Curr Opin Struct Biol 10:60-68

Rudd PM, Endo T, Colominas C, Groth D, Wheeler SF, Harvey DJ, Wormald MR, Serban H, Prusiner SB, Kobata A, Dwek RA (1999) Glycosylation differences between the normal and pathogenic prion protein isoforms. Proc Natl Acad Sci USA 96:13044-13049

Saborio GP, Permanne B, Soto C (2001) Sensitive detection of pathological prion protein by cyclic amplification of protein misfolding. Nature 411:810-813

Safar J, Roller PP, Gajdusek DC, Gibbs CJ Jr (1993) Thermal stability and conformational transitions of scrapie amyloid (prion) protein correlate with infectivity. Protein Sci 2:2206-2216

Sakudo A, Lee DC, Yoshimura E, Nagasaka S, Nitta K, Saeki K, Matsumoto Y, Lehmann S, Itohara S, Sakaguchi S, Onodera T (2004) Prion protein suppresses perturbation of cellular copper homeostasis under oxidative conditions. Biochem Biophys Res Commun 313:850-855

Santoso A, Chien P, Osherovich LZ, Weissman JS (2000) Molecular basis of yeast prion species barrier. Cell 100:277-288

Schlumpberger M, Wille H, Baldwin MA, Butler DA, Herskowitz I, Prusiner SB (2000) The prion domain of yeast Ure2p induces autocatalytic formation of amyloid fibers by a recombinant fusion protein. Prot Sci 9:440-451
Schoun J, Lacroute F (1969) Etude physiologique d'une mutation permettant l'incorporation d'acide ureidosuccinique chez la levure. C R Acad Sci (Paris) 269:1412-1414
Schubert U, Anton LC, Gibbs J, Norbury CC, Yewdell JW, Bennink JR (2000) Rapid degradation of a large fraction of newly synthesized proteins by proteasomes. Nature 404:770-774
Schlumpberger M, Prusiner SB, Herskowitz I (2001) Induction of distinct [URE3] yeast prion strains. Mol Cell Biol 21:7035-7046
Scott M, Foster D, Mirenda C, Serban D, Coufal F, Walchli M, Torchia M, Groth D, Carlson G, DeArmond SJ, Prusiner SB (1989) Transgenic mice expressing hamster prion protein produce species-specific scrapie infectivity and amyloid plaques. Cell 59:847-857
Scott MR, Will R, Ironside J, Nguyen HO, Tremblay P, DeArmond SJ, Prusiner SB (1999) Compelling transgenetic evidence for transmission of bovine spongiform encephalopathy prions to humans. Proc Natl Acad Sci USA 96:15137-15142
Serio TR, Cashikar AG, Kowal AS, Sawicki GJ, Moslehi JJ, Serpell L, Arnsdorf MF, Lindquist SL (2000) Nucleated conformational conversion and the replication of conformational information by a prion determinant. Science 289:1317-1321
Shmerling D, Hegyi I, Fischer M, Blättler T, Brandner S, Götz J, Rülicke T, Flechsig E, Cozzio A, von Mering C, Hangartner C, Aguzzi A, Weissmann C (1998) Expression of amino-terminally truncated PrP in mouse leading to ataxia and specific cerebellar lesions. Cell 93:203-214
Singh A, Helms C, Sherman F (1979) Mutation of the non-Mendelian suppressor, Psi+, in yeast by hypertonic media. Proc Natl Acad Sci USA 76:1952-1956
Sipe JD, Cohen AS (2000) History of the amyloid fibril. J Struct Biol130:88-98
Sondheimer N, Lindquist S(2000) Rnq1: an epigenetic modifier of protein function in yeast. Mol Cell 5:63-172
Sondheimer N, Lopez N, Craig EA, Lindquist S (2001) The role of Sis1 in the maintenance of the [RNQ+] prion. EMBO J 20:2435-2442
Stahl N, Borchelt DR, Hsiao K, Prusiner SB (1987) Scrapie prion protein contains a phosphatidylinositol glycolipid. Cell 51:229-240
Stahl N, Borchelt DR, Prusiner SB (1990) Differential release of cellular and scrapie prion proteins from cellular membranes by phosphatidylinositol-specific phospholipase C. Biochemistry 29:5405-5412
Stahl N, Baldwin MA, Teplow DB, Hood L, Gibson BW, Burlingame AL, Prusiner SB (1993) Structural studies of the scrapie prion protein using mass spectrometry and amino acid sequencing. Biochemistry 32:1991-2002
Stamp JT (1962) Scrapie. A transmissible disease of sheep. Vet Rec 74:357-362
Stansfield I, Jones KM, Kushnirov VV, Dagkesamanskaya AR, Poznyakovski AI, Paushkin SV, Nierras CR, Cox BS, Ter-Avanesyan MD, Tuite MF (1995)The products of the SUP45 (eRF1) and SUP35 genes interact to mediate translation termination in *Saccharomyces cerevisiae.* EMBO J 14:4365-4373
Stefani M, Dobson CM (2003) Protein aggregation and aggregate toxicity: new insights into protein folding, misfolding diseases and biological evolution. J Mol Med 81:678-699

Sunde M, Serpell LC, Bartlam M, Fraser PE, Pepys MB, Blake CC (1997) Common core structure of amyloid fibrils by synchrotron X-ray diffraction. J Mol Biol 273:729-739

Swietnicki W, Morillas M, Chen SG, Gambetti P, Surewicz WK (2000) Aggregation and fibrillization of the recombinant human prion protein huPrP90-231. Biochemistry 39:424-431

Tanaka M, Chien P, Naber N, Cooke R, Weissman JS (2004) Conformational variations in an infectious protein determine prion strain differences. Nature 428:323-328

Taylor JP, Hardy J, Fischbeck KH (2002) Toxic proteins in neurodegenerative disease. Science 296:1991-1995

Taylor KL, Cheng N, Williams RW, Steven AC, Wickner RB (1999) Prion domain initiation of amyloid formation *in vitro* from native Ure2p. Science 283:1339-1343

Telling GC, Scott M, Mastrianni J, Gabizon R, Torchia M, Cohen FE, DeArmond SJ, Prusiner SB (1995) Prion propagation in mice expressing human and chimeric PrP transgenes implicates the interaction of cellular PrP with another protein. Cell 83:79-90

Ter-Avanesyan MD, Kushnirov VV, Dagkesamanskaya AR, Didichenko SA, Chernoff YO, Inge-Vechtomov SG, Smirnov VN (1993) Deletion analysis of the SUP35 gene of the yeast *Saccharomyces cerevisiae* reveals two non-overlapping functional regions in the encoded protein. Mol Microbiol 7:683-692

Ter-Avanesyan MD, Dagkesamanskaya AR, Kushnirov VV, Smirnov VN (1994) The SUP35 omnipotent suppressor gene is involved in the maintenance of the non-Mendelian determinant [psi+] in the yeast *Saccharomyces cerevisiae*. Genetics 137:1339-1343

Thual C, Komar AA, Bousset L, Fernandez-Bellot E, Cullin C, Melki R (1999) Structural characterization of *Saccharomyces cerevisiae* prion-like protein Ure2. J Biol Chem 274:13666-13674

Thual C, Bousset L, Komar A A, Walter S, Buchner J, Cullin C, Melki R (2001) Stability, folding, dimerization, and assembly properties of the yeast prion Ure2p. Biochemistry 40:1764-1773

Torrent J, Alvarez-Martinez MT, Harricane MC, Heitz F, Liautard JP, Balny C, Lange R (2004) High pressure induces scrapie-like prion protein misfolding and amyloid fibril formation. Biochemistry 43:7162-7170

True HL, Lindquist SL (2000) A yeast prion provides a mechanism for genetic variation and phenotypic diversity. Nature 407:477-483

Tuite MF (2000) Yeast prions and their prion-forming domain. Cell 100:289-292

Tuite MF, Cox BS (2003) Propagation of yeast prions. Nat Rev Mol Cell Biol 4:878-889

Tuite MF, Mundy CR, Cox BS (1981) Agents that cause a high frequency of genetic change from [psi+] to [psi-] in *Saccharomyces cerevisiae*. Genetics 98:691-711

Turcq B, Deleu C, Denayrolles M, Bégueret J (1991) Two allelic genes responsible for vegetative incompatibility in the fungus *Podospora anserina* are not essential for cell viability. Mol Gen Genet 228:3-6

Turk E, Teplow DB, Hood LE, Prusiner SB (1988) Purification and properties of the cellular and scrapie hamster prion proteins. Eur J Biochem 176:21-30

Uptain SM, Sawicki GJ, Caughey B, Lindquist S (2001) Strains of [PSI+] are distinguished by their efficiencys of prion-mediated conformational conversion. EMBO J 20:6236-6245

Viles JH, Cohen FE, Prusiner SB, Goodin DB, Wright PE, Dyson HJ (1999) Copper binding to the prion protein: structural implications of four identical cooperative binding sites. Proc Natl Acad Sci USA 96:2042-2047

Wadsworth JDF, Joiner S, Hill AF Campbell TA, Desbruslais M, Luthert PJ, Collinge J (2001) Tissue distribution of protease resistant protein in variant CJD using a highly sensitive immunoblotting assay. Lancet 358:171-180

Weissmann C, Raeber AJ, Shmerling D, Aguzzi A, Manson JC (1999) In Prion Biology and Diseases, SB Prusiner ed. (Cold Spring Harbor, NY). Cold Spring Harbor Laboratory Press 229-272

Westermark P, Benson MD, Buxbaum JN, Cohen AS, Frangione B, Ikeda S, Masters CL, Merlini G, Saraiva MJ, Sipe JD (2002) Amyloid fibril protein nomenclature - 2002. Amyloid 9:197-200

Wickner RB (1994) Evidence for a prion analog in *S. cerevisiae*: the [URE3] non-Mendelian genetic element as an altered URE2 protein. Science 264:566-569

Xu S, Bevis B, Arnsdorf MF (2001) The assembly of amyloidogenic yeast Sup35 as assessed by scanning (atomic) force microscopy: An analogy to linear colloidal aggregation? Biophys J 81:446-454

Yehiely F, Bamborough P, Da Costa M, Perry BJ, Thinakaran G, Cohen FE, Carlson GA, Prusiner SB (1997) Identification of candidate proteins binding to prion protein. Neurobiol 3:339-355

Zahn R, Liu A, Luhrs T, Riek R, von Schroetter C, Lopez Garcia F, Billeter M, Calzolai L, Wider G, Wuthrich K (2000) NMR solution structure of the human prion protein. Proc Natl Acad Sci USA 97:145-150

Zhang Y, Swietnicki,W, Zagorski MG, Surewicz WK, Sonnichsen F (2000) Solution structure of the E200K variant of human prion protein. Implications for the mechanism of pathogenesis in familial prion diseases. J Biol Chem 275:33650-33654

Zhouravleva G, Frolova L, Le Goff X, Le Guellec R, Inge-Vechtomov S, Kisselev L, Philippe M (1995) Termination of translation in eukaryotes is governed by two interacting polypeptide chain release factors, eRF1 and eRF3. EMBO J 14:4065-4072

Zurdo J, Guijarro JI, Dobson CM (2001) Preparation and characterization of purified amyloid fibrils. J Am Chem Soc 123:8141-8142

Bousset, Luc

Laboratoire d'Enzymologie et Biochimie Structurales, CNRS, 91198 Gif-sur-Yvette Cedex, France.

Present address: EMBL, 6 rue Jules Horowitz, BP181, 38042 Grenoble Cedex 9, France.

Fay, Nicolas

Laboratoire d'Enzymologie et Biochimie Structurales, CNRS, 91198 Gif-sur-Yvette Cedex, France.

Melki, Ronald
Laboratoire d'Enzymologie et Biochimie Structurales, CNRS, 91198 Gif-sur-Yvette Cedex, France.
melki@lebs.cnrs-gif.fr

The Hsp60 chaperonins from prokaryotes and eukaryotes

M. Giulia Bigotti, Anthony R. Clarke, and Steven G. Burston

Abstract

The Hsp60 molecular chaperones (the chaperonins) are essential proteins throughout biology. They can be separated into two evolutionary classes: the Group I chaperonins from eubacteria and their endosymbiotic counterparts in eukaryotic cells, and the Group II chaperonins from archaea and the eukaryotic cytosol. While the two classes have some similarity to each other in structural and functional characteristics, they also have a number of important distinctions implying that they may have some significant differences in their modes of action. In this review we first examine our current understanding of the Group I class, typified by GroE from *Escherichia coli*, before looking at the recent developments in the much less well-studied Group II chaperonins, including the archeal thermosome and eukaryotic CCT.

1 The Group I chaperonins

1.1 Introduction

The Group I chaperonins were discovered from three diverse sources during the 1970's and 1980's. Costa Georgopoulos and colleagues, who were studying bacteriophage morphogenesis, isolated temperature-sensitive (ts) mutants of the essential *groE* operon in *Escherichia coli* that were unable to support the growth of bacteriophage λ (Georgopoulos et al. 1973; Hohn and Georgopoulos 1978). The gene product was isolated and found to be an ATPase with subunit M_r ~60kDa, which we now know to be the Group I chaperonin, GroEL. Further investigation demonstrated that it was not just bacteriophage λ growth which was affected by mutants in the *groEL* gene but also bacteriophages T4 and T5, and that mutation resulted in an accumulation of aggregated phage proteins within the cells, which had not correctly assembled into mature phage particles (Tilly et al. 1981). This led to the suggestion that GroEL was somehow involved as an "assembly factor" to assist the assembly of the phage capsid (Georgopoulos and Tilly 1981). As George Lorimer has pointed out (Lorimer 2001), the lack of suggestion that GroEL was involved in protein folding was not surprising given that Anfinsen had been awarded the Nobel Prize in 1972 for his dogma that correct protein folding depended entirely upon the information contained within the primary amino acid sequence; although it should be noted that Anfinsen himself had predicted the existence of proteins to assist with protein folding (Anfinsen 1973). Costa Georgopoulos and his colleagues moved on to look for suppressors of the ts-*groEL* mutants by mutating the other gene in the *groE* operon, *groES*. Indeed suppressors in *groES* were found and when the 10kDa GroES protein was purified it was found to inhibit the ATPase activity of GroEL to some degree (Chandrasekhar et al. 1986).

During the same period, John Ellis and his colleagues were studying protein synthesis in intact chloroplasts via the incorporation of radiolabelled amino acids during translation (Barraclough and Ellis 1980). The most abundant protein synthesised in chloroplasts is Rubisco and as they followed the production of Rubisco they noted that the large Rubisco subunit was associated with another protein before becoming incorporated into the holo-enzyme. This protein was termed the Rubisco-subunit binding protein and found to be part of a large oligomeric complex. Ellis continued to study this protein complex (now known to be the chloroplast chaperonin) eventually coining the term "molecular chaperone" to describe its function, which he suggested was to assist the folding and assembly of Rubisco.

The third discovery of the chaperonins was made by Art Horwich and Ulrich Hartl via studies of ts-mutants of the mitochondrial hsp60 in yeast, which resulted in the accumulation of misfolded conformations of newly imported mitochondrial proteins under non-permissive conditions (Cheng et al. 1989). Again their initial conclusion was that accumulation of defective proteins was due to a problem with the assembly of newly-imported proteins into multi-subunit protein complexes, although it was later recognised to be a problem with protein folding rather than assembly (Ostermann et al. 1989).

The seminal experiment demonstrating chaperonin function *in vitro* was performed by George Lorimer and colleagues (Goloubinoff et al. 1989). They examined the effect of purified *E. coli* GroEL and GroES on the refolding of dimeric Rubisco from the purple bacterium *Rhodospirillus rubrum*. This enzyme refolds very inefficiently when diluted from denaturant into renaturing conditions, typically yielding less than 1% active protein. However, when refolding was initiated in an aqueous buffer containing GroEL, GroES, and Mg-ATP a massive increase in refolding efficiency was observed with the return of approximately 80% of enzymatic activity. This pioneering study sparked off feverish activity amongst the protein folding community and soon work was published showing that GroEL binds protein folding intermediates and suppresses non-productive aggregation as part of its function in enhancing the efficiency of protein folding (Martin et al. 1991; Badcoe et al. 1991; Buchner et al. 1991; Fisher 1992). Having established the exact function of the GroE chaperonin the scene was then set for elucidating its mechanism of action.

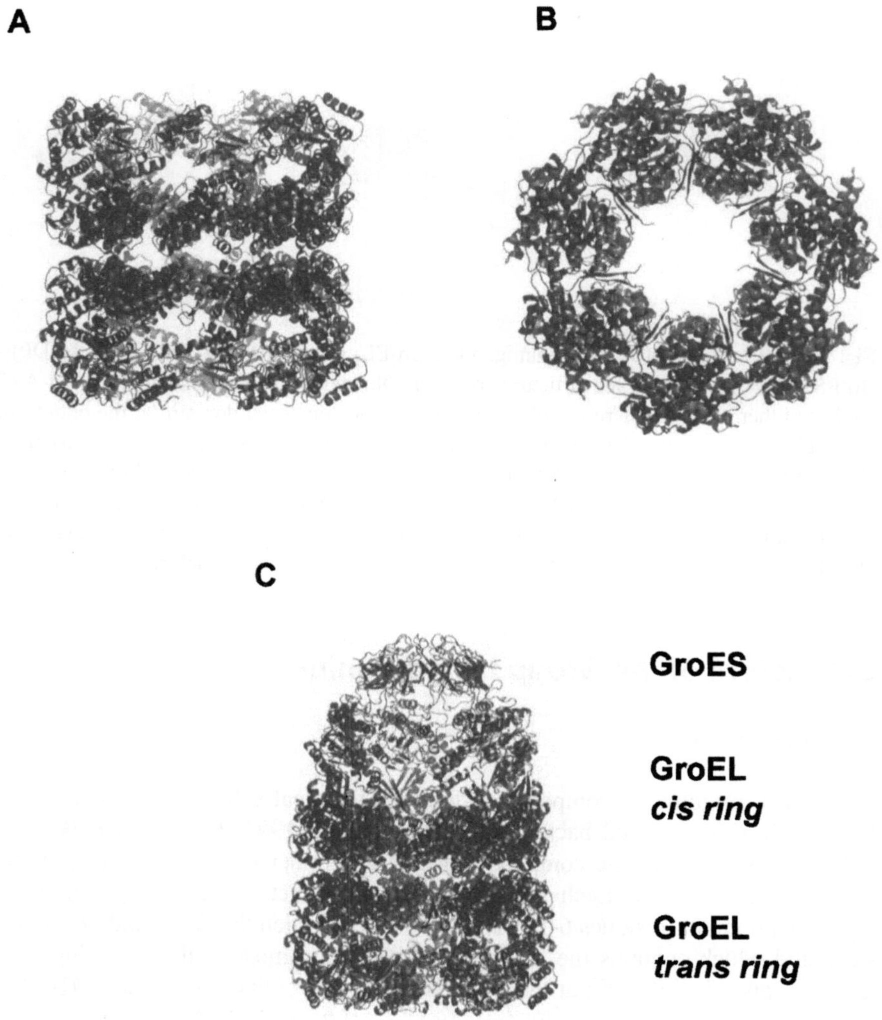

Fig. 1. (A) A side view of the structure of the GroEL chaperonin (Braig et al. 1994) showing the two heptameric rings stacked back-to-back. (B) The top view of the GroEL chaperonin showing the seven-fold axis of symmetry running through the centre of the GroEL cylinder. (C) GroEL-$(ADP)_7$-GroES (Xu et al. 1997) now showing the GroES co-chaperonin capping one ring of the GroEL structure. The apical domains of the *cis* ring (adjacent to GroES) have also extended upwards to interact with GroES and open up a large central cavity underneath GroES. The *trans* ring (opposite GroES) has a similar structure to the unliganded GroEL.

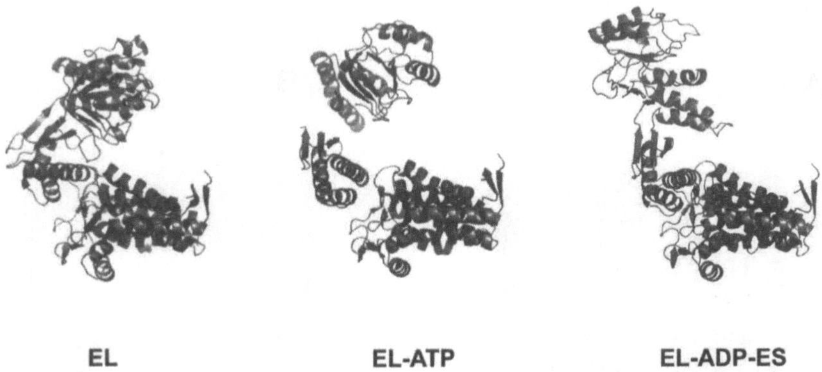

Fig. 2. GroEL subunits from the unliganded GroEL, GroEL-$(ATP)_7$ and GroEL-$(ADP)_7$-GroES x-ray crystallographic structures revealing the extent of the movement as first ATP binds and then GroES. There is little movement in the equatorial domain (at the bottom of each subunit), however, ATP induces a small opening of the subunit and a 25° counter-clockwise rotation of the apical domain (at the top of each subunit structure) relative to the unliganded GroEL. The GroES bound structure shows the subunit to have extended upwards to a much greater extent and the apical domain has rotated 90° clockwise relative to the unliganded GroEL (115° relative to the GroEL-(ATP_7) subunit structure).

2 Structure of the Group I chaperonins

2.1 GroEL structure

The GroEL oligomer is composed of fourteen identical subunits arranged as two heptameric rings stacked back-to-back (Braig et al. 1994; Fig. 1A and B). The overall architecture of the complex is that of a cylinder which is 146Å in length and 137Å in diameter. Each subunit has three distinct domains (Fig. 2): (i) an equatorial domain (residues 6-133 and 409-523) in which the ATP binding site resides and which contains the inter-ring contacts and most of the intra-ring contacts, (ii) an apical domain at the ends of the GroEL cylinder (residues 191-376), and (iii) an intermediate domain (residues 134-190 and 377-405) which acts as a flexible linker for the other two domains. Each heptameric ring encloses a central cavity whose opening is 45Å in diameter and which is lined with apical residues predominantly hydrophobic in character and which provides the binding site for substrate protein and also GroES. The central cavity has been shown to be the location of folding polypeptides (Braig et al. 1993; Weissman et al. 1995; Mayhew et al. 1996).

2.2 GroES structure

The x-ray crystallographic structure of the GroES co-protein has also been determined (Hunt et al. 1996; see Fig. 1C). The overall architecture contains seven subunits arranged in a ring, thus preserving the symmetry existing in GroEL. Each subunit contains an eight-stranded β-barrel at its core with an upwards extension towards the seven-fold axis giving the GroES oligomer a dome-shaped appearance. At the top of the dome is a cluster of acidic residues, which presumably repel each other perhaps giving the oligomer some degree of flexibility. A highly mobile loop region (residues 17-34) extends downwards from each subunit and interacts with the GroEL apical domain in the presence of adenine nucleotides (Landry et al. 1993; Xu et al. 1997).

One interesting discovery made was that the morphogenesis of bacteriophage T4 requires a specialised GroES homologue, encoded for by the phage, called gp31. It was found that the folding of a 56kDa major capsid protein, gp23, cannot be encapsulated in the GroEL cavity beneath GroES, but can when GroES is replaced with gp31. The x-ray crystallographic structure of gp31 has been solved and was found to have similar tertiary and quaternary structure to GroES, despite the fact that they only share 14% sequence identity. However, gp31 has a number of features which help to enlarge the GroEL cavity to accommodate the large capsid protein: a longer mobile loop, deletion of the conserved aromatic residue on the lower rim within the dome and deletion of the roof β-hairpin covering the top of the dome (Hunt et al. 1997; Bakkes et al. 2005).

2.3 The structure of the GroEL-ATP complex

The difficulty in determining the structure of a GroEL-ATP complex is due to the problem posed by the rapid hydrolysis of ATP to ADP during sample preparation or crystallisation. The first attempt to view the effect of ATP binding on the GroEL structure was made using negative stain electron microscopy and three-dimensional image reconstruction (Saibil et al. 1993). Although the resulting reconstruction was of a low resolution (~30Å) it clearly showed that there were significant movements in quaternary structure, and in particular of the apical domains relative to the equatorial domains when compared to the reconstruction of GroEL alone. Following the solution of the x-ray structure of GroEL, the structure of the complex liganded with the non-hydrolysable ATP analogue ATPγS was also solved (Boisvert et al. 1996). Unlike the EM reconstructions this structure showed little difference when compared to the structure of apo-GroEL. The reason for this is unclear but may be due to lattice constraints preventing the incorporation of the more open structure. A recent attempt to revise this structure using the same data proposes much larger domain motions (Wang and Boisvert 2003).

The problem was eventually overcome by using a hydrolysis-deficient mutant of GroEL (D398A) and rapidly freezing in liquid ethane after mixing with ATP to embed the sample in vitreous ice before recording cryo-electron micrographs (Ranson et al. 2001). Three-dimensional image reconstructions using the high resolution (~10Å) data revealed that the GroEL cylinder had been elongated and there was also distinct asymmetry between the two heptameric rings. The intermediate domain had rotated downwards to interact with the ATP binding site while the apical domains extended upwards and rotated 25° counter-clockwise relative to their position in the apo-GroEL structure (Fig. 2B).

2.4 The structure of the GroEL-GroES complexes

GroES is able to associate with GroEL in the presence of ATP or ADP. Low resolution EM images of the GroEL-GroES complex (Saibil et al. 1991; Roseman et al. 1996) show GroES to be binding to one end of the GroEL cylinder giving the resultant structure a "bullet" shape. Under other conditions a complex has been observed in which a GroES complex is bound to either end of the GroEL cylinder giving rise to an American "football"-shaped molecule (Azem et al. 1995). It is unclear whether these "football"-shaped intermediates have any functional role; however, they are not obligatory for efficient chaperonin-assisted protein folding.

The GroEL-ADP_7-GroES x-ray structure has GroES bound to the ADP-liganded ring of GroEL (Xu et al. 1997; Fig. 1C). As with the GroEL-ATP cryoEM structure described above, the intermediate domain has rotated downwards to interact with the nucleotide while the apical domains have extended upwards, in this case to interact with the GroES mobile loops, thus opening up the central cavity within the heptameric GroEL ring to which GroES is attached. However, in contrast to the GroEL-ATP structure, the apical domains have now rotated 90° clockwise compared to the position in the apo-GroEL structure. Since GroEL-ATP_7 is an obligatory intermediate on the way to GroEL-ADP_7-GroES then it would suggest that upon association of GroES to the GroEL-ATP_7 the apical domains must rotate through a large 115° angle (Fig. 2C). This structural rearrangement also has the effect of changing the physicochemical nature of the central cavity lining from hydrophobic to hydrophilic.

While GroEL-ATP_7-GroES is able to support the folding of a stringent protein substrate within the GroEL central cavity GroEL-ADP_7-GroES is unable to do so. In an attempt to understand the role of the γ-phosphate Art Horwich and colleagues showed that the ATP analogues GroEL-$(ADP{\cdot}AlF_3)_7$-GroES and GroEL-$(ADP{\cdot}BeF_3)_7$-GroES could also support refolding. Surprisingly however, upon solving the structure of the GroEL-$(ADP{\cdot}AlF_3)_7$-GroES complex there were no significant differences when compared to the GroEL-ADP_7-GroES structure (Chaudhry et al. 2003) indicating that hydrolysis and loss of the γ-phosphate exerts an effect on the energetic potentials within the system, (i.e. strain) rather than causing a change in the structure *per se*.

3 Interaction between Group I chaperonins and protein substrate

Amongst the earliest observations was that inactivating the yeast mitochondrial Hsp60 resulted in the misfolding and aggregation of a large number of newly-imported polypeptides (Cheng et al. 1989), while the *E. coli* GroEL could bind ~40% of all chemically denatured cytoplasmic proteins (Viitanen et al. 1992). This points to an essential characteristic of the Group I chaperonins that is the apparent lack of substrate specificity. However, since it is known that *GroE* is an essential gene for *E. coli* viability at all temperatures, workers have attempted to identify which polypeptide substrates have an absolute requirement for GroEL and GroES in order to fold. When an *E. coli* strain expressing a ts-mutant of GroEL was shifted to a higher temperature the rate of translation was seen to decline and the folding of a number of cytoplasmic proteins was affected (Horwich et al. 1993). More recently, a combination of pulse-chase radiolabelling *in vivo*, immunoprecipitation and 2D-PAGE was used to screen for *E. coli* substrates of GroEL (Houry et al. 1999). The identified substrates are a diverse set of proteins with differing functions and structures.

The possibility that GroEL is able to recognise a particular structural motif has also been addressed. One early indication, based on NMR studies of a bound peptide, suggested that GroEL might specifically recognise amphipathic α-helices within a folding intermediate (Landry et al. 1991), however, GroEL is also able to assist the refolding of an all β-sheet protein (Schmidt et al. 1992). Substrate polypeptides bind to GroEL in a multivalent manner but do not necessarily need all seven binding sites provided by the subunits within a ring (Farr et al. 2003). Mutational analysis (Fenton et al. 1994), calorimetry (Lin et al. 1995) and the structure of GroEL with a bound peptide (Chen et al. 1999) suggests that the most important factor is the spatial clustering of hydrophobic amino acids on the apical domain and that there is significant structural plasticity in this region to accommodate a range of structural motifs from a diverse set of proteins. Recent structural data also suggests that the binding of substrate polypeptides may also induce structural rearrangements in the polypeptide in order to optimise the binding affinity (Wang et al. 1999; Falke et al. 2003; Wang and Chen 2003) and that this is an important aspect of the overall allostery of the complex.

4 Allostery and asymmetry in nucleotide binding to GroEL

GroEL binds and hydrolyses ATP in a K^+-dependent manner (Gray and Fersht 1991; Jackson et al. 1993; Todd et al. 1993). ATP binding exhibits positive cooperativity within a heptameric ring and negative cooperativity between rings, revealing an apparent asymmetry within the GroEL complex (Yifrach and Horovitz 1994, 1995; Burston et al. 1995) which has also been observed within the structure using cryoEM (Ranson et al. 2001). This behaviour has been described using a

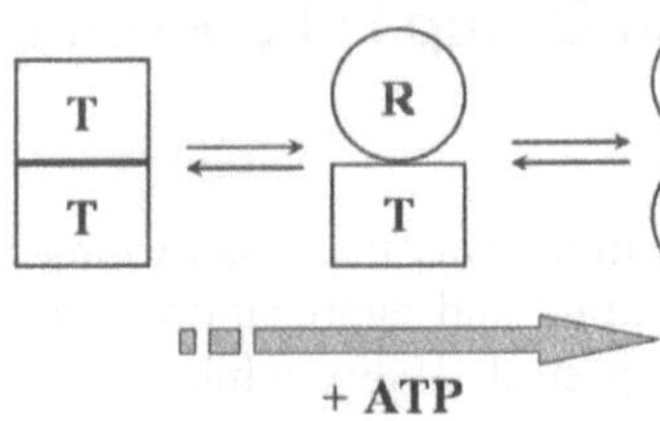

Fig. 3. GroEL shows nested cooperativity with respect to ATP binding. The binding of ATP is positively cooperative within a heptameric ring and induces a T to R "all-or-none"-type allosteric transition. However, negative cooperativity between the two rings means that the conversion of the rings from T to R occurs sequentially as ATP binds.

nested model of cooperativity (Yifrach and Horovitz 1995; Fig. 3) in which each ring can be considered to switch between a T-conformation (apo-GroEL), which has high affinity for unfolded polypeptide (Staniforth et al. 1994), and an R-conformation (GroEL-ATP), which has a much lower affinity for substrate polypeptide. This inter-conversion takes place in an "all-or-none" manner described by the MWC model of cooperativity (Monod et al. 1965). However, because of the negative cooperativity between the rings then each ring can be said to interconvert between the T- and R-conformations in a sequential manner, typical of the KNF model of cooperativity (Koshland et al. 1966). Of course any mechanistic model of GroEL action must also account for the product ADP, which exhibits only a negligible degree of positive cooperativity when binding within a heptameric ring (Cliff et al. 1999; Inobe et al. 2001) while the binding of ADP to all 14 nucleotide binding sites is so unfavourable that at high concentrations of ADP the heptameric rings are forced to dissociate (D. Poso, A. R. Clarke, and S. G. Burston, unpublished data).

The cooperativity observed in ATP binding is due to the large structural rearrangements which have been observed using cryoEM (Roseman et al. 1996; Ranson et al. 2001). The kinetics of these structural rearrangements have been measured by monitoring the intrinsic fluorescence of engineered single tryptophan probes (Yifrach and Horovitz 1998; Cliff et al. 1999; Inobe et al. 2003). This revealed the presence of a number of intermediate species along the pathway from the T- to the R-conformations. This kinetic mapping of the allostery has also been made easier by using the engineered single-ring version of GroEL (Weissman et al. 1995) in which the intra-ring cooperativity could be studied in the absence of the negative cooperativity between the rings. Surprisingly, there was little difference between the single-ring GroEL and the wild type double-ring version (Poso et al. 2004a; Amir and Horovitz 2004) except in the rate-limiting step of the ATP hydrolytic cycle in which the presence of the second ring increases the activation enthalpy from 42 to 94 kJ/mol demonstrating that the rings are conformationally coupled at this point but are remarkably autonomous otherwise (Poso et al. 2004b). The roles of some of the conformational intermediates on the allosteric pathway have also been assigned more recently, including the GroES acceptor conformation and the subsequent conformation from which polypeptide is displaced into the central cavity (M. Cliff, C. Limpkin, S. G. Burston, and A. R. Clarke, unpublished data). One other advantage of mapping the kinetic pathway is that the effect of mutations can be analysed quantitatively to construct a structural description of the changes taking place during each conformational transition from the T- to R-states. One method of doing this is to use the ϕ-value analysis, which has been used so successfully both in physical organic chemistry and in protein folding studies (Horovitz et al. 2002; Inobe and Kuwajima 2004). These ongoing studies will allow the mapping of the ligand-induced structural changes in GroEL and also the pathway of allosteric communication.

The effect of GroES is to diminish the rate of ATP hydrolysis by GroEL to approximately 35% of the rate in the absence of GroES, and apparently increase the degree of positive cooperativity within a heptameric ring (Gray and Fersht 1991; Jackson et al. 1993; Todd et al. 1993). In addition, the GroEL-GroES interaction is highly dynamic (Todd et al. 1994). GroES associates rapidly with GroEL-$(ATP)_7$ to form a stable ternary complex (Burston et al. 1995; Rye et al. 1999). Hydrolysis of ATP in the *cis* ring (the ring to which GroES is associated) to ADP reduces the strength of the GroEL-GroES interaction while the subsequent binding of seven ATP molecules to the *trans* ring (opposite GroES) results in GroES (and any folding polypeptide underneath GroES) being obligatorily dissociated from GroEL (Weissman et al. 1994; Burston et al. 1996; Rye et al. 1997). Simultaneous association of a polypeptide to the *trans* ring results in much faster enforced dissociation of GroES and any encapsulated polypeptide (Rye et al. 1999). As mentioned previously (Section 2.4) GroEL-GroES and either ATP, or the non-hydrolysable analogues, ADP-(AlF_3) or ADP-(BeF_3) are able to support the refolding of a protein substrate, while ADP cannot, despite the fact that the x-ray structures of GroEL-$(ADP)_7$-GroES and GroEL-(ADP-$(AlF_3)_7$)-GroES have no significant structural differences (Chaudhry et al. 2003). However, when the GroEL-nucleotide-GroES structure is made using ATP rather than ADP the presence of the γ-phosphate provides an additional 180 kJ/mol of energy (Motojima et al. 2004).

5 Reaction cycle of the Group I chaperonins

The chaperonin reaction cycle can be thought of as consisting of two half-reactions occurring sequentially on each heptameric ring in the manner of a two-stroke motor. Each half-reaction can be broken down into four stages (Fig. 4): (i) substrate binding, (ii) substrate encapsulation, (iii) priming for the release of substrate and GroES ejection, and (iv) the ejection step itself.

5.1 Substrate binding to GroEL

GroEL must be able to recognise unfolded or misfolded polypeptides. These polypeptide species generally have much more exposed hydrophobic surface than a folded protein, which would normally buried it in the interior of the protein. The

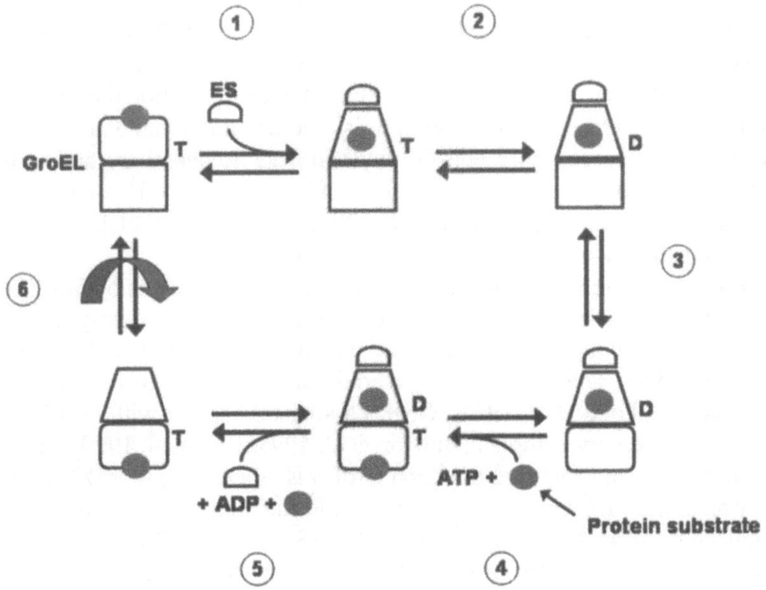

Fig. 4. A schematic model of the GroE chaperonin half-reaction. (Step 1) ATP (T) and GroES bind to a GroEL-protein substrate binary complex. This displaces the protein substrate into the central cavity underneath GroES where it begins to refold. (Step 2) ATP in the *cis*-ring now hydrolyses to ADP (D) with a half-time ~6s. This weakens the GroEL-GroES affinity priming the complex for GroES release. (Step 3) A slow conformational change takes place in the *trans*-ring. (Step 4) ATP and further protein substrate bind to the *trans* ring which (Step 5) induces the ejection of GroES and the encapsulated polypeptide, whether folded or not, from the *cis*-ring. (Step 6) The complex is now ready to start a new half-reaction on the opposite ring. The curved arrow indicates that the GroEL molecule is rotated through 180°. Polypeptide that does not refold during the time of encapsulation will rebind and proceed through further cycles of binding, encapsulation, and release.

hydrophobic binding sites on the GroEL apical domains bind to this surface stabilising the polypeptide and preventing it from irreversible aggregation. It has also been proposed that a protein substrate may unfold to some extent as it anneals to the hydrophobic binding sites (Jackson et al. 1993; Todd et al. 1994; Walter et al. 1996). Recently an elegant study has been performed which addresses this possibility using fluorescence resonance energy transfer (FRET) to monitor intramolecular distances within the substrate Rubisco. When refolding was initiated at low temperatures, the protein formed a kinetically-trapped, non-aggregating, non-native, monomeric intermediate state, which required GroEL, GroES, and ATP to refold. FRET analysis revealed that the polypeptide had to proceed through chaperonin-induced structural changes, including unfolding, to proceed to the native state (Lin and Rye 2004).

If substrate is binding to a *trans* ring (as with Fig. 4 Step 4), along with ATP, at the end of a previous half-cycle (i.e. the ejection step) then it has been observed that the presence of the substrate protein enhances the rate at which the ligands are ejected from the *cis* side. The explanation for this may come from a recent structure of GroEL with a bound substrate, glutamine synthetase, which has been investigated using cryoEM (Falke et al. 2005). The binding of the substrate appears to induce structural movements within the ring of GroEL to which it is bound, albeit less dramatic than those induced by nucleotide. Importantly however, it also causes a tilting inwards of the apical domains in the ring *opposite* the one to which the unfolded protein has bound. If GroES were bound to the opposite ring (as is the case in the ejection step) then effectively forcing the apical domains back towards their apo-GroEL positions would help GroES to dissociate.

5.2 Encapsulation and the initiation of protein folding

ATP binds to a GroEL-polypeptide complex with positive cooperativity, inducing a large structural rearrangement as the intermediate domain moves down over the ATP binding site and the apical domains move upwards and twist. GroES can rapidly associate to stabilise the opening of the central cavity and induce further twisting of the apical domains. As ATP and GroES bind the polypeptide binding sites within the ring open like an iris and become occluded, displacing the polypeptide into the enlarged central cavity of GroEL underneath GroES where refolding is initiated (Fig. 4 Step 1). The residues lining the central cavity are now predominantly hydrophilic providing a more polar environment, similar to bulk aqueous solvent, in which the folded state is energetically favoured. One study suggested that the polypeptide may undergo a degree of forced unfolding as the apical domains twist and the GroES associates (Shtilerman et al. 1999), however, this experiment has been repeated recently with a much higher signal-to-noise ratio and the result does not support this kind of forced unfolding at this stage (Park et al. 2004). Additionally GroEL-bound MDH only showed a very mild degree of deprotection of amide upon binding GroES and ATP, and the deprotection was spread randomly across the substrate suggesting this was due to the loss of interaction with the binding sites on the apical domains (Chen et al. 2001). Since the folding polypeptide is now sequestered underneath GroES it is able to refold without the possibility of inter-molecular association resulting in irreversible aggregation. There is a degree of controversy over the role of the central cavity in the chaperonin-assisted refolding reaction. On one level, it may be acting merely as a "cage of infinite dilution" to suppress the possibility of aggregation. However, there has been a suggestion that the central cavity is able to change the energy landscape of a folding protein (Brinker et al. 2001). One way in which this might be achieved is due to the spatial confinement of the polypeptide in the central cavity which could stabilise compact folding intermediates and might result in an acceleration in the rate of refolding within the chaperonin when compared to the intrinsic rate of refolding of the substrate protein (Takagi et al. 2003; Thirumalai et al. 2003). Indeed this rate enhancement has been observed for several proteins including bacterial Rubisco and MDH. It might well be imagined that this is the case if the central cavity can stabilise intermediate conformations, which are normally of a higher free energy than the unfolded conformational ensemble. However for protein substrates where the compact intermediate ensemble is already more stable than the unfolded ensemble then no rate enhancement would be observed. Pertinently, it has been noted that the mobility of a substrate polypeptide refolding inside the GroEL cavity is reduced when compared to free solution (Weissman et al. 1996), indicating that the polypeptide must be interacting with the walls of the cavity.

5.3 Priming the complex for the release of GroES and polypeptide substrate

ATP hydrolysis in the *cis* ring has a half-time of ~6s (Burston et al. 1995) and the release of the inorganic phosphate product is very fast (Cliff et al. 1999) leaving ADP bound to the *cis* ring. The hydrolysis of the γ-phosphate of ATP results in the loss of 180 kJ/mol of energy, which was used to stabilise the GroEL-GroES interaction (Chaudhry et al. 2003; Motojima et al. 2004). The result is that the complex is now primed for the discharge of GroES and polypeptide from the *cis* ring (Fig. 4 Step 2). However, although the energetic potential of the system has changed, the structure stays the same and the encapsulated protein can continue to refold (Chaudhry et al. 2003). The time that is available to the substrate to refold in the cavity is therefore dictated by both the rate of ATP hydrolysis in the *cis* ring and the rate at which the ejection signal is received from the *trans* ring, which is ~15-20s.

5.4 Ejection of the substrate and GroES from the *cis* ring

The signal for ejection of the ligands from the *cis* ring comes from the opposite *trans* ring. A slow conformational change in the *trans* ring (the rate-limiting step in the half-cycle; Fig. 4 Step 3) which immediately precedes (or is concomitant with) ATP binding to the *trans* ring (Fig. 4 Step 4) provides the signal which is communicated across the ring-ring interface forcing the *cis* apical domains to collapse releasing the bound GroES and liberating the polypeptide substrate (Fig. 4 Step 5) (Rye et al. 1997 and 1999). At this time, a further unfolded polypeptide substrate may also bind to the *trans* ring in preparation for the next half-reaction (as described in Section 5.1 above; Fig. 4 Step 4). The effect of this is to speed up the *trans* conformational change and, hence, the ejection of ligands from the *cis* ring. The substrate protein is ejected from the *cis* ring regardless of whether it has refolded or not (Weissman et al. 1994; Smith and Fisher 1995; Burston et al. 1996; Ranson et al. 1997). Proteins, which have not folded, can migrate either to another chaperonin molecule or back to the same chaperonin molecule for another round of assistance.

6 The Group I chaperonin-assisted protein folding reaction

How does GroE enhance the efficiency of protein refolding and why does it need to be enhanced anyway? While many small proteins are able to refold rapidly and efficiently under renaturing conditions many larger and more complex proteins refold much more slowly, falling prey to inter-molecular association and hence irreversible loss of material through aggregation. The reason for this is that as the complexity of the protein rises then the conformational energy landscape also becomes more complex. The presence of local energy minima can lead to the kinetic trapping of intermediate conformations or even off-pathway misfolded conformations. These conformational species are relatively long-lived and expose internal hydrophobic surface leaving them vulnerable to self-association and eventual loss. As mentioned in the discussion of the GroE reaction cycle above there are a number of characteristics that GroEL exploits to prevent this loss.

Firstly, by sequestering an unfolded protein into the GroEL central cavity underneath GroES then the polypeptide is protected from detrimental aggregation reactions (Buchner et al. 1991). This has been termed the “Anfinsen cage” and can be thought of as a chamber of infinite dilution.

Secondly, the substrate polypeptide may be unfolded to some degree as it anneals to the structurally plastic binding sites on the apical domains of GroEL (Jackson et al. 1993; Todd et al. 1994; Walter et al. 1996; Lin and Rye 2004). This may relieve any misfolding or just allow refolding to reinitiate from a higher energy conformation. Ranson et al. (1995) found that the misfolding of MDH involves early aggregation steps as the protein is presumably forming low-order aggregates, although the presence of a unimolecular misfolding step prior to chain-chain association could not be distinguished kinetically. Unlike the formation of large aggregates, these early steps are reversible, even though the equilibrium favours the aggregate formation. Surprisingly, GroE was able to actively reverse these processes, even at sub-stoichiometric levels. Using cycles of binding and release, material, which is being unproductively lost to the aggregation pathway can be restored to the productive folding pathway and, in the case of MDH, results in an enhancement in the apparent rate of refolding.

Finally, the cavity itself alters the conformational energy landscape of the folding protein (Brinker et al. 2001). This may be due to the confinement of the polypeptide within the cavity (Takagi et al. 2003; Thirumalai et al. 2003) or some other, as yet unknown, interaction between the cavity and the folding protein. The importance of the properties of the central cavity have been elegantly demonstrated by the experiment in which directed evolution was used to optimise the central cavity for the refolding of the green fluorescent protein (GFP) (Wang et al. 2002). This was achieved by mutations altering both the rate of ATP hydrolysis and the inherent allostery of the system, and also changing the polarity of the central cavity. It may be that GroEL has evolved a balance of the ATPase rate and central cavity properties in order to make it a generalised chaperone.

One final aspect of the central cavity concerns its volume. The volume in the GroEL-$(ADP)_7$-GroES crystal structure is 175,000 $Å^3$, which has been estimated to be large enough to accommodate the folding intermediate of a protein up to 60kDa in size. However, it has been observed that the yeast mitochondrial aconitase (82kDa) was also dependent upon the homologous chaperonin, mitochondrial Hsp60. *In vitro* studies revealed that, although the protein cannot become encapsulated underneath GroES in the GroEL cavity, it is able to bind to the *trans* ring where it passes through cycles of binding and release until it has efficiently refolded to the holoenzyme (Chaudhuri et al. 2001). GroEL-GroES complexes were constructed in which the substrate protein could fold either only in the *cis* cavity underneath GroES or only on the *trans* ring opposite GroES. Surprisingly, when these were tested with bacterial Rubisco and MDH it was found that they could both fold productively from the *trans* ring, albeit at a slower rate than refolding in the *cis* cavity.

One important aspect of all of the above mechanistic aspects of GroE is that none of them contradict Anfinsen's Dogma that the three-dimensional structure of the protein in a particular solvent is dictated entirely by its primary amino acid sequence. This explains why the Group I chaperonins are able act on such a wide variety of proteins from diverse sources.

7 The Group II chaperonins

7.1 Introduction

The first member of this family to be identified was the chaperonin from the hyperthermophilic archaeon *Pyrodictum occultum*, when the lysis of accidentally heat-shocked *P. occultum* cells released several toroidal particles composed of two stacked rings of eight subunits each enclosing a central cavity (Phipps et al. 1991). The name "thermosome" was chosen to highlight the extreme thermal stability of this heat-inducible protein complex. Furthermore, the overall particle shape, the subunit organization, as well as the ATPase activity and the heat-shock inducibility of this protein were reminiscent of the Group I chaperonin family, suggesting the possibility for the thermosome to represent an archaebacterial chaperonin. Subsequently, another protein complex (TF55, for thermophilic factor of 55kDa subunits) was found to accumulate in the cytosol of the thermophilic archaebacterium *Sulfolobus shibatae* in response to heat-shock (Trent et al. 1991). Electron microscopy analysis revealed TF55 to be a bi-toroidal complex composed by 18 subunits arranged as two nine-membered rings, closely resembling the seven-membered-ring complexes of the eubacterial chaperonins. The evidence for this protein to be also a weak ATPase and to bind unfolded polypeptides *in vitro* corroborated the hypothesis for a role of the thermosome as an archaebacterial chaperonin. Similar protein complexes have been identified since then in all classes of archaea: thermophiles (e.g. *Thermoplasma acidophilum*, Waldmann et al.1995a), methanogens (e.g. *Methanopyrus kandleri*, Andra et al. 1996; Minuth et al. 1999) and halophiles (e.g. *Haloferax volcanii*, Kuo et al. 1997; and *Haloarcula marismortui*, Franzetti et al. 2001). The term “thermosome”, initially coined for the chaperonin from *P. occultum*, is now used to indicate all these multimeric protein complexes, and has been adopted as a generic name for archaebacterial chaperonins.

Further investigation of TF55 revealed another interesting feature: although its sequence was not highly homologous with the eubacterial chaperonins (e.g. GroEL), it shared a high degree of homology with the sequence of the eukaryotic protein t-complex polypeptide-1 (TCP-1, Trent et al. 1991). TCP-1 is a weak ATPase ($M_r \sim 60$ kDa), originally isolated from murine testes (Silver et al. 1979), but later found to be constitutively expressed in other mammalian cell types (Silver et al. 1987; Willison et al. 1987) as well as in *Drosophila melanogaster* (Ursic and Ganetzky 1988) and *Saccharomyces cerevisiae* (Ursic and Culbertson 1991). TCP-1 exists within a double-toroidal hetero-oligomeric complex, now known as TRiC (TCP-1 Ring Complex) or CCT (Chaperonin Containing TCP-1), and before its sequence homology with the thermosome was pointed out, it had already been tentatively identified as the chaperonin of the eukaryotic cytosol (Ellis 1990; Ahmad and Gupta 1990). Supporting the hypothesis of its role as a eukaryotic chaperonin was the evidence of its involvement in the biogenesis of actin and tubulin (Frydman et al. 1992; Gao et al. 1992; Lewis et al. 1992; Yaffe et al. 1992). CCT has been proposed to be present in the cytosol of all eukaryotic cells (Horwich and Willison 1993; Kubota et al. 1995) where, in contrast to the biosynthesis of the other chaperonins, its expression can neither be induced nor enhanced by heat-shock (Ursic and Culbertson 1992).

Since the identification of the first archaebacterial and eukaryotic chaperonins, many others have emerged and are still emerging in both phylogenetic domains, thanks to the completion of the relative genome sequencing projects. In fact, in 1999 the SwissProt database contained 78 sequences of Group II members (20 archaeal and 58 eukaryotic sequences; Gutsche et al. 1999), while currently (2005) the number has increased up to 442, of which 76 sequences are from archaebacteria and 366 from eukaryotic organisms (source: cpnDB, the chaperonin sequence database, see Hill et al. 2004), such that complete gene sequences for the chaperonin systems are now available for a wide number of archaea (*Haloferax volcanii, Paracoccus furiosus, Sulfolobus sulfataricus, Thermococcus sp.* ks) and eukaryotes such as *Caenorabditis elegans, Mus musculus, Homo sapiens, Xenopus laevis*. Contrary to the case of eubacteria, no GroES-like co-chaperonin has been identified in archaea or in the eukaryotic cytosol.

Finally, it’s worth noting that the occurrence of Group I chaperonins in eubacteria as well as in mitochondria and chloroplasts, and of Group II chaperonins in eukaryotes as well as in archaea has been brought up as a crucial evidence to support the endosymbiotic theory for the evolution of eukaryotic cells (Gupta 1995), stating that certain eukaryotic organelles may have originated from primordial eubacteria engulfed by archaebacterial host cells (Margulis 1971; for a review, see Lopez-Garcia and Moreira 1999).

8 Group II chaperonin subunit composition and organization

The subunit composition of Group II chaperonins is generally more complex and heterogeneous than that of their eubacterial counterpart. Some of the thermosomes are homo-oligomeric (e.g. those from methanogens, but also P45 from the halophile *Haloarcula Marismortui*, Franzetti et al. 2001), but the majority consists of two subunit types, named α and β, alternating in each eight-membered ring (Nitsch et al. 1997). The ancestral Group II chaperonin was homo-oligomeric, as the Group I *ur*-chaperonin, from which it has evolved independently for more than two million years (Gupta 1990; Kubota et al. 1995). Gene duplication events in most of the archaeal species subsequently led to hetero-oligomeric αβ-thermosomes, with the two subunits sharing 55% sequence identity. In at least one species (*Sulfolobus)* the *ur*-α-subunit has undergone a further duplication into α and γ-paralogs, leading to a thermosome composed by three subunit types (Gutsche et al. 1999; Archibald et al. 2001; Archibald and Roger 2002). On the other hand, the third Group II gene (Hsp60-3) identified in archaea of the *Methanosarcinae* species is not homologous to the γ-subunit of the *Sulfolobus* species. The hypothesis that it may have arisen early and evolved independently from the α and β genes in the *Methanosarcinae* lineage is supported by a very recent study which reported the presence of further two Group II chaperonin genes (Hsp60-4 and Hsp60-5) in the archaeon *Methanosarcina acetivorans*, whose origin and evolution could be similar to the one of Hsp60-3 (Maeder et al. 2005). As the authors suggest, the presence of five different genes should imply the possibility of more than a single type of thermosome in the cytosol of *M. acetivorans*, composed of different combinations of the corresponding five subunits. This kind of scenario, that requires further investigation, is of particular interest in light of the previous finding that in *M. acetivorans*, as well as in *M. mazei* and *M. barkeri*, Group I chaperonin genes co-exist with the Group II ones, and are simultaneously expressed to similar levels in the cytosol (Klunker et al. 2003). In *M. mazei*, while the thermosome consists of the three paralogous subunits, α, β, and γ described above, which assemble preferentially at a molar ratio of 2:1:1 in a nucleotide-dependent manner, the Group I proteins have the same structural features as their bacterial counterparts. The co-occurrence of both chaperonin families in the *Methanosarcina* species has led to the proposition that each chaperonin system assists in the folding of a specific subset of cytosolic proteins. Another hypothesis allows for the possibility that the presence of the Group I chaperonins facilitated the specialization of the *M. mazei* thermosome (and possibly, in the light of the results by Maeder et al. (2005) reported above, of its various isoforms) in the folding of proteins that cannot be handled by GroEL (Klunker et al. 2003; Figuereido et al. 2004).

Nevertheless, the complexity of the subunit composition within the Group II chaperonins reaches its maximum in the CCT complex, whose two superimposed rings are each constituted by eight different, though homologous, subunits (CCT α, β, γ, δ, ε, ζ, η, θ; CCT1 - CCT8 in yeast; Kubota et al. 1994 and 1995). Although the range of possible combinations of such a number of different subunits in a single ring would be very large, the presence of preferential inter-subunits contacts introduces sufficient combinatorial bias for a unique topology to ensue in which each subunit occupies a well defined position in the 8-membered ring (Liou and Willison 1997). The clockwise order is proposed to be $\alpha(1)\rightarrow \varepsilon(5)\rightarrow \zeta(6)\rightarrow \beta(2)\rightarrow \gamma(3)\rightarrow \theta(8)\rightarrow \delta(4)\rightarrow \eta(7)$; a specific subunit arrangement which seems to be shared by all the eukaryotic CCT. Each subunit is encoded by a unique gene in all the organisms and tissues studied so far (Willison and Grantham 2001), the only exception being that of mammalian testis, which contain a tissue-specific CCTζ2 (codified by the gene Cctz-2), whose sequence is more than 80% identical to the one of the CCTζ1 subunit expressed in the other tissues (Hynes et al. 1995; Kubota et al. 1997). It is interesting to note that different subunits of CCT from the same species have closer sequence homology to the thermosome subunits (35-40% identity; Waldmann et al. 1995a) than to one another (25-30% identity; Kubota et al. 1995).

9 Structure of the Group II chaperonins

The first evidence of a substantial similarity in the subunit architecture of the Group I and Group II chaperonins came from the comparison of the relative sequences (Lewis et al. 1992; Kim et al. 1994; Waldman et al. 1995b) which were identified in the former as an apical domain and an equatorial domain, connected by an intermediate hinge domain (see Fig. 5 and Section 2 above). The equatorial domain shows the highest degree of homology with that of GroEL, while the sequences of the apical domains are more divergent; conclusions confirmed by the crystal structure of the thermosome apical domain (Klumpp et al. 1997), revealing the presence in the thermosome of a helix-turn-helix motif that is absent in GroEL. Subsequently the crystal structure of the intact thermosome from *Thermoplasma acidophilum* (Ditzel et al. 1998) was resolved at 2.6 Å, showing the assembly of the two subunits (α and β) in two eight-membered rings, stacked back-to-back to form a double toroidal cylinder with a diameter of 150-160 Å and a height of 160-180 Å (Fig. 5). Although it shares with GroEL an almost identical pattern of intra-ring contacts, most of which provided by the equatorial domain, the thermosome presents very different inter-ring (between subunits in opposite rings) contacts, all concentrated in the equatorial domain. If in fact in GroEL two helices (D and Q) of a subunit in one ring interact with those of two different subunits in the facing ring (Braig et al. 1994: Xu et al. 1997), in the thermosome the two rings are in register, every subunit in one ring interacting with only one subunit in the opposite ring, in a pattern of α–α and β–β contacts. The same overall structure has been determined for CCT at 25-30 Å resolution by cryo-electron microscopy (Llorca et al. 1999a, 2000), the only available at present, due to the absence of high resolution information on the intact CCT complex. Also in this case, every subunit in one ring interacts with only one subunit on the opposite ring (Llorca et al. 1999b). This important difference in the inter-ring packing opens the question as to

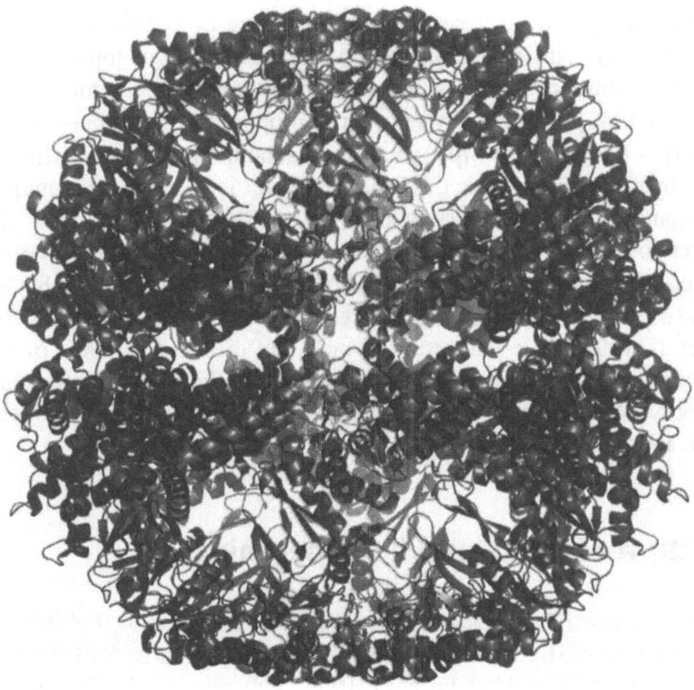

Fig. 5. The structure of the archeal Group II chaperonin, the thermosome. Similarities can be seen with GroEL in that it is a double-ring structure, and each subunit is composed of the same three domains as for Group I chaperonins. However, in this case each ring is a hetero-octamer of composition $(\alpha\beta)_4$ and the complex has no equivalent of GroES.

whether the communication between the rings may be similar to that characterised for GroEL, or whether it proceeds by means of a different allosteric behaviour, unique to the Group II chaperonins.

As mentioned above, another striking difference relative to the Group I chaperonins is the presence of a 28 residue insertion, strictly conserved in all the Group II apical regions, which forms a unique helical protrusion 25 Å long. In the closed, compact form of thermosome, which has been crystallised (Ditzel et al. 1998), the helical protrusions are projected towards the central axis to form a lid domain, occluding the central cavity and thus controlling the access to the folding compartment without need for a detachable co-chaperonin like GroES. The sealing of the ring cavity, operated by the lid following large nucleotide-induced conformational changes in the apical domains, has been visualised by cryo-electron microscopy both for chaperonins from archaea (Schoehn et al. 2000a, 2000b) and for CCT (Llorca et al. 1998, 1999b). Furthermore, the helical protrusions have been proposed to have a role in substrate recognition (Klumpp et al. 1997; Ditzel et al. 1998; Heller et al. 2004). In the Group I chaperonins, the polypeptide-binding surface, composed by a cluster of hydrophobic residues, has been located on one side of the apical domain exposed to the central cavity in the apo-GroEL structure (Fenton et al. 1994; Chen and Sigler 1999). The corresponding region in the Group II chaperonins has a mostly hydrophilic character (Klumpp et al. 1997) and since hydrophobic interactions have been implicated in substrate binding to both Group I and Group II chaperonins (Guagliardi et al. 1994; Dobrzynski et al. 1996), the identification of a hydrophobic core in the lid segment of the apical domain has been considered as consistent with a role of the helical protrusion in substrate binding to the thermosome (Ditzel et al. 1998; Gutsche et al. 1999) and to CCT (Rommelaere et al. 1999). This view is supported by the analysis of an open state of the thermosome, visualised by a electron-tomographic 3D reconstruction of vitreous ice-embedded thermosomes (Nitsch et al. 1998), in which these hydrophobic patches are organised to form surface areas facing the wide central channel, where the substrate could bind. However, it has been shown that deletion mutants of the thermosome from *Thermococcus* sp. strain KS-1 lacking different portions or even the whole of the helical protrusions are still capable of binding and protecting them from thermal aggregation, although they significantly affected their ability to mediate folding of unfolded substrates (Iizuka et al. 2004). These results would confirm the involvement of the helical protrusions in the active folding of substrates, but question their importance in substrate recognition and binding. On the other hand, in contrast with the work of Rommelaere et al. (1999) cited earlier, electron microscopy studies on CCT:substrate complexes have identified in the substrate-binding regions of the apical domains some charged residues likely to be responsible for the specific interactions (Llorca et al. 1999a, 2000). This evidence would point to an important difference between the archaebacterial and the eukaryotic chaperonin in the nature of their interaction with substrates.

10 Nucleotide-induced structural rearrangements in the Group II chaperonins

The activity of the Group II chaperonins, like that of the Group I proteins, is dependent upon ATP. However, although the ATP reaction cycle of the Group II complexes still lacks a detailed characterization, some important differences, as well as analogies, in the structural rearrangements undergone by the two groups during the reaction cycle have been identified. As described above, the tips of the apical domains appear to substitute for the function of GroES in closing the chamber of the cis-ring, and their ability in forming a built-in lid depend on significant ATP-induced structural rearrangements involving the equatorial and apical domains. Although the crystal structure of the thermosome in the absence of ATP reveals the complex in a closed state (Ditzel et al. 1998), cryo-electron microscopy and small angle neutron scattering (SANS) analyses on CCT and on the thermosome have shown that in their apo-form Group II chaperonins adopt an open conformation (Gutsche et al. 2000a, 2001; Llorca et al. 1999b), and that the addition of ATP induces the closure of the complex (Gutsche et al. 2000a; Llorca et al. 2001). These results suggest that the crystallization conditions (i.e. high salt concentration) or lattice formation may have trapped the chaperonin in the closed state normally observed in solution only upon addition of ATP, producing an image of the central chamber of the thermosome that closely resembles the cis-ring of the asymmetric complex MgATP-GroEL /GroES, identified as the folding-active state (Weissman et al. 1996; Mayhew et al. 1996). To reinforce the analogy between the two groups of chaperonins, the structure of the thermosome in the presence of the transition state analogue Mg-ADP-AlF_3 has revealed a pattern of interactions between the nucleotide and residues from the intermediate and equatorial domains which is nearly identical to the one present in the cis-ring of the folding-active GroEL/GroES complex, suggesting a common mechanism of ATP hydrolysis for the two groups. Furthermore, the ATP-CCT structure from cryoEM analysis (Llorca et al. 1999) is that of an asymmetric complex, where one of the rings is in a conformation similar to that of apo-CCT, and the other appears more open. These differences are consistent with an ATP-dependent rotation of the apical domains, which in the apo-form contacts adjacent subunits, to a position in which the tips point towards the axis of the complex, accompanied by a concomitant upward movement of the equatorial domain towards the axis of the particle. These combined movements of apical and equatorial domains in CCT-ATP generate in a different inner wall surface, and a closed cavity slightly larger than that of the apo-form (Carrascosa et al. 2001). Small angle x-ray scattering (SAXS) and SANS studies (Meyer et al. 2003; Gutsche et al. 2000b) have demonstrated for CCT and for the thermosome, respectively, that binding of ATP to the fairly open apo-complex leads to a further expansion, that has been proposed to be the substrate binding form. A fast rearrangement upon ATP binding has also been observed by tryptophan fluorescence both for CCT (Kafri and Horovitz 2003) and the thermosome (Bigotti and Clarke 2005). The requirement for ATP in order for the chaperonin-substrate interaction to occur has been recently been suggested by the finding that the apo-form of the *T. acidophilum* thermosome does not appear to bind unfolded protein substrate, either by being unable to form hydrophobic contacts with exposed non polar groups in the substrate and/or not allowing its entry in the cavity (Bigotti and Clarke 2005). As hydrolysable ATP has been shown to be required both for the lid to close (Szpikowska et al. 1998; Meyer et al. 2003) and for substrates to fold (Meyer et al. 2003), subsequent ATP hydrolysis would trigger the lid closure, allowing the substrate encapsulation and the folding reaction to proceed. The events involved in the opening of the ring to complete and restart the cycle have still to be determined. For the thermosome it has been proposed, on the basis of ATPase activity measurements and of SANS analysis (Gutsche et al. 2000c, 2000b) that the rate-limiting step of the reaction cycle (the release of either ADP and/or P_i product from the nucleotide pocket) would trigger the opening of the lid.

11 Allostery in the Group II chaperonins

The mechanism by which the structural rearrangements occurring during the hydrolytic cycle of the Group I chaperonins coordinate to drive the folding machine is reasonably well described. On the contrary, the mechanism of energy transduction in the Group II chaperonins and its coupling to the binding and folding of protein substrates remains obscure. In one respect, the type II system is simpler, since there is no need for a co-protein that is ejected and re-bound during each complete hydrolytic cycle. In contrast however, it is more complicated in that the double-ring complexes are hetero-oligomers, so that there is the added difficulty of understanding the properties of the differing subunits and how these relate to the integrated mechanism. As reported above, much of the information available on the Group II class regarding the effects of nucleotide binding and hydrolysis on the gross conformation of the complex has been elucidated by scattering methods and by cryo-EM. More recently, transient kinetics experiments have been used to provide quantitative comparisons of allosteric communication amongst the subunits within and between rings. In GroEL, the two-tier cooperativity of the structure (i.e. positive within the ring and negative between rings) is strikingly manifest by the time-resolved response to ATP, and when a similar analysis was performed on the CCT complex, it revealed a similar pattern of behaviour (Kafri et al. 2001; Kafri and Horovitz 2003), implying at least some commonality in the mechanism of nested cooperativity within the chaperonins from the two groups. In contrast with the concerted mechanism characterised for GroEL however, the same studies, based on the evidence that the rates of the ATP-dependent transitions of CCT are considerably lower than those of GroEL, suggest that CCT may undergo sequential allosteric transitions within a ring (confirming what had already been proposed on the basis of genetic evidence; Lin and Sherman 1997), and that this may depend on differences in the intrinsic affinities for ATP among different subunits in a ring. The same kind of conclusions have been drawn in a recent electron microscopy study of CCT complexes labelled with monoclonal antibodies specific for distinct subunits (Rivenzon-Segal et al. 2005) which detected preferential patterns by which the ATP-induced conformational changes spread around the ring in a sequential manner. The authors proposed that this should allow the sequential folding and release of single domains of multidomain protein substrates. These conclusions would add structural evidence to the view that domain-by-domain co-translational folding is more common in eukaryotes than prokaryotes (Netzer and Hartl 1997). Transient-kinetics experiments on the thermosome from *T. acidophilum* (Bigotti and Clarke 2005) revealed that the rate constant of the fast rearrangement observed upon ATP-binding follows a two-phase saturation profile, as it does for GroEL and CCT. This result, in keeping with these precedents, reveals that the thermosome is also a negatively cooperative system with respect to inter-ring communications, in spite of its apparent lack of positive intra-ring cooperativity in ATP-binding and hydrolysis (Gutsche et al. 2000c; Bigotti and Clarke 2005). Also, as in the case of GroEL, the loading of the second ring is weakened

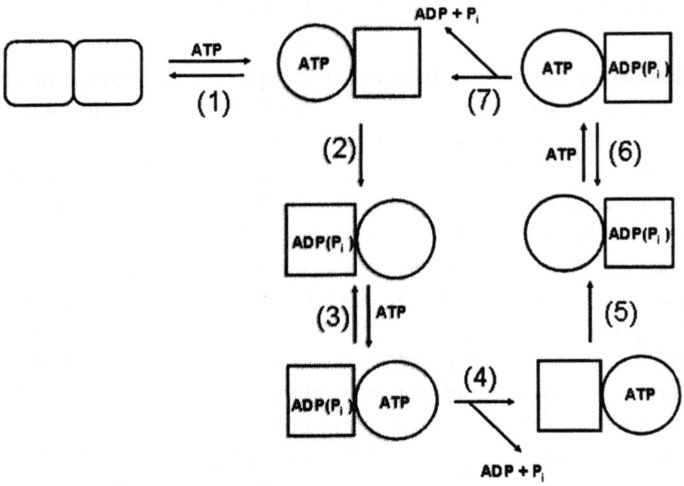

Fig. 6. The schematic of the proposed model shows three conformations for the octameric ring of the thermosome; the apo (rounded square), the ATP state (circle) and the hydrolytic product state (square). At low nucleotide concentrations the cycle is primed by the binding of ATP to one ring of the complex (Step 1), which lowers the affinity of the second ring. The hydrolytic reaction (Step 2) then leads to the formation of the product state, thus, switching the conformation of the opposite ring and allowing it to bind ATP with high affinity (Step 3). The rate-limiting step of the cycle is the release of products (Step 4). Steps 5, 6, and 7 are repeats of Steps 2, 3, and 4 on the opposite ring. In this cycle the mixed ATP/ADP.Pi complexes would be predominant in the steady state with, presumably, the ATP-ring being an open acceptor state where polypeptide substrates have access to the cavity and the product state being closed and capable of encapsulating the substrate.

by ADP, implying that asymmetric ATP/ADP complexes are favoured over symmetric ones: i.e. the preferred state in the presence of mixed nucleotides being an ATP_8:ADP_8 complex. Based on the fact that such a species, according to previous cryo-EM work (Gutsche et al. 2000a), is likely to have one open and one closed ring, a model has been tentatively proposed that might unify the experimental findings regarding the functional cycle of the thermosome (Bigotti and Clarke 2005; see Fig. 6). Despite the difference in co-protein involvement in the type I and II chaperonins, the results reported either for CCT and for the thermosome show that negative cooperativity is a common feature in inter-ring communication in all chaperonins thus far examined, and imply at least some commonality with the reciprocating encapsulation mechanism shown for the GroE chaperonin.

12 Interaction between the Group II chaperonins and protein substrates

In recent years many efforts have been made to characterise the nature of the folding substrates of Group II chaperonins and of the relative interactions involved in the folding process. While unfortunately no data are yet available on the identity of the natural substrates of the archaebacterial chaperonins, the number of known substrates for CCT is steadily increasing (Siegers et al. 2003; for a review see Dunn et al. 2001). Although this chaperonin was initially thought to fold specifically only actin and tubulin, an activity which is essential for cell viability, pulse-chase analysis indicates that at least 9-15% of the newly synthesised proteins transit through CCT (Thulasiraman et al. 1999). Among others, firefly luciferase (Frydman et al. 1994), G-α-transducin (Farr et al. 1997), the von Hippel-Lindau tumour suppressor protein (VHL, Feldman et al. 1999) and cyclin E (Won et al. 1998) are chaperoned by CCT. No clear structural or sequence features are shared by all the CCT substrates, but it has been identified that a family of proteins, the WD40 domain proteins, containing the WD repeat (a weakly conserved sequence motif of ~31 amino acids ending with conserved tryptophan and aspartate residues) and sharing a β-propeller fold, constitute the first class of chaperonin substrates defined by a common fold (Siegers et al. 2003). The dependence of such proteins on CCT for folding has been tentatively connected to their high β-sheet content, which may make them particularly prone to aggregation. The heterogeneity of folding substrates has been linked to subunit diversity in CCT, implying that each subunit may contribute to the recognition of specific motifs within the substrates (Feldman et al. 1999; McCallum et al. 2000). As an example, cryo-EM of actin-CCT complexes revealed the binding of actin to the chaperonin in a specific 1,4 order, one of the actin-binding subunits being δ, the other one β or ε, depending on the orientation of the interaction (Llorca et al. 1999). Specific interactions of the elongating polypeptide with CCT individual subunits have also been detected using photo-crosslinking experiments (McCallum et al. 2000). The same kind of analysis, however, has recently demonstrated that actin and luciferase nascent chains contact multiple CCT subunits during their elongation, proposing a model of dynamic, as opposed to specific, interactions between the substrate and the chaperonin (Etchells et al. 2005). CCT substrates also vary in size, with molecular weights spanning from 20 kDa to > 100 kDa. Since the volume of the central chamber, as determined on the basis of the crystal structure of the thermosome (Ditzel et al. 1998), in its closed state could accommodate a polypeptide up to 50-60 kDa, it has been proposed that for larger proteins CCT could promote their folding in individual domains (Netzer et al. 1997). In this respect, it is important to note that, unlike its eubacterial counterpart, CCT binds nascent polypeptides co-translationally, as they emerge from the ribosome (Frydman et al. 1994; McCallum et al. 2000; Siegers et al. 2003). The binding of the elongating chains to CCT is mediated by upstream chaperones, which capture and transfer them to the chaperonin for functional folding (Hansen et al. 1999; Siegers et al. 2003). Of the two known chaperone systems, GimC, or prefoldin, seems to be specific for actin and tubulin, while Hsp70 seems to bind generally other classes of substrates (Siegers et al. 2003), although their functions partially overlap. Prefoldins have been found and characterised also in archaea (Leroux et al. 1999), and the crystal structure of GimC from *Methanobacterium thermoautotrophicum* (Siegert et al. 2000), together with electron microscopy of the eukaryotic one (Martin-Benito et al. 2000) have shown that prefoldin is assembled in a double β-barrel with six coiled coils protruding from it, giving it the appearance of a jellyfish. The protrusions are thought to bind to the substrate in a concerted manner (Martin-Benito et al. 2002). Furthermore, electron microscopy of prefoldin-CCT complexes showed an interaction between the outer regions of the GimC tentacles and the inner region of the apical domains of the chaperonin (Martin-Benito et al. 2002). Support for the possible handoff mechanism of substrates from prefoldin to the chaperonin has come from a recent study on prefoldin from hyperthermophilic archaea, which reveals a dynamic equilibrium between binding and release of substrates from GimC, the release being facilitated in the presence of the thermosome (Zako et al. 2005). As to the state of the substrate bound, it is still under debate if the substrate polypeptide binds to the chaperonin in a quasi-native or unstructured conformation. Biochemical data on the binding of actin (Meyer et al. 2003) or VHL (Feldman et al. 2003) to CCT seem to indicate that the substrates are bound in an unfolded state. However, the conformation of CCT-interacting actin and tubulin, although determined at low resolution by cryo-EM, appears native-like, a state that the two kind of proteins may have attained before interacting with the chaperonin (Llorca et al. 2000). Finally, it's worth pointing out what seems to be a main difference in the folding mechanism of the Group II chaperonins relative to the Group I. As opposed to the passive folding mechanism described for GroEL, it seems that CCT displays an active mechanism, in which the ATP-dependent movements of the apical domains force the change of the bound substrates actin and tubulin from an open conformation to a compact one, that are not liberated in the cavity, but remain bound to the chaperonin (Llorca et al. 2001). CCT would thus use the conformational changes that seal the ring to force the folding of its substrates.

13 Future perspectives

Although a great deal of progress has been made in our understanding of the mechanism of the Group I chaperonins there still remains some significant gaps in our knowledge. There are many details concerning how GroEL affects the folding pathway of a protein substrate, which need to be clarified. To this end one interesting development is the use of NMR techniques, previously unavailable to large protein complexes, to study the interaction between polypeptides and the chaperonins (Fiaux 2002). Additionally, there still remains a great deal to do to understand the allosteric mechanism, the underlying structural changes and their roles in the assisted folding reaction. Meanwhile, the Group II chaperonins are starting to come of age in terms of our understanding of their mechanisms. Previous difficulties in over-production and isolation have been overcome and this opens up the opportunity for the kind of biochemical, biophysical and structural studies that have been performed on the Group I proteins. One significant gap in our knowledge though is the absence of any known natural substrate for the thermosome. This makes the use of any non-natural substrate difficult to interpret in the physiological context. While the Group I proteins may be ahead in our understanding there is clearly much work still to be done before we can completely understand the mechanisms of both types of chaperonin at the molecular level.

References

Ahmad S, Gupta RS (1990) Cloning of a Chinese hamster protein homologous to the mouse t-complex protein TCP-1: structural similarity to the ubiquitous 'chaperonin' family of heat-shock proteins. Biochim Biophys Acta 1087:253-255

Amir A, Horovitz A (2004) Kinetic analysis of ATP-dependent inter-ring communication in GroEL. J Mol Biol 338:979-988

Andra S, Frey G, Nitsch M, Baumeister W, Stetter KO (1996) Purification and structural characterization of the thermosome from the hyperthermophilic archaeon *Methanopyrus kandleri.* FEBS Lett 379:127-131

Anfinsen CB (1973) Principles that govern the folding of protein chains. Science 181:223-230

Archibald JM, Roger AJ (2002) Gene duplication and gene conversion shape the evolution of archaeal chaperonins. J Mol Biol 316:1042-1050

Archibald JM, Blouin C, Doolittle WF (2001) Gene duplication and the evolution of Group II chaperonins: implications for structure and function. J Struct Biol 135:157-169

Azem A, Diamant S, Kessel M, Weiss C, Goloubinoff P (1995) The protein-folding activity of chaperonins correlates with the symmetrical $GroEL_{14}(GroES_7)_2$ heterooligomer. Proc Natl Acad Sci USA 92:12021-12025

Badcoe IG, Smith CJ, Wood S, Halsall DJ, Holbrook JJ, Lund P, Clarke AR (1991) Binding of a chaperonin to the folding intermediates of lactate dehydrogenase. Biochemistry 30:9195-9200

Bakkes PJ, Faber BW, van Heerikhuizen H, van der Vies SM (2005) The T4-encoded co-chaperonin, gp31, has unique properties that explain its requirement for the folding of the T4 major capsid protein. Proc Natl Acad Sci USA 102:8144-8149

Barraclough R, Ellis RJ (1980) Protein synthesis in chloroplasts IX Assembly of newly-synthesized large subunits into ribulose bisphosphate carboxylase in isolated pea chloroplasts. Biochim Biophys Acta 607:19-31

Bigotti MG, Clarke AR (2005) Cooperativity in the thermosome. J Mol Biol 348:13-26

Boisvert DC, Wang J, Otwinowski Z, Horwich AL, Sigler PB (1996) The 2.4Å crystal structure of the bacterial chaperonin GroEL complexed with ATPγS. Nat Struct Biol 3:170-177

Braig K, Otwinowski Z, Hegde R, Boisvert DC, Joachimiak A, Horwich AL, Sigler PB (1994) The crystal structure of the bacterial chaperonin GroEL at 2.8Å. Nature 371:578-586

Brinker A, Pfeifer G, Kerner MJ, Naylor DJ, Hartl FU, Hayer-Hartl MK (2001) Dual function of protein confinement in chaperonin-assisted protein folding. Cell 107:223-233

Buchner J, Schmidt, M, Fuchs M, Jaenicke R, Rudolph R, Schmid FX, Kiefhaber T (1991) GroE facilitates refolding of citrate synthase by suppressing aggregation. Biochemistry 30:1586-1591

Burston SG, Ranson NA, Clarke AR (1995) The origins and consequences of asymmetry in the chaperonin reaction cycle. J Mol Biol 249:138-152

Burston SG, Weissman JS, Farr GW, Fenton WA, Horwich AL (1996) Release of both native and non-native proteins from a cis-only GroEL-ternary complex. Nature 383:96-99

Carrascosa JL, Lorca O, Valpuesta JM (2001) Structural comparison of prokaryotic and eukaryotic chaperonins. Micron 32:43-50

Chandrasekhar GN, Tilly K, Woolford C, Hendrix R, Georgopoulos C (1986) Purification and properties of the groES morphogenetic protein of *Escherichia coli.* J Biol Chem 261:12414-419

Chaudhry C, Farr GW, Todd MJ, Rye HS, Brunger AT, Adams PD, Horwich AL, Sigler PB (2003) Role of the γ-phosphate of ATP in triggering protein folding by GroEL-GroES: function, structure and energetics. EMBO J 22:4877-4887

Chaudhuri TK, Farr GW, Fenton WA, Rospert S, Horwich AL (2001) GroEL/GroES-mediated folding of a protein too large to be encapsulated. Cell 107:235-246

Chen L, Sigler PB (1999) The crystal structure of a GroEL/peptide complex: plasticity as a basis for substrate diversity. Cell 99:757-768

Chen J, Walter S, Horwich AL, Smith D (2001) Folding of malate dehydrogenase inside the GroEL-GroES cavity. Nat Struct Biol 8:721-728

Cheng MY, Hartl FU, Martin J, Pollock RA, Kalousek F, Neupert W, Hallberg RL, Horwich AL (1989) Mitochondrial heat-shock protein hsp60 is essential for assembly of proteins imported into yeast mitochondria. Nature 337:620-625

Cliff MJ, Kad NM, Hay N, Lund PA, Webb MR, Burston SG, Clarke AR (1999) A kinetic analysis of the nucleotide-induced allosteric transitions of GroEL. J Mol Biol 293:667-684

Ditzel L, Lowe J, Stock D, Stetter KO, Huber H, Huber R, Steinbacher S (1998) Crystal structure of the thermosome, the archaeal chaperonin and homolog if CCT. Cell 93:125-138

Dobrzynski JK, Sternlicht ML, Farr GW, Sternlicht H (1996) Newly synthesized β-tubulin demonstrates domain-specific interactions with the cytosolic chaperonin. Biochemistry 35:15870-15882

Dunn AY, Melville MW, Frydman J (2001) Review: cellular substrates of the eukaryotic chaperonin TriC/CCT. J Struct Biol 135:176-184

Ellis RJ (1990) The molecular chaperone concept. Semin Cell Biol 1:1-9

Falke S, Fisher MT, Gogol EP (2001) Structural changes in GroEL effected by binding of a denatured protein substrate. J Mol Biol 308:569-577

Falke S, Tama F, Brooks III CL, Gogol EP, Fisher MT (2005) The 13Å structure of a chaperonin GroEL-protein substrate complex by cryo-electron microscopy. J Mol Biol 348:219-230

Farr GW, Scharl EC, Schumacher RJ, Sondek S, Horwich AL (1997) Chaperonin-mediated folding in the eukaryotic cytosol proceeds through rounds of release of native and non-native forms. Cell 89:927-937

Farr GW, Furtak K, Rowland MB, Ranson NA, Saibil HR, Kirchausen T, Horwich AL (2000) Multivalent binding of nonnative substrate proteins by the chaperonin GroEL. Cell 100:561-573

Farr GW, Fenton WA, Chaudhuri TK, Clare DK, Saibil HR, Horwich AL (2003) Folding with and without encapsulation by *cis*- and *trans*-only GroEL-GroES complexes. EMBO J 22:3220-3230

Feldman DE, Thulasiraman V, Ferreyra RG, Frydman J (1999) Formation of the VHL-elongin BC tumor soppressor complex is mediated by the chaperonin TriC. Mol Cell 4:1051-1061

Feldman DE, Spiess C, Howard DE, Frydman J (2003) Tumorigenic mutations in VHL disrupt folding *in vivo* by interfering with chaperonin binding. Mol Cell 12:1213-1224

Fenton WA, Kashi Y, Furtak K, Horwich AL (1994) Residues in chaperonin GroEL required for polypeptide binding and release. Nature 371:614-619

Figuereido L, Klunker D, Ang D, Naylor DJ, Kerner MJ, Georgopulos C, Hartl FU, Hayer-Hartl M (2004) Functional characterization of an archaeal GroEL/GroES chaperonin system. J Biol Chem 279:1090-1099

Fisher MT (1992) Promotion of the *in vitro* renaturation of dodecameric glutamine synthetase from *Escherichia coli* in the presence of GroEL (chaperonin-60) and ATP. Biochemistry 31:3955-3963

Franzetti B, Schoehn G, Ebel C, Gagnon J, Ruigrok RW, Zaccai G (2001) Characterization of a novel complex from halophilic archaeabacteria, which displays chaperone-like activities *in vitro*. J Biol Chem 276:29906-29914

Frydman J, Nimmesgern E, Erdjument-Bromage H, Wall JS, Tempst P, Hartl FU (1992) Function in protein folding of TRiC, a cytosolic ring complex containing TCP-1 and structurally related subunits. EMBO J 11:4767-4778

Frydman J, Nimmesgern E, Ohtsuka K, Hartl FU (1994) Folding of nascent polypeptide chains in a high molecular mass assembly with molecular chaperones. Nature 370:111-117

Gao Y, Thomas JO, Chow RL, Lee GH, Cowan NJ (1992) A cytoplasmic chaperonin that catalyzes β-actin folding. Cell 69:1043-1050

Georgopoulos C, Hendrix RW, Casjens SR, Kaiser AD (1973) Host participation in bacteriophage lambda head assembly. J Mol Biol 76:45-60

Georgopoulos CP, Hohn B (1978) Identification of a host protein necessary for bacteriophage morphogenesis (the GroE gene product). Proc Natl Acad Sci USA 75:131-135

Georgopoulos C, Tilly K (1981) Bacteriophage-host interactions in assembly. Prog Clin Biol Res 64:21-34

Goloubinoff P, Christeller JT, Gatenby AA, Lorimer GH (1989) Reconstitution of active dimeric ribulose bisphosphate carboxylase from an unfolded state depends on two chaperonin proteins and Mg-ATP. Nature 342:884-889

Gray TE, Fersht AR (1991) Cooperativity in ATP hydrolysis by GroEL is increased by GroES. FEBS Lett 292:254-258

Guagliardi A, Cerchia L, Bartolucci S, Rossi M (1994) The chaperonin from the archaeon *Sulfolobus solfataricus* promotes correct refolding and promotes thermal denaturation *in vitro*. Protein Sci 3:1436-1443

Gupta RS (1990) Sequence and structural homology between a mouse t-complex proteinTCP-1 and the 'chaperonin' family of bacterial (GroEL, 60-65 kDa heat shock antigen) and eukaryotic proteins. Biochem Int 20:833-841

Gupta RS (1995) Evolution of the chaperonin families (Hsp60, Hsp10 and Tcp-1) of proteins and the origin of the eukaryotic cells. Mol Microbiol 15 1-11

Gutsche I, Essen L-O, Baumeister W (1999) Group II chaperonins: new TriC(k)s and turns of a protein folding machine. J Mol Biol 293:295-312

Gutsche I, Mihalache O, Hegerl R, Typke D, Baumeister W (2000a) ATPase cycle controls the conformation of an archaeal chaperonin as visualized by cryo-electron microscopy. FEBS Lett 477:278-282

Gutsche I, Holzinger J, Rossle M, Heumann H, Baumeister W, May RP (2000b) Conformational rearrangements of an archaeal chaperonin upon ATPase cycling. Curr Biol 10:405-408

Gutsche I, Mihalache O, Baumeister W (2000c) ATPase cycle of an archaeal chaperonin. J Mol Biol 300:187-196

Gutsche I, Holzinger J, Rauh N, Baumeister W, May RP (2001) ATP-induced structural change of the thermosome is temperature-dependent. J Struct Biol 135:139-146

Hansen WJ, Cowan NJ, Welch WJ (1999) Prefoldin nascent chain complexes in the folding of cytoskeletal proteins. J Cell Biol 145:265-277

Hill JE, Penny SL, Crowell KG, Goh SH, Hemmingsen SM (2004) cpnDB: a chaperonin sequence database. Genome Res 14:1669-1675

Horovitz A, Amir A, Danziger O, Kafri G (2002) Phi-value analysis of heterogeneity in pathways of allosteric transitions: Evidence for parallel pathways of ATP-induced conformational changes in a GroEL ring. Proc Natl Acad Sci USA 99:14095-14097

Horwich AL, Willison KR (1993) Protein folding in the cell: functions of two families of molecular chaperone, hsp 60 and TF55-TCP-1. Philos Trans Roy Soc Lond B Biol Sci 339:313-326

Horwich AL, Low KB, Fenton WA, Hirshfield IN (1993) Folding *in vivo* of bacterial cytoplasmic proteins: role of GroEL. Cell 74:909-917

Houry WA, Frishman D, Eckerskorn C, Lottspeich F, Hartl FU (1999) Identification of *in vivo* substrates of the chaperonin GroEL. Nature 402:147-154

Hunt JF, Weaver AJ, Landry SJ, Gierasch L, Deisenhofer J (1996) The crystal structure of the GroES co-chaperonin at 2.8Å. Nature 379:37-45

Hunt JF, van der Vies SM, Henry L, Deisenhofer J (1997) Structural adaptations in the specialized bacteriophage T4 co-chaperonin Gp31 expand the size of the Anfinsen cage. Cell 90:361-371

Hynes G, Kubota H, Willison KR (1995) Antibody characterisation of two distinct conformations of the chaperonin-containing TCP-1 from mouse testis. FEBS Lett 358:129-132

Iizuka R, So S, Inobe T, Yoshida T, Zako T, Kuwajima K, Yohda M (2004) Role of the helical protrusion in the conformational change and molecular chaperone activity of the archaeal Group II chaperonin. J Biol Chem 279:18834-18839

Inobe T, Makio T, Takasu-Ishikawa E, Terada TP, Kuwajima K (2001) Nucleotide binding to the chaperonin GroEL: non-cooperative binding of ATP analogs and ADP, and cooperative effect of ATP. Biochim Biophys Acta 1545:160-173

Ionobe T, Arai M, Nakao M, Ito K, Kamagata K, Makio T, Amemiya Y, Kihara H, Kuwajima K (2003) Equilibrium and kinetics of the allosteric transition of GroEL studied by solution X-ray scattering and fluorescence spectroscopy. J Mol Biol 327:183-191

Jackson GS, Staniforth RA, Halsall DJ, Atkinson T, Holbrook JJ, Clarke AR, Burston SG (1993) Binding and hydrolysis of nucleotides in the chaperonin catalytic cycle: implications for the mechanism of assisted protein folding. Biochemistry 32:2554-2563

Kafri G, Willison KR, Horovitz A (2001) Nested allosteric interactions in the cytoplasmic chaperonin containing TCP-1. Protein Sci 10:445-449

Kafri G, Horovitz A (2003) Transient kinetic analysis of ATP-induced allosteric transitions in the eukaryotic chaperonin containing TCP-1. J Mol Biol 326:981-987

Kim S, Willison KR, Horwich AL (1994) Cytosolic chaperonin subunits have a conserved ATPase domain but diverged polypeptide-binding domains. Trends Biochem Sci 19:543-548

Klumpp M, Baumeister W, Essen LO (1997) Structure of the substrate binding domain of the thermosome, an archaeal Group II chaperonin. Cell 91:263-270

Klunker D, Haas B, Hirtreiter A, Figueiredo L, Naylor DJ, Pfeifer G, Muller V, Deppenmeier U, Gottschalk G, Hartl FU, Hayer-Hartl M (2003) Coexistence of Group I and Group II chaperonins in the archaeon *Methanosarcina mazei*. J Biol Chem 278:33256-33267

Koshland DE, Nemethy G, Filmer D (1966) Comparison of experimental binding data and theoretical models in proteins containing subunits. Biochemistry 5:365-385

Kubota H, Hynes G, Willison K (1995) The chaperonin containing t-complex polypeptide 1 (TCP-1): Multisubunit machinery assisting in protein folding and assembly in the eukaryotic cytosol. Eur J Biochem 230:3-16

Kubota H, Hynes G, Carne A, Ashworth A, Willison K (1994) Identification of six Tcp-1-related genes encoding divergent subunits of the TCP-1-containing chaperonin. Curr Biol 4:89-99

Kubota H, Hynes GM, Kerr SM, Willison KR (1997) Tissue-specific subunit of the mouse cytosolic chaperonin-containing TCP-1. FEBS Lett 402:53-56

Kuo YP, Thompson DK, St Jean A, Charlebois RL, Daniels CJ (1997) Characterization of two heat shock genes from *Haloferax volcanii*: a model system for transcription regulation in the Archaea. J Bacteriol 179:6318-6324

Landry SJ, Gierasch L (1991) The chaperonin GroEL binds a polypeptide in an α-helical conformation. Biochemistry 30:7359-7362

Landry SJ, Zeilstra-Ryalls J, Fayet O, Georgopoulos C, Gierasch L (1993) Characterization of a functionally important mobile domain of GroES. Nature 364:255-258

Leroux MR, Fandrich M, Klunker D, Siegers K, Lupas AN, Brown JR, Schiebel E, Dobson CM, Hartl FU (1999) MtGimC, a novel archaeal chaperone related to the eukaryotic chaperonin cofactor GimC/prefoldin. EMBO J 18:6730-6743

Lewis VA, Hynes GM, Zheng D, Saibil H, Willison K (1992) T-complex polypeptide-1 is a subunit of a heteromeric particle in the eukaryotic cytosol. Nature 358:249-252

Lin Z, Schwarz FP, Eisenstein E (1995) The hydrophobic nature of GroEL-substrate binding. J Biol Chem 270:1011-1014

Lin P, Sherman F (1997) The unique hetero-oligomeric nature of the subunits in the catalytic cooperativity of the yeast CCT chaperonin complex. Proc Natl Acad Sci USA 94:10780-10785

Lin Z, Rye HS (2004) Expansion and compression of a folding intermediate by GroEL. Mol Cell 16:23-34

Liou AK, Willison KR (1997) Elucidation of the subunit orientation in CCT (chaperonin containing TCP1) from the subunit composition of CCT micro-complexes. EMBO J 16:4311-4316

Llorca O, Smyth MG, Marco S, Carrascosa JL, Willison KR, Valpuesta JM (1998) ATP binding induces large conformational changes in the apical and equatorial domains of the eukaryotic chaperonin containing TCP-1 complex. J Biol Chem 273:10091-10094

Llorca O, McCormack EA, Hynes G, Grantham J, Cordell J, Carrascosa JL, Willison KR, Fernandez JJ, Valpuesta JM (1999a) Eukaryotic type II chaperonin CCT interacts with actin through specific subunits. Nature 402:693-696

Llorca O, Smyth MG, Carrascosa JL, Willison KR, Radermacher M, Steinbacher S, Valpuesta JM (1999b) 3D reconstruction of the ATP-bound form of CCT reveals the asymmetric folding conformation of a type II chaperonin. Nat Struct Biol 6:639-642

Llorca O, Martin-Benito J, Ritco-Vonsovici M, Grantham J, Hynes GM, Willison KR, Carrascosa JL, Valpuesta JM (2000) Eukaryotic chaperonin CCT stabilizes actin and tubulin folding intermediates in open quasi-native conformations. EMBO J 19:5971-5979

Llorca O, Martin-Benito J, Grantham J, Ritco-Vonsovici M, Willison KR, Carrascosa JL, Valpuesta JM (2001) The 'sequential allosteric ring' mechanism in the eukaryotic chaperonin-assisted folding of actin and tubulin. EMBO J 20:4065-4075

Lopez-Garcia P, Moreira D (1999) Metabolic symbiosis at the origin of eukaryotes. Trends Biochem Sci 24:88-93

Lorimer GH (2001) A personal account of chaperonin history. Plant Physiol 125:38-41

Maeder DL, Macario AJL, Conway de Macario E (2005) Novel chaperonins in a Prokaryote. J Mol Evol 60:409-416

Margulis L (1971) Symbiosis and evolution. Sci Am 225:48-57

Martin J, Langer T, Boteva R, Schramel A, Horwich AL, Hartl FU (1991) Chaperonin-mediated protein folding at the surface of GroEL through a 'molten-globule'-like intermediate. Nature 352:36-42

Martin-Benito J, Boskovic J, Gomez-Puertas P, Carrascosa JL, Simons CT, Lewis SA, Bartolini F, Cowan NJ, Valpuesta JM (2002) Structure of eukaryotic prefoldin and of its complexes with unfolded actin and the cytosolic chaperonin CCT. EMBO J 21:6377-6386

Mayhew M, da Silva AC, Martin J, Erdjument-Bromage H, Tempst P, Hartl FU (1996) Protein folding in the central cavity of the GroEL-GroES chaperonin complex. Nature 379:420-426

McCallum CD, Do H, Johnson AE, Frydman J (2000) The interaction of the chaperonin tailless complex polypeptide 1 (TCP1) ring complex (TRiC) with ribosome-bound nascent chains examined using photo-cross-linking. J Cell Biol 149:591-601

Meyer AS, Gillespie JR, Walther D, Millet IS, Doniach S, Frydman J (2003) Closing the folding chamber of the eukaryotic chaperonin requires the transition state of ATP hydrolysis. Cell 113:369-381

Minuth T, Henn M, Rutkat K, Andra S, Frey G, Rachel R, Stetter KO, Jaenicke R (1999) The recombinant thermosome from the hyperthermophilic archaeon *Methanopyrus kandleri*: *in vitro* analysis of its chaperone activity. Biol Chem 380:55-62

Monod J, Wyman J, Changeux J-P (1965) On the nature of allosteric transitions: a plausible model. J Mol Biol 12:88-118

Motojima F, Chaudhry C, Fenton WA, Farr GW, Howich AL (2004) Substrate polypeptide presents a load on the apical domains of the chaperonin GroEL. Proc Natl Acad Sci USA 101:15005-15012

Netzer WJ, Hartl FU (1997) Recombination of protein domains facilitated by co-translational folding in eukaryotes. Nature 388:343-349

Nitsch M, Klumpp M, Lupas A, Baumeister W (1997) The thermosome: alternating α and β-subunits within the chaperonin of the archaeon *Thermoplasma acidophilum*. J Mol Biol 267:142-149

Nitsch M, Walz J, Typke D, Klumpp M, Essen LO, Baumeister W (1998) Group II chaperonin in an open conformation examined by electron tomography. Nat Struct Biol 5:855-857

Ostermann J, Horwich AL, Neupert W, Hartl FU (1989) Protein folding in the mitochondria requires complex formation with hsp60 and ATP hydrolysis. Nature 341:125-130

Phipps BM, Hoffmann A, Stetter KO, Baumeister W (1991) A novel ATPase complex selectively accumulated upon heat-shock is a major cellular component of thermophilic archaeabacteria. EMBO J 10:1711-1722

Poso D, Clarke AR, Burston SG (2004a) A kinetic analysis of the nucleotide-induced allosteric transitions in a single-ring mutant of GroEL. J Mol Biol 338:969-977

Poso D, Clarke AR, Burston SG (2004b) Identification of a major inter-ring coupling step in the GroEL reaction cycle. J Biol Chem 279:38111-38117

Ranson NA, Dunster NJ, Burston SG, Clarke AR (1995) Chaperonins can catalyse the reversal of early aggregation steps when a protein misfolds. J Mol Biol 250:581-586

Ranson NA, Burston SG, Clarke AR (1997) Binding, encapsulation and ejection: substrate dynamics during a chaperonin-assisted folding reaction. J Mol Biol 266:656-664

Ranson NA, Farr GW, Roseman AM, Gowen B, Fenton WA, Horwich AL, Saibil HR (2001) ATP-bound states of GroEL captured by cryo-electron microscopy. Cell 107:869-879

Rivenzon-Segal N, Wolf SG, Shimon L, Willison KR, Horovitz A (2005) Sequential ATP-induced allosteric transitions of the cytoplasmic chaperonin containing TCP-1 revealed by EM analysis. Nat Struct Mol Biol 12:233-237

Rommelaere H, De Neve M, Melki R, Vandekerckhove J, Ampe C (1999) The cytosolic class II chaperonin CCT recognizes delineated hydrophobic sequences in its target proteins. Biochemistry 38:3246-3257

Roseman AM, Chen S, White HE, Braig K, Saibil HR (1996) The chaperonin ATPase cycle: mechanism of allosteric switching and movements of substrate-binding domains in GroEL. Cell 87:241-251

Rye HS, Burston SG, Fenton WA, Beechem JM, Xu Z, Sigler PB, Horwich AL (1997) Distinct actions of cis and trans ATP within the double ring of the chaperonin GroEL. Nature 388:792-797

Rye HS, Roseman AM, Chen S, Furtak K, Fenton WA, Saibil HR, Horwich AL (1999) GroEL-GroES cycling: ATP and nonnative polypeptide direct alternation of folding-active rings. Cell 97:325-338

Saibil H, Dong Z, Wood S, auf der Mauer A (1991) Binding of chaperonins. Nature 353:25-26

Saibil, HR, Zheng D, Roseman AM, Hunter AS, Watson GM, Chen S, auf der Mauer A, O'Hara BP, Wood SP, Mann NH, Barnett LK, Ellis RJ (1993) ATP induces large quaternary rearrangements in a cage-like chaperonin structure. Curr Biol 3:265-273

Schmidt M, Buchner J (1992) Interaction of GroE with an all β-protein. J Biol Chem 267:16829-16833

Schoehn G, Quaite-Randall E, Jimenez JL, Joachimiak A, Saibil HR (2000a) Three conformations of an archaeal chaperonin, TF55 from *Sulfolobus shibatae.* J Mol Biol 296:813-819

Schoehn G, Hayes M, Cliff M, Clarke AR, Saibil HR (2000b) Domain rotations between open, closed and bullet-shaped forms of the thermosome, an archaeal chaperonin. J Mol Biol 301:323-332

Shtilerman M, Lorimer GH, Englander SW (1999) Chaperonin function: folding by forced unfolding. Science 284:822-825

Siegers K, Bolter B, Schwarz JP, Bottcher UM, Guha S, Hartl FU (2003) TRiC/CCT cooperates with different upstream chaperones in the folding of distinct protein classes. EMBO J 22:5230-5240

Siegert R, Leroux MR, Scheufler C, Hartl FU, Moarefi I (2000) Structure of the molecular chaperone prefoldin: unique interaction of multiple coiled coil tentacles with unfolded proteins. Cell 103:621-632

Silver LM, Artzt K, Bennet D (1979) A major testicular cell protein specified by a mouse T/t complex gene. Cell 17:275-284

Silver LM, Kleene KC, Distel RJ, Hecht NB (1987) Synthesis of mouse t complex proteins during haploid stages of spermatogenesis. Dev Biol 119:605-608

Staniforth RA, Burston SG, Atkinson T, Clarke AR (1994) Affinity of chaperonin-60 for a protein substrate and its modulation by nucleotides and chaperonin-10. Biochem J 300:651-658

Szpikowska BK, Swiderek KM, Sherman MA, Mas MT (1998) MgATP binding to the nucleotide-binding domains of the eukaryotic cytoplasmic chaperonin induces conformational changes in the putative substrate-binding domains. Protein Sci 7:1524-1530

Thulasiraman V, Yang CF, Frydman J (1999) *In vivo* newly translated polypeptides are sequestered in a protective folding environment. EMBO J 18:85-95

Tilly K, McKittrick N, Georgopoulos C, Murialdo H (1981) Studies on *Escherichia coli* mutants which block bacteriophage morphogenesis. Prog Clin Biol Res 64:35-45

Todd MJ, Viitanen PV, Lorimer GH (1993) Hydrolysis of adenosine 5'-triphosphate by *Escherichia coli* GroEL: effects of GroES and potassium ion. Biochemistry 32:8560-8567

Todd MJ, Viitanen PV, Lorimer GH (1994) Dynamics of the chaperonin ATPase cycle: implications for facilitated protein folding. Science 265:659-666

Trent JD, Nimmesgern E, Wall JS, Hartl FU, Horwich AL (1991) A molecular chaperone from a thermophilic archaeabacterium is related to the eukaryotic protein t-complex polypeptide-1. Nature 354:490-493

Ursic D, Culbertson MR (1992) Is yeast TCP1 a chaperonin? Nature 356:392

Ursic D, Culbertson MR (1991) The yeast homolog to mouse Tcp-1 affects microtubule-mediated processes. Mol Cell Biol 11:2629-2640

Ursic D, Ganetzky B (1988) A *Drosophila melanogaster* gene encodes a protein homologous to the mouse t-complex polypeptide 1. Gene 68:267-274

Waldmann T, Nimmesgern E, Nitsch M, Peters J, Pfeifer G, Muller S, Kellermann J, Engel A, Hartl FU, Baumeister W (1995a) The thermosome of *Thermoplasma acidophilum* and its relationship to the eukaryotic chaperonin TRiC. Eur J Biochem 227:848-856

Waldmann T, Lupas A, Kellermann J, Peters J, Baumeister W (1995b) Primary structure of the thermosome from *Thermoplasma acidophilum*. Biol Chem Hoppe-Seyler 376:119-126

Wang J, Boisvert DC (2003) Structural basis for GroEL-assisted protein folding from the crystal structure of $(GroEL\text{-}KMgATP)_{14}$ at 2.0Å resolution. J Mol Biol 327:843-855

Wang J, Chen L (2003) Domain motions in GroEL upon binding of an oligopeptide. J Mol Biol 334:489-499

Wang JD, Herman C, Tipton KA, Gross CA, Weissman JS (2002) Directed evolution of substrate-optimized GroEL/S chaperonins. Cell 111:1027-1039

Weissman JS, Kashi Y, Fenton WA, Horwich AL (1994) GroEL-mediated protein folding proceeds by multiple rounds of binding and release of nonnative forms. Cell 78:693-702

Weissman JS, Hohl CM, Kovalenko O, Kashi Y, Chen S, Braig K, Saibil HR, Fenton WA, Horwich AL (1995) Mechanism of GroEL action: productive release of polypeptide from a sequestered position under GroES. Cell 83:577-587

Weissman JS, Rye HS, Fenton WA, Beechem JM, Horwich AL (1996) Characterization of the active intermediate of a GroEL-GroES-mediated protein folding reaction. Cell 84:481-490

Willison K, Kelly A, Dudley K, Goodfellow P, Spurr N, Groves V, Gorman P, Sheer D, Trowsdale J (1987) The human homologue of the mouse t-complex gene, TCP1, is located on chromosome 6 but is not near the HLA region. EMBO J 6:1967-1974

Willison KR, Grantham J (2001) In: Lund P (Ed), Molecular Chaperones: Frontiers in Molecular Biology. Oxford University Press, Oxford, pp 90-118

Won KA, Schumacher RJ, Farr GW, Horwich AL, Reed SI (1998) Maturation of human cyclin E requires the function of eukaryotic chaperonin CCT. Mol Cell Biol 18:7584-7589

Yaffe MB, Farr GW, Miklos D, Horwich AL, Sternlicht ML, Sternlicht H (1992) TCP1 complex is a molecular chaperone in tubulin biogenesis. Nature 358:245-248

Zako T, Iizuka R, Okochi M, Nomura T, Ueno T, Tadakuma H, Yohda M, Funatsu T (2005) Facilitated release of substrate protein from prefoldin by chaperonin. FEBS Lett 579:3718-3724

Bigotti, M. Giulia

Department of Biochemistry, University of Bristol, School of Medical Sciences, Bristol BS8 1TD, UK

Burston, Steven G.

Department of Biochemistry, University of Bristol, School of Medical Sciences, Bristol BS8 1TD, UK
s.g.burston@bristol.ac.uk

Clarke, Anthony R.

Department of Biochemistry, University of Bristol, School of Medical Sciences, Bristol BS8 1TD, UK

Index

α-1,2-mannosidase, 41, 197
α-casein degradation, 127
α-galactosidase A, 210
α1-antitrypsin deficiency, 91, 207
α_1-AT. *See* α1-antitrypsin
αB-crystallin, 14
β-glucosidase, 210
β-oxidation of fatty acids, 150
γ-crystallins, 18
σ^{32} degradation, 3
σ^{32} factor, 1-3
ΔF508CFTR, 207

[Het-s] phenotype, 227
[PIN], 225
[*PSI*$^+$], 77, 79, 224
  Hsp104, 79
[PSI+] prion, 78, 83
  role in translation termination, 78
[*PSI*$^+$] stability
  Hsp104, 80
[*URE3*], 78, 224

“bullet” shape, 256
“football”-shape, 256
“seeded” conversion, 224

1,2-mannosidase, 97
14-3-3ε, 9
19S cap, 190
1-desoxy-galactonojirimycin, 210

20S proteasome, 190
26S proteasome, 134, 190
28S rRNA, 49

3-ketoacyl-CoA thiolase, 157

9G8, 21

AAA domain, 126, 128
  ATP-dependent conformational changes, 134
  chaperone activity, 132
  pore, 135
AAA family, 120
AAA module
  architecture, 71
  Hsp104, 72
  nucleotide status, 73
  positive cooperativity, 73
AAA proteases, 119, 123, 126, 127, 131-133
  assembly, 134
  domain structure, 128
  inner mitochondrial membrane, 127
  internal cavity, 134
  mechanism, 135
  membrane topology of substrate, 133
  mitochondria, 125
  oligomeric status, 127
  structure, 135
  substrate degradation, 133
  substrate dislocation, 135
  substrate recognition, 132, 133
  substrate specificity, 133
AAA superfamily, 70
AAA-ATPase, 194
AARE, 45
AB toxins, 196, 208
acceleration
  prion assembly, 82
accumulation of aggregated protein, 82
aconitase, 124
actin, 265, 273
activation domain
  HSF1, 6
AD. *See* activation domain
ADA5, 38
ADP
  GroEL, 254
ADP-ATP exchange, 169
Afg3, 128
AFG3L1, 130
AFG3L2, 130
aggregated GFP refolding, 76
aggregates, 74, 76, 81
  heat-induced, 74
  mitochondria, 131
  structure, 77
  toxicity, 84

aggregation, 221, 263
polyglutamine proteins, 83
aging, 20
AGT, 160
alanine:glyoxylate aminotransferase 1, 160
ALR, 101
Alzheimer's disease, 65, 82, 94, 207
amino acid responsive element, 45
amino acid starvation, 39
aminopterin, 160
amyloid, 230
fibrils, 231
precursor protein, 94
amyloid-like aggregation, 83
amyloid-like fibrils, 82
amyloidogenic nuclei, 81
amyloidoses, 207
Anfinsen cage, 263
ankylosing spondylitis, 91
antibodies, 92
antibody secretion, 53
antigen presentation, 94
AP1 (Activating Protein-1) transcription factor, 49
apical insertion model
AAA proteases, 134, 135
apical sorting, 93
Apo-lipoprotein B, 104
apoptosis
HSF1, 15
UPR mediated, 47
APP, 94
archaebacterial chaperonin, 264
ASC-2, 11
assembly
Lon, 124
assembly of respiratory chain complexes
prohibitins, 131
astrocytes, 137
ATF4, 43-45, 46, 52
ATF6, 35, 40-43, 46, 48, 51-53
BiP, 50
isoforms, 43
transport to the Golgi, 51
ATF6 signaling, 50
ATP, 100
GroEL, 254
ATP hydrolysis, 262
AAA proteins, 71, 127
Hsp104, 73
Atp7, 134
ATPase activity, 75
Hsp104, 73, 80
mtHsp70, 121
ATPase associated with various activities. *See* AAA-ATPase
ATPase/protease complex, 75
ATPases, 72, 100, 162, 170, 252, 264
ATPases Associated with a a variety of cellular Activities, 70, *See* AAA proteins
ATP-dependent proteases, 120, 122, 136
in mitochondria, 120, 122
autocatalytic cleavage
presequence, 124
auxillin, 165
axial channel
Hsp100/Clp, 75

bacteriophage λ, 251
BAG-1, 106, 187, 193
BAP, 100
baso-lateral sorting, 93
B-cells, 48, 53
bichaperone network, 70, 75, 76
BiP, 35, 39, 41, 42, 47, 92, 95, 100, 106, 166
ATF6, 42, 50
binding to unfolded proteins, 50
IRE1, 50
PERK, 50
BiP/Kar2p, 195
BiP-associated Protein, 100
bovine viral diarrhea virus, 50

$C(x)_nC$, 101
Ca^{2+}/calmodulin-dependent kinase II, 10
CAG expansion, 83
cage of infinite dilution, 261
Cajal bodies, 21
calcium, 99
calcium release
in UPR, 47
calnexin, 35, 39, 96, 98, 106, 195
calcium, 99
cycle, 94, 97, 99
P-domain, 98
transmembrane segments, 99
calnexin/calreticulin-cycle, 197
calreticulin, 35, 39, 96, 98, 195

calcium, 99
CaMK II, 10
catalase, 150, 165
cataract, 18
CBP, 21
CCAAT binding factor, 42
CCAAT box, 4
Ccp1, 133, 134, 136
CCT, 251, 265
ATP, 269
size of cavity, 273
structure, 267
substrate binding, 269
subunit composition, 266
CD82, 99
Cdc48p, 194, 205
Cdc48p/p97, 185, 204, 206
Cdc48p/Ufd1p/Npl4p-complex, 199
Cdk9-CyclinT1, 11
cell cycle arrest, 19, 44, 45, 48
cell cycle arrest in UPR, 40
cell growth
HSF in, 14
cell stress, 199
cell viability
HSF in, 14
Cer1p, 197
CFTR, 199, 200, 207, 209
chaperone network, 105
mitochondria, 123
chaperones, 68, 185, 197
endoplasmic reticulum, 35, 91
in prion assembly, 235
in prion stability, 234
in quality control, 186
chaperonin, 2, 121, 251
archaebacterial, 265
Group I, 251
Group II, 251
Chaperonin Containing TCP-1. *See* CCT
chemical chaperone, 65, 209
CHIP, 13, 106, 191, 200, 208
chloroplasts, 252
cholera toxin, 102, 208
cholesterol biosynthesis, 43
CHOP, 45, 46, 137
CHOP/GADD153, 44
chromatin, 21
chromatin-remodeling complex, 11
chromosome puffs, 3
chronic wasting disease, 221
CIRCE elements, 2
CJD, 221
clathrin, 165
Clp, 119
Clp adaptor proteins, 76
Clp proteases
mitochondrial, 126
ClpA, 71, 72, 75
substrate recognition, 74
ClpB, 70, 73, 122
protein unfolding/threading, 75
substrate specificity, 73, 74
thermotolerance, 74
ClpB/ClpP complex, 75, 76
ClpB/Hsp100, 122
ClpP protease, 75, 125, 126, 137
oligomeric status, 127
ClpX, 71, 72, 75, 127
substrate recognition, 74
ClpXP protease, 127
ClpY, 71
CNX. *See* calnexin
COB, 126
co-chaperones, 121, 194
coiled-coil, 7, 72, 127
complex I deficiencies, 130
conformational diseases, 221
Congo red, 229
copper homeostasis, 223
Cox1, 126, 130
Cox2, 130
Cpn10, 137
Cpn60, 137
CPVT, 95
CPY*, 199, 206
CREB, 44
CREB/ATF family transcription factor, 40, 42
Creutzfeldt Jacob's disease, 82, 91, 221
CRN, 194
cross-ß structure, 230
CRT. *See* calreticulin
Cta1p, 165
Cue1p, 199
curing of [*PSI*$^{+}$], 79
CWD, 221
CxxC motif, 101, 105
CxxCxxC motif, 101
CYC4, 194
cyclophilin, 95
cyclophilin B, 104, 105

cyclophilin-60, 194
cyclosporin A, 95
cystathionine β-synthetase, 43
cystic fibrosis, 36, 91, 207
cystic fibrosis transmembrane receptor. *See* CFTR
cytochrome *c* peroxidase, 132
cytochrome oxidase, 130, 137

DAF-16, 20
Dal5p, 227
DAXX, 13
DBD. *See* DNA-binding domain
degradation, 120, 192, 198
  from ER, 97, 195
  integral membrane proteins, 131, 134, 199
  misfolded mitochondrial proteins, 122
  misfolded proteins, 124, 136
  mitochondria, 121
  mtHsp70, 122
  of aggregated protein, 76
degradation capacity
  mitochondria, 136
dentorubral and pallidoluysian atrophy, 83
Der1p, 198, 199, 201
Derlin-1, 201
detoxification, 102
development
  HSF in, 14
  IRE1α, 52
developmental delay, 83
DGJ, 210
DHFR, 160
dioxin receptor, 171
disaggregation, 70, 74
diseases
  degradation, 206
dislocation, 200, 204
  during AAA proteolysis, 133, 134
  membrane proteins, 134
dislocon, 200
disulfide bond formation, 35, 43, 100
disulfide bonds
  in regulation, 101
disulfide isomerisation, 103
dithiol-disulfide exchange, 101
Dlp1p, 155
DMSO, 209
DNA polymerase γ, 126
DNA-binding domain
  HSF, 15
  HSF1, 6
DnaJ, 2, 3, 46, 99, 121, 164, 168
DnaJc7, 194
DnaK, 2, 3, 76, 164
  ClpB, 70
  in protein refolding, 76
DnaK/DnaJ, 76, 150
Doa10p, 200, 203, 204
dolichol phosphate, 97
dopaminergic neurons, 208
DRPLA, 83
Dsk2p, 206
DT40 cells, 17
dTRAP80, 11
DTT
  and UPR, 52
dynamin, 155

E1, 190
E2, 190, 199
E2 Hepatitis C virus envelope glycoprotein, 50
E3, 202
E3 ligase, 190, 199, 200
E3 ubiquitin ligases, 202
E4, 194
EDEM, 41, 46, 97, 197
eIF2α, 39, 41, 44, 46, 49, 52
  phosphorylation by PERK, 45
EKN1, 193
emphysema, 36
EndoPDI, 105
endoribonuclease, 35, 37, 38, 49
endosymbiotic, 265
epigenetic, 237
Eps1p, 103, 198
ER, 35, 47, 91, 100, 126, 224
  dislocation, 134
  membrane proliferation, 47
  protein folding capacity, 35, 36, 39
  redox proteins, 101
  retention signal, 92
  stress, 19, 35
  stress response element. *See* ERSE
ER associated degradation. *See* ERAD
ER degradation enhancing α-mannosidase like protein. *See* EDEM

ERAD, 36, 46, 52, 91, 195, 100, 102, 106,
UPR, 37, 39
ERdJ3, 105
ERdJ4, 46
ERdJ5, 105
ERK, 10
Ero1p, 101
ERp18, 105
ERp28, 104, 105
ERp29, 105
ERp44, 105
ERp46, 105
ERp57, 96, 98, 103, 195
ERp72, 103, 105, 106
ER-peroxisome connection, 155
ERSE, 42, 46, 52
Erv2p, 101
euchromatin, 21
Eug1p, 103
exosome, 11

F_1F_O-ATP synthase, 130
Fabry's disease, 210
FAD
in Ero1, 101
fatal familial insomnia, 221
fertility, 15
FFI, 221
fiber depolymerization, 82
fibril formation, 81, 82
Hsp104, 81
fibrillar aggregates
Sup35p, 78
fibrillogenesis, 82
fibrils, 229, 230
PrP, 223
fidelity of translation termination, 77
FK506 binding-protein, 95
FKB23, 95
FKBP, 95
FKBP51, 13
FKBP52, 13, 171
flavin-containing mono-oxygenase, 102
flavopiridol, 11
FMO, 102
folding diseases, 186
endoplasmic reticulum, 91
folding intermediates, 236
FRET, 92, 260
FtsH, 2, 3, 127, 131, 132, 134

GADD153, 44
GADD45, 46
ganglioside
in UPR, 47
Gaucher's disease, 207, 210
GC box, 4
Gcn2, 39, 45
Gcn4p, 39
Gdn, 68, 77, 79
geldanamycin, 12, 171
Gerstmann-Sträussler-Scheinker syndrome, 221
GFP
fused to polyglutamine, 83
refolding of, 76, 263
giant mitochondria, 130
GimC, 273
Gln3p, 227
glucocorticoid receptor, 191, *See* GR
glucose limitation, 39
glucose regulated proteins, 92
glucose starvation, 10
glucosidase I, 97, 197
glucosidase II, 97, 197
glucosylceramide-β-glucosidase, 207
glutamine repeats, 83
glutathione, 103
glycerol, 209
glycogen synthase kinase-3, 10
glycoprotein fate, 97
glycoprotein folding, 36, 94, 96, 103
glycoproteins, 91, 96
glycosylation, 35
glycosylphosphatidyl inositol (GPI) linked protein, 92
glyoxisome, 168
gp23, 255
gp31, 255
gp96, 104, 105, 106
GR, 171
green fluorescent protein. *See* GFP
GroE, 2, 251
central cavity, 261
half-reaction, 260
kinetic pathway, 259
protein folding, 261
release of substrate, 262
GroEL, 2, 251
allostery in binding, 257
apical domain, 254

asymmetry in binding, 257
central cavity, 254
conformations, 258
cooperativity, 258
equatorial domain, 254
nucleotide, 256
protein substrate, 257
Substrate binding, 259
GroEL-GroES complex, 256
GroES, 2, 252
mobile loop region, 255
release of, 262
Group I chaperonins, 263
Reaction cycle, 259
Group II chaperonins, 264
allostery, 271
ATP, 269
nucleotide, 269
structure, 267
substrate binding, 269, 273
subunit composition, 266
GRP170, 105, 197
GRP78, 35, 41, 42, 47
GRP94, 42, 105
GRP98, 35, 47
GrpE, 2, 3, 121
GSH, 103
GSK-3, 10
GSS, 221
GSSG, 103
GT, 97, 197
GTP-hydrolysis, 204
guanidinium, 68, 77, 79, 80, 227

HA, 96
Hac1p, 37, 39, 48
HAP, 21
HCMV, 208
HCV, 94
HDAC, 43
heat shock, 8, 13, 17, 19, 21, 22, 65, 66, 136
heat shock element, 1, *See* HSE
heat shock elements, 4
heat shock factor. *See* HSF
heat shock protein, *See* Hsp
heat shock response, 1, 9, 12, 19
bacterial, 2
eukaryotic, 3
prokaryotic, 3
heat shock transcription factor. *See* HSF
heat stress, 122, 123, 124
HECT, 189, 203
HEDJ, 46
helicases, 135
heme-regulated kinase, 45
Hepatitis C virus, 94
heptad repeat
HSF, 6, 15
hereditary spastic paraplegia, 130
HERP, 46
HET-s, 230
Hip, 13, 171
histondeacetylase, 43
histone acetylation, 38
histone acetyltransferase CREB-binding protein, 21
histones, 21
HLA, 94
Hlj1p, 199
Hmg2p, 209
HMG-CoA reductase, 48
homocysteine, 43
homocysteinemia, 43
Hop, 13, 169-171
HR. *See* heptad repeat
HrcA repressor, 2, 3
Hrd1p, 199, 203
Hrd1p/Hrd3p, 204
Hrd3p, 203
HRI, 45
Hsc70, 106, 164, 200
Hsc73, 167
HSE, 1, 4, 7, 19, 21
HSF, 1, 4, 7
alternative splicing, 4
DNA-binding activity, 8
DNA-binding domain, 7, 15
fertility, 17
function independent of Hsps, 14
heptad repeats, 15, 16
oligomeric status, 7
oogenesis, 14
phosphorylation, 7
physiological functions, 14
structure, 5
sumoylation, 7
target genes, 18
trimerization, 15
HSF1, 6, 7, 9, 11, 12, 20
and transcription, 11
as global regulator, 19

as repressor, 18
disulfide bonds, 8
heterochromatin, 21
Hsp70 interaction, 13
Hsp90 interaction, 12, 13
localization, 21, 22
longevity, 20
oligomeric status, 13
phosphorylation, 8, 9, 10, 13
pro-apoptotic function, 15
regulation, 8, 12
regulation of activity, 10
satellite III repeats, 21
structure, 6, 7
transcriptional regulation, 18
trimerization, 8, 13
HSF2, 15-17
apoptosis, 17
erythroid differentiation, 15
fertility, 17
hemin-induced upregulation, 16
HSF1 cooperation, 17
oligomeric status, 16
ubiquitin-mediated proteasomal degradation, 16
HSF2-α, HSF2-β, 16
HSF3, 17
HSF4, 18
HSF4a, 5
HsfA1, 6
HsfA2, 6
HsfB1, 6
HSFs, 5, 22
novel functions in transcription, 18
plant, 5
HslU, 71, 72, 75
HslV protease, 75
Hsp10, 2
Hsp100, 122
Hsp100/Clp ATPases, 70
classes, 71
structure, 72
Hsp100-dependent proteases, 75
Hsp100s
adaptor proteins, 74
axial channel, 75
substrate recognition, 74
Hsp104, 122, 199, 234
Hsp70/Hsp40, 70
lag phase of fibril assembly, 81
Hsp104p, 65, 66, 77, 83, 84
[PSI+], 79, 80
allosteric control, 73
as chaperone, 66
dispersion of aggregates, 68
expressed in mammalian cells, 83
Hsp70, 76
molecular chaperone activity, 69
nucleotide-dependent assembly, 73
oligomeric status, 74
refolding activity, 68
remodeling of aggregates, 81
Sti1p, 77
structure, 72
substrate recognition, 74
substrate specificity, 73
substrates, 68
Sup35 interaction, 82
thermotolerance, 65, 67, 70
tryptophan fluorescence, 75
Hsp104p orthologs, 83
Hsp104p/ClpB, 77
oligomeric status, 74
Hsp104p-dependent protein refolding, 82
Hsp104p-mediated acceleration of fibril formation, 81
Hsp25, 14
Hsp26, 11
Hsp40, 2, 46, 105, 164, 171, 187, 197
Hsp60, 2, 121, 251, 266
mitochondrial, 252
protein substrate, 257
Hsp60/Hsp10, 123
Hsp70, 2, 12, 84, 92, 100, 162, 164, 167-172, 186-189, 195, 197, 199, 208, 234
chaperone cycle, 164
duplications, 165
HSF1 interaction, 13
Hsp104, 76
in polyglutamine protein aggregation, 84
in protein refolding, 76
mitochondria, 121
peroxisomal protein import, 168
protein import, 166
regulatory proteins, 163
hsp70 promoter, 8, 11
footprinting analysis, 19
hsp70 transcription, 9
hsp70.1 promoter, 4

Hsp70/Hsp40, 68, 69, 76, 150
  Hsp104, 70
Hsp78, 122, 123
  chaperone activity, 122
Hsp90, 12, 105, 106, 167, 186, 189, 194
  chaperone cycle, 169
  HSF1 interaction, 12, 13
  interaction with IRE1α, 49
  peroxisomal protein import, 170, 172
Hsp90 chaperone machinery, 13
HspBP1, 106, 193
HSPs, 1, 2, 4, 23, 65
Htm1p, 197
human cytomegalovirus, 208
huntingtin, 83
  in *C. elegans*, 83
Huntingtin, 228
huntingtin-GFP fusion, 84
Huntington's disease, 20, 83, 228
hydrogen peroxide, 150
HYOU1, 105
hyperphosphorylation
  PolII, 11
hypophosphorylation
  PolII, 11
hypoxia, 105

IκBα, 206
i-AAA proteases, 120, 125, 128-130, 132
IAP-1, 129
Ilv5, 122
immunoglobulin, 103
immunophilins, 13, 171
import-competent state, 166
inclusion bodies, 237
Influenza virus hemagglutinin. *See* HA
inheritance, 78
  [*PSI*+] prion, 80
inner membrane proteins
  mitochondria, 131
INO1, 47
Ino2/4, 47
inositol, 47
insulin, 36, 53
insulin signaling pathway, 20
internal cavity
  mitochondrial proteases, 123
intra-membrane protease, 94
IRE1, 35, 40, 41, 44, 46, 48, 51, 53, 100
  BiP, 50
  isoforms, 40
  substrate specificity, 49
IRE1 isoforms, 49
IRE1 signaling, 49
IRE1/PERK heterodimerization, 51
Ire1p, 37, 38
  oligomerization, 38
  substrates, 39
isocitrate lyase, 172

J chain, 103
J-domain, 105
Jem1p, 197, 204
JNK, 10, 49
JPDI, 105
J-proteins, 121, 123, 164
  mitochondria, 121

Kar2p, 100, 204
Kar2p/BiP, 197
karmellae, 48
KDEL ER retention signal, 92
Keap1, 49
kinases, 10, 171
  UPR, 100
kinetic partitioning
  mtHsp70, 121
kuru, 221

lag phase
  polymerization, 79
lateral opening model
  AAA proteases, 135
LDL-R, 103, 105
leucine zipper, 39
  HSF, 7
leucine zipper transcription factor, 44
Lewy-bodies, 208
Lhs1p, 197
lipid composition of the membrane
  heat shock, 65
lipids
  as chaperones, 99
  in quality control, 99
Lon proteases, 119, 122, 125, 137
  functions, 124
  mitochondria, 124
  nucleic acid binding, 125
  oligomeric status, 124
  structure, 124
longevity, 20

Lon-like PIM1 protease, 122
loss-of-function, 207
low-density lipoprotein receptor, 103, 105
LPS, 14
luciferase, 66, 69, 75, 83
lysines in ubiquitin, 190
lysosomal enzymes, 210
lysosomal storage diseases, 207
lysosome
  protein import, 167

m-AAA protease, 122, 125, 126, 128, 130-136
  as processing peptide, 132
  complex with prohibitins, 131
maintenance of mitochondrial DNA
  PIM1 protease, 125
Major Histocompatibility Complex. *See* MHC
MAP kinases, 10
MAP-1, 128
MAPK, 10
Mdh3p, 160
Mdj1, 121, 123
mediator coactivator complex, 11
Mendelian inheritance, 78
Mendelian recessive trait, 77
metallopeptidases, 125, 127, 136
methotrexate, 160
methyl-β-cyclodextrin, 209
Mge1, 121, 123
Mgm1, 136
MGP, 46
MHC, 94, 98
MHC class I, 201, 208
misfolded proteins, 185, 192, 195, 197
misfolding, 263
misfolding diseases, 237
mitochondria, 119, 167
  degradation, 119
  folding state, 167
  protein import, 166, 167
  protein turnover, 120
  proteome, 120
  quality control, 119, 136
  stability, 119
  stress response, 136, 137
  thermotolerance, 122
mitochondrial AAA proteases
  functions, 128
  substrate specificity, 128
mitochondrial biogenesis
  PIM1 protease, 126
mitochondrial chaperone network, 123
mitochondrial chaperones, 137
mitochondrial import motor, 121
mitochondrial matrix, 121
mitochondrial morphology, 130
  prohibitins, 131
mitochondrial proteases, 125
mitochondrial proteolysis, 121
molecular crowbar model
  Hsp104p/ClpB, 74
molecular ratchet model
  Hsp104, 75
monoglucosylated intermediate, 97
mono-ubiquitination, 189
Mpd1p, 103
Mpd2p, 103
mPTS, 155
Msc2, 100
mtDNA, 125
mtDNA helicase, 126
mtDnaJ, 137
mtHsp70, 121, 123
  co-chaperones, 121
  protein import, 121
multi-membrane spanning protein, 92
Myo2p, 155

N-(*n*-nonyl)desoxynojirimycin, 210
NAC, 150
nascent chain associated complex, 150
nascent chains, 92, 150, 225, 273
neurodegeneration, 130
neurodegenerative diseases, 20, 221
NF-Y, 42
nGAAn sequences, 21
nitrogen-dependent repression, 227
NLS
  in ATF6, 42
NN-DGJ, 210
N-octyl-β-valienamine, 210
non-Mendelian inheritance, 77, 224
nonsense suppression, 224
Npl4 zinc finger, 205
Npl4p, 194, 204, 206
Nrf2, 46, 49
nSBs, 21, 22
  HSF1 in, 21
nuclear fragmentation, 84

nuclear stress bodies, 21, 22
nuclear stress granules, 21, 22
nucleotide-release factor, 121
NZF, 205

off pathway aggregates, 81, 82
oligosaccharide transfer complex, 96
Oma1, 125, 136
omnipotent suppression, 77, 79
  [PSI+] prion, 78
oogenesis
  HSF in, 14
Opi1, 47
organelle specific stress responses, 137
ORP150, 105
Oxa1, 132, 136
oxidative damage
  degradation, 124
oxidative ER protein folding, 105
oxidative stress, 14, 19, 130, 136
oxidised glutathione. *See* GSSG

p23, 13, 169, 170, 189
P5, 46, 105
p58^{IPK}, 46
p62-E1, 94
p97, 127, 128, 129
p97/Cdc48, 134
PaeI-R, 208
Pam16, 121, 123
Pam18, 121, 123
PAMP4, 46
paraplegin, 130
Parkin, 208
Parkinson's disease, 207
Pcp1, 133, 136
PDI, 39, 96, 99, 101-106
PDI homologues, 103, 105, 195, 198
Pdi1p, 101, 102
Pdi1p homologues, 103
PDILT, 105
PEK, 44
Pelizaeus-Merzbacher disease
  X-linked, 99
peptidases
  mitochondrial inner membrane, 136
peptidyl-prolyl cis/trans isomerases, 95, 194
Per100p, 197
PERK, 35, 40-53, 100
  BiP, 50
  bovine viral diarrhea virus, 50
  E2 Hepatitis C virus envelope glycoprotein, 50
  phosphorylation, 44, 45
PERK signaling, 49
permeability
  of Sec61 translocon, 92
peroxins
  complexes, 152
  interactions, 152
  list of, 151
peroxisomal fission, 155
peroxisomal protein complexes, 159
peroxisomal protein import, 149, 153, 154, 158
  ATP, 153
  Hsp70, 162, 168
  Hsp90, 170, 172
peroxisomal proteins
  folding, 160
peroxisomal targeting signal 1. *See* PTS1
peroxisomal targeting signal 2. *See* PTS2
peroxisomes, 149
  ATP, 153
  docking proteins, 159
  formation, 155
  gold particle import, 161
  pH, 153
  piggy-back import, 162
  protein import, 160
  receptor recycling, 159
  translocating proteins, 159
Pertussis toxin, 208
Pex, 151
Pex13p, 157, 158
Pex14p, 157
Pex15, 48
Pex17p, 157
Pex18p, 157
Pex18p/Pex21p, 154
Pex19p, 155
Pex20p, 154, 157
Pex21p, 157
Pex3p, 155
Pex5L, 157
Pex5p, 154, 158, 168
  extended shuttle, 156
  shuttle, 156
Pex7p, 154, 157

shuttle, 157
Pex8p, 159
pH
in prion conversion, 229
pharmacological chaperones, 209
phase II detoxifying enzymes, 49
Phb1, 131
Phb2, 131
phosphatases, 10
phosphoethanolamine, 48
phosphoinositol regulation, 47
phospholipid synthesis, 47
phospholipids
in UPR, 47
phosphorylation, 11
HSF, 7
HSF1, 10
in UPR, 48
PERK, 44
photo cross-linking, 93
piggy-back translocation, 160, 162
piggy-back import, 160, 162
Pim1
oligomeric status, 124
PIM1, 122, 124, 125
PIM1 protease, 126, 130
maintenance of mitochondrial DNA, 125
mtDNA metabolism, 125
pipe, 104
PKC, 10
PKR, 45
plant peroxisomes, 168
protein import, 168
plasma B cells, 48, 92
plasma cell differentiation, 47
P-loop ATPase domain
in AAA ATPases, 123
PLP, 99
Pma1p, 103
PML bodies, 21
PMP, 155
PMP61, 168
PMP73, 168
Pol II, 11, 21
polyadenylation factors, 11
polyglutamine
diseases, 84, 237
expansions, 20
fused to GFP, 83
repeat, 83
polyglutamine/asparagine-containing region, 236
poly-L-lysine
Hsp104, 73
polymerization, 232
polyQ, 20
polyQ repeat, 83
polyribosomes
HAC1, 37
polytopic membrane protein degradation, 136
poly-ubiquitination, 185, 190, 205
PPIases, 95
prefoldin, 273
prefoldin-CCT complexes, 274
preimplex, 161
presenilin, 94
primary hyperoxaluria 1, 160
prion, 68
assembly, 234, 238
conformational isoform, 232
conversion, 232
diseases, 207, 221
domain, 227
fibrils, 235
hypothesis, 78, 222, 225
infectious form, 232
inheritance, 237
nucleus, 233
protein, 221, 222
resistance to proteolysis, 223
seeds, 80, 233
species barrier, 224
strains, 224, 233
structure, 228
transmission, 78, 221, 232
variants, 233
yeast, 225, 237
prion propagation, 221, 225, 231-233
Hsp104, 82
mechanisms, 231
prion stability, 79
Hsp104, 79, 82
prionogenic protein, 79
PRNP, 222
proapoptotic signaling, 84
Hsp104, 84
procaspase-12, 45
processing enzyme
mitochondria, 136
processive unfolding, 123

by AAA ATPases, 123
prohibitins, 125, 131
assembly, 131
complex with *m*-AAA protease, 131
respiratory chain assembly, 131
prointerleukin-1β, 18
proliferation of the ER, 48
proline isomerisation, 95
pro-region
Lon protease, 124
proteases, 119
proteasomal degradation, 106, 185
proteasome, 199, 205
architecture, 190
protein aggregates, 20
mitochondria, 123, 124
protein aggregation, 65, 66, 122
protein aggregation disease, 82
protein disaggregase, 65
protein disaggregation, 73, 77, 82, 83
Hsp104, 76
protein disulfide isomerase. *See* PDI
protein folding, 65, 91, 186, 251, 263, 273, 274, *See*
mitochondria, 121
protein folding capacity, 47
endoplasmic reticulum, 47, 53
mitochondria, 136
protein folding stress, 44
protein import
mtHsp70, 121
protein misfolding, 221
protein refolding, 75, 186, 252
Hsp104, 74, 82
mitochondria, 123
protein sorting, 93
inner nuclear membrane proteins, 93
protein synthesis
mitochondrial, 122
protein t-complex polypeptide-1. *See* TCP-1
protein translocase in the inner membrane. *See* TIM complex
protein unfolding/threading
ClpB, 75
proteolipid protein, 99
proteolysis
mitochondria, 121
PrP, *See* prion protein
prp genes, 194
PrP(C), 209, 222
function, 223
PrP(Sc), 209, 223
PRP19, 194
P-TEFb, 11
PTS1, 153, 154
PTS2, 154, 157

Q and N rich domains, 228
Q6, 101
QC. *See* quality control
QSOX, 101
quality control, 95, 119, 185
endoplasmic reticulum, 36, 91, 93, 98, 102, 195
mitochondria, 120, 124-127, 131, 136, 137
regulation, 136
quiescin-sulfhydryl oxidase, 101

Rad23p, 206
RAP, 106
Rbd1, 133, 136
Rca1, 128
RD. *See* regulatory domain
read-through of nonsense codons, 77
recycling receptor import model, 156
redox potential, 102
reduced glutathione. *See* GSH
reduction
in ERAD, 102
refolding
Hsp70/Hsp40, 74
Hsp100s, 75
re-glycosylation, 36
regulation, 1
post-translational, 5
regulatory domain
HSF1, 6
remodeling protein aggregates
Hsp104, 81
respiratory chain
assembly, 131
biogenesis, 126
response
heat shock, 1
retrograde translocation, 36, 97
retrotranslocation, 36, 97
rhomboid-like peptidase, 133, 136
ribosome, 164
ricin toxin, 102, 208
RING-domain, 202

RING-finger E3, 189
RNA polymerase, 3
RNA polymerase II, 11
RNA processing, 37
RNA splicing, 37
RNase L, 38
Rnq1p, 226
rpoH gene, 2
Rrp46p, 68
Rubisco, 252
ryanodine receptors, 95

S1P, 42, 43, 47, 50
  ATF6 cleavage, 42
S2P, 42, 43, 47, 50
SAGA, 38
Sam68, 21
sarcoplasmic reticulum, 95, 99
satellite III repeats, 22, 23
Sbh1p, 200
SBMA, 83
SCA genes, 83
scaffold
  fibril formation, 81
SCAP, 42, 50
Scj1p, 99, 197, 204
scrapie, 221
Scs2, 48
Sec11, 94
Sec12p, 199
Sec18p, 199
Sec61 complex, 92, 200
Sec62p, 200
Sec63, 166, 200
Sec71p, 200
Sec72p, 200
SecA, 156
secretory pathway, 35, 37, 91
SecYEG complex, 201
seed
  for aggregation, 78
seeded polymerization, 79
seeded-polymerization model, 232
segregase, 205
SEL-1, 203
SEL-1-like repeats, 203
Semliki Forest virus, 94, 96
SERCA2b, 99
serine/threonine kinase, 35, 44
SF2/ASF, 21
SFV. *See* Semliki Forest virus
Shiga toxin, 208
Shp1p/p47, 204
sHsps, 20
signal peptidase complex, 93
signal peptide cleavage, 93
signal peptide peptidase, 94
signal sequence
  endoplasmic reticulum, 92
Sil1p, 197
Sis1p, 189
Site 1 Protease. See S1P
Site 2 Protease. See S2P
SKL, 154
SLP, 203
Sls1p, 197
small heat shock proteins, 20
Snf1, 10
sorting
  of glycoproteins, 93
spastic paraplegia, 99
SPC, 93
sphingolipidose, 207
spinal and bulbar muscular atrophy, 83
spinocerebellar ataxia, 83
splicing
  HAC1, 37, 41
  Ire1p mediated, 37
  XBP1, 41
splicing complexes
  Hsp104p, 67
splicing factors, 21
spongiform encephalopathy, 78
SPP, 94
Spt23p, 205
SREBP, 42, 43, 50
SRp30c, 21
S-S bridge, 100
Ssa proteins, 164, 165
Ssa1/Ydj1, 68, 69
Ssa1p, 199
Ssa2p, 168
Ssb proteins, 164, 165
Ssc1, 121, 123
Sse proteins, 164
Ssi1p, 197
Sss1, 126, 200
Ssz1p, 164
sterol response element, 43
sterol response element binding protein,
  42
Sti1p, 76, 77

stop codons, 225
storage diseases
  endoplasmic reticulum, 106
stress, 10, 11, 14, 136
stress response, 39
  mitochondrial, 136
structure
  GroEL, 254
  GroES, 255
substrate recognition
  AAA proteases, 132
sulfhydryl, 100
SUMO-1 modification, 10, 11
  HSF, 7, 10, 16
sumoylation, 10, 11
  HSF, 7, 10, 16
Sup35p, 68, 77, 78, 79, 224, 236
  fibrils, 80
  Hsp104, 77
  oligomeric status, 78
  oligomerization, 82
  aggregate size, 80
Sup35p/Hsp104p interaction, 82
Sup45p, 78, 225
suppressors, 252
SWI/SNF, 11
symplekin, 11, 12
systemic amyloidosis, 83

TATA-box, 4
*T*ClpB, 71, 72
TCP-1, 265
TCP-1 Ring Complex. *See* TRiC and CCT
Tef3p, 68
template
  fibril formation, 81
template-assistance model, 232
tetraspanin, 99
tetratricopeptide repeats. *See* TPR
TF, 150
TF55, 264
TFIIF, 18
thermal stress, 65, 136
thermolysin, 127
thermosome, 264, 267, 269, 272
thermotolerance, 65, 66, 74, 75, 80
  ClpB, 74
  Hsp104, 65, 67, 70
  mitochondrial, 122
thiolase, 160
thioredoxin domain, 102, 105
threading model
  AAA proteases, 135
thyroglobulin, 104-106
TIM complex, 121
Tim14, 121, 123
Tim16, 121, 123
TIM23 translocase, 134
Tim44, 121
tissue-type plasminogen activator, 103
TMX3, 105
TNFα, 18
TOM complex, 123
toxins, 206, 208
tPA, 103
TPR, 154, 189, 191, 203
TRAF2, 49
TRAM, 93
transcription factors, 1, 4, 39
transcriptional activators, 6
transcriptional regulation, 4
  HSF1, 11
  mitochondrial chaperones and proteases, 136
translation elongation, 77
translation termination, 77, 226, 238
  [PSI+] prion, 78
translational repression in UPR, 40
translocase of inner membrane. *See* TIM
translocase of outer membrane. *See* TOM
translocation, 167, 171
  AAA proteases, 135
  mitochondria, 121
translocon, 92
  endoplasmic reticulum, 126
transmembrane serine/threonine kinase/endoribonuclease, 35
trehalose, 65
TRiC, 265
trigger factor, 150
trimethylamine-*N*-oxide, 209
tRNA ligase, 37
TSE, 223
tubulin, 265, 273
tunicamycin, 52
Twinkle, 126
type I diabetes, 36

UBA domain, 206
Ubc, 190, 199, 202

Ubc6p, 203
ubiquitin, 190
ubiquitin ligase, 106, 185
ubiquitin proteosome system, 36
ubiquitin-associated domain, 206
ubiquitination, 202, 204, 205
ubiquitination pathway, 189
ubiquitin-conjugating enzymes, 190, 202
ubiquitin-proteasome system, 185, 189
U-box domain, 194
U-box ligases, 106, 191, 194, 206
UDP-glucose-glycoprotein glucosyltransferase, 97, 197
Ufd1p, 194, 204, 206
Ufd2p, 194, 206
UIP5, 194, 206
unfoldase activity
  of AAA protease, 134
unfolded protein response, 195, *See* UPR
unfolded protein response element. *See* UPRE
unfolding of preproteins, 123
UPR, 35, 37, 40, 41, 43, 44, 46-48, 51, 53, 137, 195
  and ER lipid content, 47
  cell cycle arrest, 48
  cell fate, 48
  in yeast, 37
  phospholipids, 47
  physiological roles, 52
  three signaling branches, 46
  translation repression, 48
  Zinc depletion, 100
UPR signaling branches, 51
UPR signaling pathway, 41
UPRE, 39, 41, 42, 46, 52
Ure2p, 224, 227, 236
URE3, 227
ureidosuccinate usage, 227
US11, 201, 208
US2, 208

Walker A motif, 72, 127
Walker B motif, 72, 127
Walker-type ATPases, 120
vasopressin V_2 receptor, 209
WCGHC motif, 102
VCP/p97, 201
WD40 repeats, 154, 194
VID vesicles, 167
wind, 104
Vps1p, 155
WW domain, 203

XAP1, 171
XBP1, 40, 46, 47, 48, 49, 52, 53
X-linked Pelizaeus-Merzbacher disease, 99

Ydj1p, 189, 199
yeast prions, 77, 225, 228
yellow-green birefringence in polarized light, 229
ykl056c, 68
Yme1, 127-129, 132
YME1, 130
Yme1 AAA domains
  structure, 129
Yme2, 133
Yta10, 128, 130, 134
Yta12, 128, 130, 134

Zinc finger motif in ClpX, 72
Zn^{2+}, 99
Zn^{2+} transporter
  endoplasmic reticulum, 100
Zuo1p, 164

MIX
Papier aus verantwortungsvollen Quellen
Paper from responsible sources
FSC® C105338

If you have any concerns about our products,
you can contact us on
ProductSafety@springernature.com

In case Publisher is established outside the EU,
the EU authorized representative is:
Springer Nature Customer Service Center GmbH
Europaplatz 3, 69115 Heidelberg, Germany

Printed by Libri Plureos GmbH
in Hamburg, Germany